Andreas Heinrich, Heiko Pleines (eds.)

Export Pipelines from the CIS Region

Geopolitics, Securitization,
and Political Decision-Making

CHANGING EUROPE

Edited by Dr. Sabine Fischer, Dr. Heiko Pleines and
Prof. Dr. Hans-Henning Schröder

ISSN 1863-8716

3   Daniela Obradovic, Heiko Pleines (eds.)
    The Capacity of Central and East European Interest Groups to
    Participate in EU Governance
    ISBN 978-3-89821-750-7

4   Sabine Fischer, Heiko Pleines (eds.)
    Crises and Conflicts in Post-Socialist Societies
    The Role of Ethnic, Political and Social Identities
    ISBN 978-3-89821-855-9

5   Julia Kusznir, Heiko Pleines (eds.)
    Trade Unions from Post-Socialist Member States in EU Governance
    ISBN 978-3-89821-857-3

6   Sabine Fischer, Heiko Pleines (eds.)
    The EU and Central & Eastern Europe
    Successes and Failures of Europeanization in Politics and Society
    ISBN 978-3-89821-948-8

7   Sabine Fischer, Heiko Pleines (eds.)
    Civil Society in Central and Eastern Europe
    ISBN 978-3-8382-0041-5

8   Zdenka Mansfeldová, Heiko Pleines (eds.)
    Informal Relations from Democratic Representation to Corruption
    Case Studies from Central and Eastern Europe
    ISBN 978-3-8382-0173-3

9   Leonid Kosals, Heiko Pleines (eds.)
    Governance Failure and Reform Attempts after the
    Global Economic Crisis of 2008/09
    Case Studies from Central and Eastern Europe
    ISBN 978-3-8382-0336-2

10 Andreas Heinrich, Heiko Pleines (eds.)
    Export Pipelines from the CIS Region
    Geopolitics, Securitization, and Political Decision-Making
    ISBN 978-3-8382-0639-4

Andreas Heinrich, Heiko Pleines (eds.)

EXPORT PIPELINES FROM THE CIS REGION

Geopolitics, Securitization,
and Political Decision-Making

ibidem-Verlag
Stuttgart

Bibliografische Information der Deutschen Nationalbibliothek
Die Deutsche Nationalbibliothek verzeichnet diese Publikation in der
Deutschen Nationalbibliografie; detaillierte bibliografische Daten sind im
Internet über http://dnb.d-nb.de abrufbar.

Bibliographic information published by the Deutsche Nationalbibliothek
Die Deutsche Nationalbibliothek lists this publication in the Deutsche Nationalbibliografie;
detailed bibliographic data are available in the Internet at http://dnb.d-nb.de.

∞

Gedruckt auf alterungsbeständigem, säurefreien Papier
Printed on acid-free paper

ISSN: 1863-8716

ISBN-13: 978-3-8382-0639-4

© *ibidem*-Verlag
Stuttgart 2014

Alle Rechte vorbehalten

Das Werk einschließlich aller seiner Teile ist urheberrechtlich geschützt. Jede Verwertung außerhalb der engen Grenzen des Urheberrechtsgesetzes ist ohne Zustimmung des Verlages unzulässig und strafbar. Dies gilt insbesondere für Vervielfältigungen, Übersetzungen, Mikroverfilmungen und elektronische Speicherformen sowie die Einspeicherung und Verarbeitung in elektronischen Systemen.

All rights reserved. No part of this publication may be reproduced, stored in or introduced into a retrieval system, or transmitted, in any form, or by any means (electronic, mechanical, photocopying, recording or otherwise) without the prior written permission of the publisher. Any person who does any unauthorized act in relation to this publication may be liable to criminal prosecution and civil claims for damages.

Printed in Germany

Contents

List of Figures 7

List of Tables 8

Abbreviations 9

Preface 11

Andreas Heinrich
0. Introduction: Export Pipelines in Eurasia 13

PART I. RUSSIA'S FOREIGN ENERGY POLICY

András Molnár
1. Russian Foreign Energy Policy Under Vladimir Putin: Norms, Ideas, and Determining Factors 77

Inna Chuvychkina
2. An Actor-Centred Institutionalist Approach to Russia's Pipeline Policies 91

Niels Smeets
3. Opening Up the Black Box: Russia's Energy Security Concept 107

PART II. ENERGY POLICY IN CENTRAL ASIA

Farkhod Aminjonov
4. Challenges Along the Way Towards a Maximally Secure Central Asian Gas System 131

Boris Barkanov
5. The Geo-Economics of Eurasian Gas: the Evolution of Russian–Turkmen Relations in Natural Gas (1992–2010) 149

Paolo Sorbello
6. Pipelines and Hegemonies in the Caspian: a Gramscian Appraisal 175

Part III. The Southern Energy Corridor

Rufat Rustamov
7. The Trans-Adriatic Pipeline and Nabucco West Pipeline Projects: Advantages and Disadvantages for Azerbaijan — 201

Irina Kustova
8. EU Energy Policy Towards the Caspian Region: Assessing the Southern Gas Corridor — 215

Lusine Badalyan
9. The EU's Democracy Promotion in its Eastern Neighbourhood: Consistency in the Shadow of Energy Trade and the EU's Ability — 229

Part IV. Ukraine's Energy Policy

Eric Pardo Sauvageot
10. Gas Disputes Between Russia and Ukraine: Patterns of Escalation and Russian Stakes (2006–2009) — 253

Katerina Malygina
11. The Struggle Over Ukraine's Gas Transit Pipeline Network Through the Lenses of Securitisation Theory — 277

Annex

Oil and Gas Reserves and Production — 309

Oil and Gas Pipelines in Europe and Central Asia — 313

Selected Bibliography on Eurasian Export Pipelines — 319

About the Authors — 339

List of Figures

3-1:	Coded references to SOS in Russia's energy strategies, 2003 and 2009 (NVivo 10)	126
3-2:	Coded references to SOD in Russia's energy strategies, 2003 and 2009 (NVivo 10)	127
5-1:	Evolution of Price Paid by Gazprom for Turkmen Gas Versus the Price of Russian Gas at the German Border (US$/1,000cm)	158
5-2:	Change in the Turkmen Gas Price as a Proportion of the German Border Price	159
6-1:	The Crude Oils and Their Key Characteristics	190
9-1:	Democratic Change in EaP Countries According to Freedom House and the EU	242
10-1:	Decision Tree by Zagare and Kilgour	260
10-2:	The Gas Conflict in 2006	265
10-3:	The Gas Conflict in 2008	269
10-4:	The Gas Conflict in 2009	271
A-1:	Oil Pipelines in Central Asia	313
A-2:	Oil Pipelines in Europe	314
A-3:	Gas Pipelines in Europe	316
A-4:	Gas Pipelines in Central Asia	318

List of Tables

0-1:	Soviet Oil and Gas Production (in Millions of Tonnes and Billions of Cubic Metres)	23
0-2:	Soviet Oil Export Pipelines	26
0-3:	Soviet Gas Export Pipelines	37
0-4:	Russia's Oil Export Pipelines	41
0-5:	Russia's Gas Export Pipelines	45
0-6:	Kazakhstan's Oil Export Pipelines	50
0-7:	Kazakhstan's Gas Export Pipeline	52
0-8:	Turkmenistan's Gas Export Pipelines	59
0-9:	Azerbaijan's Oil Export Pipelines	66
0-10:	Azerbaijan's Gas Export Pipelines	68
3-1:	Security of Demand Strategies (Energy Strategy 2003, 2009)	119
5-1:	Volumes of Turkmen Gas Purchased Annually by Gazprom	157
9-1:	Freedom House Democratisation Categories	240
9-2:	EU Democratisation Assessment Units and Definitions	241
9-3:	Ordinary Least Squares (OLS) Results	248
11-1:	Ukraine's Debate over GTS Privatisation in 1990–2009	302
A-1:	Proven Crude Oil Reserves (in Million Tonnes)	309
A-2:	Crude Oil Production (in Million Tonnes)	310
A-3:	Proven Natural Gas Reserves (in Trillion Cubic Metres)	311
A-4:	Natural Gas Production (in Billion Cubic Metres)	312

Abbreviations

€	Euro
ACG	Azeri–Chirag–Guneshli oil fields
AMG	AqtobeMunaiGaz
ASSR	Autonomous Soviet Socialist Republic
bcm	Billion cubic metres
BG	British Gas
BPS	Baltic Pipeline System
BTC	Baku–Tbilisi–Ceyhan oil pipeline
BTE	Baku–Tbilisi–Erzurum gas pipeline (also known as South Caspian Pipeline, SCP)
BYuT	Bloc Yulia Timoshenko (Blok Yulii Timoshenko)
CEO	Chief executive officer
cm	Cubic metres
CMEA	Council for Mutual Economic Assistance
CNPC	China National Petroleum Corporation
EaP	Eastern Partnership
EC	European Commission
ENP	European Neighbourhood Policy
ESPO	East Siberia–Pacific Ocean
ESS	European Security Strategy
EU	European Union
FH	Freedom House
FSU	Former Soviet Union
GDP	Gross domestic product
GDR	German Democratic Republic
GTS	Gas transit system
HPP	Hydropower plant(s)
IAP	Ionian–Adriatic Pipeline
IEA	International Energy Agency
IMF	International Monetary Fund
ITGI	Interconnector Turkey–Greece–Italy
KCP	Kazakhstan–China pipeline
km	Kilometre(s)
KMG	KazMunaiGaz

kWh	Kilowatt hour(s)
LNG	Liquefied natural gas
mt	Million tonnes
NATO	North Atlantic Treaty Organization
NEL	Nordeuropäische Erdgasleitung (North European Gas Pipeline)
NOC(s)	National oil companies
NPE	Normative Power Europe
NUNS	Our Ukraine–People's Self-Defence (Nasha Ukraina–Narodna Samooborona)
OLS	Ordinary Least Squares
OPAL	Ostsee–Pipeline–Anbindungs-Leitung (Baltic Sea Pipeline Link)
OPEC	Organization of Petroleum Exporting Countries
PKK	Kurdistan Workers' Party
QCA	Quantitative Content Analysis
SEEP	South East Europe Pipeline
SOCAR	State Oil Company of the Azerbaijan Republic
SOD	Security of demand
SOS	Security of supply
SSR	Soviet Socialist Republic
TANAP	Trans–Anatolia gas pipeline
TAP	Trans–Adriatic Pipeline
TNC(s)	Transnational companies
US$	US dollar
US/ USA	United States of America
USSR	Union of Soviet Socialist Republics (i.e., Soviet Union)
WTO	World Trade Organization

Preface

This book presents the papers discussed at the 8th Changing Europe Summer School on 'Export pipelines from the CIS region: National debates, political decision-making and geopolitics' held in Almaty in August 2013. Organized by the Central Asian Foundation for the Development of Democracy, Almaty, and the Research Centre for East European Studies at the University of Bremen with financial support from the Volkswagen Foundation, the Summer School brought together young academics from different disciplines of the social sciences and the humanities to share their research on energy issues in the post-Soviet region. Participants were selected by means of an anonymous review process that is kindly supported by the members of our international review panel.

The results of each Summer School are published in this book series. It goes without saying that this book would not have been possible without ample support. First of all, our thanks go to the participants themselves, whose enthusiasm and knowledge made the Summer School a truly worthwhile event. We would also like to thank all the referees who aided us in the selection process for appropriate participants and all the discussants who enriched the debates in Almaty, namely Margarita Balmaceda, Julia Kusznir and Indra Øverland.

The Summer School was organized within the research project on 'Domestic debates and foreign policy-making in the Caspian region: The case of export pipelines from Azerbaijan, Kazakhstan and Turkmenistan', which is also funded by the Volkswagen Foundation. Our project partners Kenan Aslanli and Tolganay Umbetalieva also greatly contributed to the success of the event.

We are especially grateful to those who helped to organize the Summer School and the book production, namely Nozima Akhrarkhodjaeva (organizational support), Malika Elkhan (organizational support), and Matthias Neumann (layout).

Bremen, January 2014
The Editors

Andreas Heinrich

0. Introduction: Export Pipelines in Eurasia

In the 1990s, the end of the Cold War—with the dissolution of the Soviet Union—was the major topic for social science scholars, especially those in International Relations. Interpretations focused on the emergence of a multi-polar world, and the Caspian Sea region (comprising the Southern Caucasus and countries of Central Asia) was considered one of the most important regions in which the global balance of power would be re-adjusted, especially after the discovery of considerable oil and gas reserves. The competition among foreign powers, primarily Russia as the successor to the Soviet Union and the United States (US), but to a lesser degree the European Union (EU) and increasingly China and regional powers such as Turkey and Iran, for access to Caspian energy resources and control over infrastructure is often called a 'new great game'. Accordingly, a large body of literature has emerged.[1]

Although relatively small on a global scale, the oil and gas reserves of the Caspian Sea basin are of major geopolitical importance, as they are not controlled by OPEC or Russia. Therefore, they offer the EU, the US and China the opportunity to diversify their energy supplies.[2]

However, Russia remains a major player in the region because it controls the Soviet pipeline network on which Eurasian oil and gas producers still rely. Russia considers this dependence to be a major foreign policy tool that it can use to maintain its influence in the so-called 'near abroad'. In reaction to the rather aggressive and unilateral use of this tool, many analysts term it Russia's 'energy weapon'.[3] Additionally, Russia also

---

1   For a comprehensive bibliography on Central Asia see the following works: Laumulin, Murat T.: Tsentral'naya Aziya v zarubezhnoy politologii i mirovoy geopolitike, five volumes, Almaty: KISI, 2005-2009; Hanks, Reuel (ed.): Routledge handbook of Central Asian politics, London: Routledge, 2010.
2   Cf. e.g., Mavrakis, Dimitrios/ Thomaidis, Fotios/ Ntroukas, Joannis: An assessment of the natural gas supply potential of the south energy corridor from the Caspian region to the EU, in: Energy Policy, 2006 (vol. 34), no. 13, pp. 1671–1680; Dorian, James P.: Central Asia: a major emerging energy player in the 21st century, in: Energy Policy, 2006 (vol. 34), no. 5, pp. 544–555; Ahrend, Rüdiger/ Tompson, William: Caspian oil in a global context, in: Transition Studies Review, 2007 (vol. 14), no. 1, pp. 163–187; Bahgat, Gawdat: Prospects for energy cooperation in the Caspian Sea, in: Communist and Post-Communist Studies, 2007 (vol. 40), no. 2, pp. 157–168; Remme, Uwe/ Blesl, Markus/ Fahl, Ulrich: Future European gas supply in the resource triangle of the former Soviet Union, the Middle East and Northern Africa, in: Energy Policy, 2008 (vol. 36), no. 5, pp. 1622–1641.
3   Cf. e.g., Stulberg, Adam N.: Well-oiled diplomacy: strategic manipulation and Russia's energy statecraft in Eurasia, Albany, NY: State University of New York Press, 2007; Nygren, Bertil: Putin's use of natural gas to reintegrate the CIS region, in: Problems of Post-Communism, 2008 (vol. 55), no. 4, pp. 3–15; Orban, Anita: Power, energy, and the new Russian imperialism, Westport, CT: Praeger, 2008; Hashim, S. Mohsin: Power-loss or power-transition? Assessing the limits of using the energy sector in reviving Russia's geopolitical stature, in: Communist and Post-Communist

employs soft power, namely personal and cultural links cultivated during the Soviet period, and its support for authoritarian rulers to improve its position in the Caspian region.[4] Thus, it succeeded, for example, in convincing Kazakhstan to construct its first post-Soviet oil export pipeline through Russian territory.[5]

Moreover, after years of negotiations and intensive lobbying by the US government, the main oil export pipeline from Azerbaijan runs via Georgia to Turkey.[6] In addition, the EU began to lobby for a pipeline transporting Caspian gas through Turkey to Central Europe.[7] This offered Turkey the opportunity to play a more prominent role

---

Studies, 2010 (vol. 43), no. 3, pp. 263-274; Smith Stegen, Karen: Deconstructing the 'energy weapon': Russia's threat to Europe as case study, in: Energy Policy, 2011 (vol. 39), no. 10, pp. 6505–6513.

4   Cf. e.g., Baev, Pavel K.: Assessing Russia's cards: three petty games in Central Asia, in: Cambridge Review of International Affairs, 2004 (vol. 17), no. 2, pp. 269–283; Tolstrup, Jakob: Studying a negative external actor: Russia's management of stability and instability in the 'near abroad', in: Democratization, 2009 (vol. 16), no. 5, pp. 922–944; Gayoso, Carmen A.: Russian hegemonies: historical snapshots, regional security and changing forms of Russia's role in the post-Soviet region, in: Communist and Post-Communist Studies, 2009 (vol. 42), no. 2, pp. 233–252; Jackson, Nicole: The role of external factors in advancing non-liberal democratic forms of political rule: a case study of Russia's influence on Central Asian regimes, in: Contemporary Politics, 2010 (vol. 16), no. 1, pp. 101–118; Melnykovska, Inna/ Plamper, Hedwig/ Schweickert, Rainer: Do Russia and China promote autocracy in Central Asia?, in: Asia Europe Journal, 2012 (vol. 10), no. 1, pp. 75–89.

5   Cf. e.g., Kovalev, Felix N.: Transportation of Caspian oil through Russia, in: Amirahmadi, Hooshang (ed.): The Caspian region at a crossroad: challenges of a new frontier of energy and development, New York: St. Martin's Press, 2000, pp. 155–162; Kubicek, Paul: Russian energy policy in the Caspian basin, in: World Affairs, 2004 (vol. 166), no. 4, pp. 207–217; Marten, Kimberly: Russian efforts to control Kazakhstan's oil: the Kumkol case, in: Post-Soviet Affairs, 2007 (vol. 23), no. 1, pp. 18–37; Nanay, Julia: Russia's role in the Eurasian energy market: seeking control in the face of growing challenges, in Perovic, Jeronim/ Orttung, Robert/ Wenger, Andreas (eds): Russian energy power and foreign relations: implications for conflict and cooperation, London: Routledge, 2009, pp. 109–131; Zhukov, Stanislav: Turkmenistan: an exporter in transition, in: Pirani, Simon (ed.): Russian and CIS gas markets and their impact on Europe, Oxford: Oxford University Press, 2009, pp. 271–315; Baev, Pavel K.: Russian energy policy and military power: Putin's quest for greatness, London: Routledge, 2009; Øverland, Indra/ Torjesen, Stina: Just good friends: Kazakhstan's and Turkmenistan's energy relations with Russia, in: Øverland, Indra/ Kjærnet, Heidi/ Kendall-Taylor, Andrea (eds): Caspian energy policy: Azerbaijan, Kazakhstan and Turkmenistan, London: Routledge, 2010, pp. 136–149.

6   Cf. e.g., Barnes, Joseph: U.S. national interests: getting beyond the hype, in: Kalyuzhnova, Yelena/ Jaffe, Amy M./ Lynch, Dov/ Sickels, Robin C. (eds.): Energy in the Caspian region: present and future, Basingstoke: Palgrave Macmillan, 2002, pp. 212–233; Starr, S. Frederick/ Cornell, Svante E. (eds): The Baku-Tbilisi-Ceyhan pipeline: oil window to the West, Stockholm/ Washington, DC: Central Asia – Caucasus Institute & Silk Road Studies Program, 2005; Cooley, Alexander: Principles in the pipeline: managing transatlantic values and interests in Central Asia, in: International Affairs, 2008 (vol. 84), no. 6, pp. 1173–1188.

7   Cf. e.g., Kalyuzhnova, Yelena: The EU and the Caspian region: an energy partnership, in: Economic Systems, 2005 (vol. 29), no. 1, pp. 59–76; Giuli, Marco: Nabucco pipeline and the Turkmenistan conundrum, in: Caucasian Review of International Affairs, 2008 (vol. 2), no. 3, pp. 124–132; Pashkovskaya, Irina: Evropeiskii Soyuz. Energeticheskaya politika v otnoshenii novykh nezavisimykh gosudarstv, Moscow: MGIMO, 2009; Bilgin, Mert: Geopolitics of European natural gas demand: supplies from Russia, Caspian and the Middle East, in: Energy Policy, 2009 (vol. 37),

0. Introduction: Export Pipelines in Eurasia   15

in the Caspian region, which it considers its special sphere of interest due to historical links.[8]

Iran, however, which would have offered the easiest export route from the Caspian Sea with its established port facilities on the Persian Gulf coast, was sidelined for chiefly political reasons (i.e., US sanctions). It also played an uncooperative role by blocking agreements on the division of the Caspian Sea shelf among the coastal states and thus hampered resource exploration and plans for a trans-Caspian pipeline.[9]

China, however, which became the world's most rapidly growing importer of energy resources at the end of the 1990s, has increasingly gained access to the Caspian region. It invested heavily in the Kazakh oil industry and was also the first of the major

---

no. 11, pp. 4482–4492; Locatelli, Catherine: Russian and Caspian hydrocarbons: energy supply stakes for the European Union, in: Europe-Asia Studies, 2010 (vol. 62), no. 6, pp. 959–971; Baev, Pavel K./ Øverland, Indra: The South Stream versus Nabucco pipeline race: geopolitical and economic (ir)rationales and political stakes in mega-projects, in: International Affairs, 2010 (vol. 86), no. 5, pp. 1075–1090.

8   Cf. e.g., Sayari, Sabri: Turkey's Caspian interests: economic and security opportunities, in: Ebel, Robert/ Menon, Rajan (eds): Energy and conflict in Central Asia and the Caucasus, Lanham, MD: Rowman and Littlefield, 2000, pp. 225–246; Cutler, Robert M.: Turkey and the geopolitics of Turkmenistan's natural gas, in: Review of International Affairs, 2001 (vol. 1), no. 2, pp. 20–33; Tayfur, M. Fatih/ Göyman, Korel: Decision making in Turkish foreign policy: the Caspian oil pipeline issue, in: Middle Eastern Studies, 2002 (vol. 38), no. 2, pp. 101–122; Winrow, Gareth: Turkish national interests, in: Kalyuzhnova, Yelena/ Jaffe, Amy M./ Lynch, Dov/ Sickels, Robin C. (eds.): Energy in the Caspian region: present and future, Basingstoke: Palgrave Macmillan, 2002, pp. 234–250; Bilgin, Mert: The emerging Caspian energy regime and Turkey's new role, in: Turkish Yearbook, 2003 (vol. 34), pp. 1–22; Bilgin, Mert: New prospects in the political economy of inner-Caspian hydrocarbons and western energy corridor through Turkey, in: Energy Policy, 2007 (vol. 35), no. 12, pp. 6383–6394; Babali, Tuncay: Turkey at the energy crossroads, in: Middle East Quarterly, 2009 (vol. 16), no. 2, pp. 25–33; Tekin, Ali/ Williams, Paul A.: EU-Russian relations and Turkey's role as an energy corridor, in: Europe-Asia Studies, 2009 (vol. 61), no. 2, pp. 337–356.

9   Cf. e.g., Miles, Carolyn: The Caspian pipeline debate continues: why not Iran?, in: Journal of International Affairs, 1999 (vol. 53), no. 1, pp. 325–346; Kemp, Geoffrey: US-Iranian relations: competition or cooperation in the Caspian Sea basis, in Ebel, Robert/ Menon, Rajan (eds): Energy and conflict in Central Asia and the Caucasus. Lanham, MD: Rowman and Littlefield, 2000, pp. 145–162; Namazi, S./ Farzin, F.: Division of the Caspian Sea: Iranian policies and concerns, in: Akiner, Shirin (ed.): The Caspian; politics, energy and security, London: Routledge, 2004, pp. 230–243; Yuldasheva, Guli: Geopolitics of Central Asia in the context of the Iranian factor, in: Caucasian Review of International Affairs, 2008 (vol. 2), no. 3, pp. 133–145; Sadegh-Zadeh, Kaweh: Iran's strategy in the South Caucasus, in: Caucasian Review of International Affairs, 2008 (vol. 2) no. 1, pp. 35–41; Pahlavi, Pierre/ Hojati, Afshin: Iran and Central Asia: politics of prudent pragmatism, in: Kavalski, Emilian (ed.): The new Central Asia: the regional impact of international actors, Singapore: World Science Publishing, 2010, pp. 215–238; Khalifa-Zadeh, Mahir: Iran and the Southern Caucasus: a struggle for influence, in: Central Asia and the Caucasus, 2011 (vol. 12), no. 1, pp. 51–61; Sadri, Houman A.: Iran and the Caucasus states in the 21st Century: a study of foreign policy goals and means, in: Journal of Balkan and Near Eastern Studies, 2012 (vol. 14), no. 3, pp. 383–396.

powers to become involved in gas production in Turkmenistan. Oil and gas pipelines from Kazakhstan and Turkmenistan provide China with considerable energy supplies.[10]

Thus, the exports of crude oil and natural gas from the Caspian Sea region and the construction of export pipelines have principally been analysed in the social sciences from a geopolitical perspective. While the geopolitical perspective is important, it oversimplifies international relations in the region and fails to consider the Caspian countries as actors in their own right. References to their foreign policy options and preferences are only made in passing; there is still no complete academic analysis focusing on the geopolitical strategies of the Caspian states.[11]

This gap in the literature became obvious when Azerbaijan developed a more critical stance towards Western projects during the debate over the Nabucco pipeline while simultaneously supporting Georgia against Russian pressure. These moves could not be explained by the rising influence of a major outside power and has recently

---

10 Cf. e.g., Xu, Xiaojie: The oil and gas links between Central Asia and China: a geopolitical perspective, in: OPEC Review, 1999 (vol. 23), no. 1, pp. 33–54; Guangcheng, Xing: China's foreign policy toward Kazakhstan, in: Legvold, Robert (ed.): Thinking strategically: the major powers, Kazakhstan and the Central Asian nexus, Cambridge, MA: MIT Press, 2003, pp. 141–164; Sír, Jan / Horák, Slavomír: China as an emerging superpower in Central Asia: the view from Ashkhabad, in: China and Eurasia Forum Quarterly, 2008 (vol. 6), no. 2, pp. 75–88; Janardhanan, Nandakumar: China's search for energy and its strategy towards Central Asia, in: International Journal of Energy Sector Management, 2009 (vol. 3), no. 2, pp. 102–107; Marketos, Thrassy: China's energy geopolitics: the Shanghai Cooperation Organization and Central Asia, London: Routledge, 2009; Lateigne, Marc: China, energy security and Central Asian diplomacy: bilateral and multilateral approaches, in: Øverland, Indra/ Kjærnet, Heidi/ Kendall-Taylor, Andrea (eds): Caspian energy policy: Azerbaijan, Kazakhstan and Turkmenistan, London: Routledge, 2010, pp. 101–115; Kennedy, Ryan: In the 'new great game', who is getting played? Chinese investment in Kazakhstan's petroleum sector, in: Øverland, Indra/ Kjærnet, Heidi/ Kendall-Taylor, Andrea (eds): Caspian energy policy: Azerbaijan, Kazakhstan and Turkmenistan, London: Routledge, 2010, pp. 116–125; Petelin, Evgeny: China's energy monologue in Central Asia, in: Security Index: A Russian Journal on International Security, 2011 (vol. 17), no. 4, pp. 29–46; Huiron, Zhao/ Hongwei, Wu: China's energy foreign policy towards the Caspian region: the case of Kazakhstan, in: Amineh, Mehdi Parvizi/ Guang, Yang (eds): Secure oil and alternative energy: the geopolitics of energy paths of China and the European Union, Leiden: Koninklijke Brill, 2012, pp. 167–196; Syroezhkin, Konstantin: China's presence in the energy sector of Central Asia, in: Central Asia and the Caucasus, 2012 (vol. 13), no. 1, pp. 20–43.
11 It is telling that in his article, Sadri primarily addresses the attitudes of the major foreign powers towards Azerbaijan and reserves only one page for Azerbaijan's foreign policy goals. Sadri, Houman: Elements of Azerbaijan foreign policy, in: Journal of Third World Studies, 2003 (vol. 20), no. 1, pp. 179–192. The same is true, for example, for Yesdauletova, Ardak: Kazakhstan's energy policy: its evolution and tendencies, in: Journal US-China Public Administration, 2009 (vol. 6), no. 4, pp. 31–39, which refers to Russia, the US, China, and the EU and does not devote a single paragraph to Kazakhstan's own policy preferences. The same applies to Baizakova, Kuralai: Energy security issues in the foreign policy of the Republic of Kazakhstan, in: American Foreign Policy Interests, 2010 (vol. 32), no. 2, pp. 103–109. A slight exception are Azizian, Rouben/ Bainazarova, Elnara: Eurasian response to China's rise: Russia and Kazakhstan in search of optimal China policy, in: Asian Politics & Policy, 2012 (vol. 4), no. 3, pp. 377–399, who deal to some extent with Kazakhstan's perception of China (and Russia's, respectively).

0. Introduction: Export Pipelines in Eurasia

caused some researchers to examine the foreign policy of Azerbaijan in its own right. Kjærnet concludes that

> Azerbaijan has exploited its new situation to position itself as an increasingly powerful and independent actor in the South Caucasus. Due to its energy independence, Azerbaijan has been able to do this without suffering any reprisal from the Russian side.[12]

However, none of these analyses attempt to explain the formation of foreign policy positions or decision-making processes, but they instead examine the impact these positions have on the geopolitical balance of power in the region.[13]

Few studies consider the domestic actors involved in policy-making processes in the Caspian states. These studies either neglect the foreign policy dimension and instead analyse informal networks and the lack of transparency in business transactions[14] or do not address energy issues.[15] Related public debates have not been included in any analysis conducted thus far. These gaps in the literature are not restricted to the Caspian region. In general, the International Relations literature, first, tends to underestimate the role of small powers,[16] and second, domestic foreign policy debates[17] and decision-making processes continue to receive much less attention than actual foreign policy and the comparative dimension is clearly underdeveloped.[18]

The aim of this collected volume is to broaden the scope of analysis by adding further perspectives to the investigation of exports of crude oil and natural gas from

---

12 Kjærnet, Heidi: Azerbaijani-Russian relations and the economization of foreign policy, in: Øverland, Indra/ Kjærnet, Heidi/ Kendall-Taylor, Andrea (eds): Caspian energy policy: Azerbaijan, Kazakhstan and Turkmenistan, London: Routledge, 2010, pp. 150–161, here p. 158.
13 See also Ipek, Pinar: The role of oil and gas in Kazakhstan's foreign policy: looking east or west?, in: Europe-Asia Studies, 2007 (vol. 59), no. 7, pp. 1179–1199; Anceschi, Luca: Turkmenistan's foreign policy: positive neutrality and the consolidation of the Turkmen regime, London: Routledge, 2008; Ipek, Pinar: Azerbaijan's foreign policy and challenges for energy security, in: Middle East Journal, 2009 (vol. 63), no. 2, pp. 227–239; Franke, Anja/ Gawrich, Andrea/ Melnykovska, Inna/ Schweickert, Rainer: The European Union's relations with Ukraine and Azerbaijan, in: Post-Soviet Affairs, 2010 (vol. 26), no. 2, pp. 149–183.
14 Guliyev, Farid/ Akhrarkhodjaeva, Nozima: The trans-Caspian energy route: cronyism, competition and cooperation in Kazakh oil export, in: Energy Policy, 2009 (vol. 37), no. 8, pp. 3171–3182; Lussac, Samuel: The state as an (oil) company? The political economy of Azerbaijan, Coventry: University of Warwick, 2010 (GARNET Working Paper 74), http://www.garnet-eu.org/fileadmin/documents/working_papers/7410.pdf.
15 Cummings, Sally: Eurasian bridge or murky waters between East and West? Ideas, identity and output in Kazakhstan's foreign policy, in: Journal of Communist Studies and Transition Politics, 2003 (vol. 19), no. 3, pp. 139–155; Bukkvoll, Tor: Astana's privatized independence: private and national interests in the foreign policy of Nursultan Nazarbayev, in: Nationalities Papers, 2004 (vol. 32), no. 3, pp. 631–650.
16 For an elaboration of this point, see Kassimeris, Christos: The foreign policy of small powers, in: International Politics, 2009 (vol. 46), no. 1, pp. 84–101.
17 An exception for the former Soviet Union is Fawn, Rick: Ideology and national identity in post-communist foreign policies, in: Journal of Communist Studies and Transition Politics, 2003 (vol. 19), no. 3, pp. 1–41.
18 For a concise summary of these deficits, see Breuning, Marijke: Bringing 'comparative' back to foreign policy analysis, in: International Politics, 2004 (vol. 41), no. 4, pp. 618–628.

the Caspian Sea region; therefore, the authors apply a wider range of theories to the issue of export pipelines and energy policy. Thus, it examines the securitisation of energy issues, includes the interests and strategies of national political and business actors in the analysis and addresses the formation of energy policies in the countries of the region.

0.1. Pipelines in General

Global conventional energy resources (crude oil and natural gas) are geographically highly concentrated in the so-called 'strategic ellipse'—running roughly from Russia over the Caspian region to the Persian Gulf. It contains approximately 70% of the world's proven oil and gas reserves.[19] In modern, industrial societies, which are generally net energy importers, energy imports have become essential for economic welfare and political stability. Thus, the cross-border oil and gas trade has grown significantly over the last 60 years; an increasing amount of these resources has been transported via pipelines.[20]

The literature has generally neglected the economic and financial drivers of pipeline construction, despite that politics and diplomacy alone cannot ensure the construction of pipelines that do not have a strong financial and business rationale. Pipelines are built if all parties involved benefit from their construction.[21]

While they require substantial upfront investments for construction, pipelines are considered a cost-effective means of transporting oil and gas in the long run. The cost structure is characterised by high fixed costs and low variable costs. Thus, total costs are largely independent of the throughput of the pipeline, and large pipe diameters create economies of scale. As throughput depends on market consumption, long-term supply contracts, especially for gas, are preferred by producers to guarantee that the pipeline will be used until the investment is amortised. Supply contracts often have

---

19 International Energy Agency: Resources to reserves 2013: oil, gas and coal technologies for the energy markets of the future, Paris: IEA, 2013, http://www.oecd-ilibrary.org/energy/resources-to-reserves-2013_9789264090705-en. Unconventional or shale oil and gas are more dispersed globally, see Energy Information Administration: Technically recoverable shale oil and shale gas resources: an assessment of 137 shale formations in 41 countries outside the United States, Washington, DC: US Department of Energy, 2013, http://www.eia.gov/analysis/studies/world shalegas/pdf/fullreport.pdf.
20 United Nations Development Programme/ World Bank: Cross-border oil and gas pipelines: problems and prospects, Washington DC: Energy Sector Management Assistance Programme, 2003, p. 18, http://siteresources.worldbank.org/INTOGMC/Resources/crossborderoilandgaspipelines.pdf.
21 Guillet, Jérôme: How to get a pipeline built: myth and reality, in: Dellecker, Adrian/ Gomart, Thomas (eds.): Russian energy security and foreign policy, London: Routledge, 2011, pp. 58–73, here pp. 58–59.

0. Introduction: Export Pipelines in Eurasia

a 'take-or-pay' clause that requires the consumer to pay for the contracted amount of gas even it is not consumed, guaranteeing an investment return for the supplier.[22]

In many instances, it will be important to differentiate between oil and gas pipelines, as their characteristics, consequences and results often differ. These differentiating factors include the following:[23]

- Natural gas can only be transported by pipeline or liquefied and transported by cryogenic tanker as liquefied natural gas (LNG), making it a much less flexible fuel than oil. The vast majority of crude oil is transported in ocean-going tankers; in addition to pipelines, it can be shipped by rail and trucks.
- Gas pipeline transportation requires very different technical conditions than oil (e.g., precautions must be taken to prevent air from entering the gas pipes, blending stations have to be constructed to harmonise the characteristics of gas from different sources, and gas appliances are geared to accept a uniform calorific value at a uniform pressure).
- The security of supply is more important for gas than for oil because a gas outage involves a much greater technical problem for reconnection.

As a result, transporting natural gas is generally far more expensive and complex than transporting crude oil.[24]

0.2. Pipelines in Eurasia

All of the contemporary, large Eurasian[25] oil and gas exporters (i.e., Azerbaijan, Kazakhstan, the Russian Federation, and Turkmenistan) were part of the Soviet Union, which ceased to exist in 1991. Thus, every analysis of Eurasian export pipelines must begin there.

In the Soviet Union, the construction of export pipelines for both crude oil and natural gas was closely connected to the development of the energy industry and the geological exploration and production cycles of these resources. In general, such

---

22 United Nations Development Programme/ World Bank: Cross-border oil and gas pipelines: problems and prospects, Washington DC: Energy Sector Management Assistance Programme, 2003, pp. 13, 15, http://siteresources.worldbank.org/INTOGMC/Resources/crossborderoilandgaspipelines.pdf; Guillet, Jérôme: How to get a pipeline built: myth and reality, in: Dellecker, Adrian/ Gomart, Thomas (eds.): Russian energy security and foreign policy, London: Routledge, 2011, pp. 58–73, here p. 67.
23 Stern, Jonathan P.: Specters and pipe dreams, in: Foreign Policy, 1982, No 48, pp. 21–36, here p. 24; United Nations Development Programme/ World Bank: Cross-border oil and gas pipelines: problems and prospects, Washington DC: Energy Sector Management Assistance Programme, 2003, pp. xiv, 1, 6, 13, http://siteresources.worldbank.org/INTOGMC/Resources/crossborderoil andgaspipelines.pdf.
24 Guillet, Jérôme: How to get a pipeline built: myth and reality, in: Dellecker, Adrian/ Gomart, Thomas (eds.): Russian energy security and foreign policy, London: Routledge, 2011, pp. 58-73, here p. 60.
25 The term 'Eurasia' as used in this chapter comprises the countries of the former Soviet Union, especially Russia and the Central Asian and Transcaucasian republics.

a production cycle includes the exploration and, if successful, production of the energy resource. To ship the produced oil and/or gas to domestic consumers, pipelines need to be constructed. Only after domestic demand for this energy resource was satisfied and the production prospects for the years to come were positive could the government consider exporting energy resources, which entails the construction of pipelines. The depletion of resources in a production region requires new exploration efforts and the entire cycle begins again. Thus, in the Soviet Union, the export pipeline infrastructure was developed out of the domestic pipeline network.[26]

As the configuration of the domestic energy system came to strongly influence subsequent export patterns,[27] the first part of this chapter traces the production cycles of the Soviet oil and gas industry from the shifting of production regions, to domestic pipeline transportation and eventually the construction of export pipelines (first to CMEA countries[28] and later to Western Europe). This part of the chapter also includes pipelines that were not export pipelines at the time of their construction, as they delivered oil and gas from the production regions to the consumption centres within the Soviet Union. However, these pipelines became export pipelines after the dissolution of the Soviet Union (such as the Bukhara-Urals and Central Asia-Centre gas pipelines or the Atyrau-Samara oil pipeline).

Thus, the Soviet pipeline system still forms the backbone of contemporary export infrastructure in Eurasia. Nevertheless, the main oil and gas exporters, Azerbaijan, Kazakhstan, Russia and Turkmenistan, have constructed numerous new export pipelines, which are discussed in the second part of the chapter.

0.3. Soviet Union[29]

As a planned economy, the planning and construction process for oil and gas pipelines in the Soviet Union was bureaucratic and inefficient; political desirability trumped economic considerations (not to speak of environmental concerns).[30] Kornai observed that

---

26   The connection between production regions and huge export pipeline projects was the strongest in the Soviet gas industry.
27   Cf. e.g., Stern, Jonathan P.: Specters and pipe dreams, in: Foreign Policy, 1982, No 48, pp. 21–36, here p. 22; Högselius, Per: Red gas: Russia and the origins of European energy dependence, New York: Palgrave Macmillan, 2013, p. 20.
28   The Soviet Union formed the Council for Mutual Economic Assistance (CMEA) in January 1949. It was envisaged that the CMEA would provide all necessary imports and exports for all member states to ensure that they would never rely on imports from other countries. In this article, the term CMEA only refers to its East European member countries, i.e., Poland, Hungary, Czechoslovakia, the German Democratic Republic, Bulgaria and Romania.
29   A brief overview of the Soviet oil and gas sector can be found in Perović, Jeronim: Russlands Aufstieg zur Energiegroßmacht. Geschichte einer gesamteuropäischen Verflechtung, in: Osteuropa, 2013 (vol. 63), no. 7, pp. 5–28.
30   For the construction of gas pipelines, see Högselius, Per: Red gas: Russia and the origins of European energy dependence, New York: Palgrave Macmillan, 2013.

0. Introduction: Export Pipelines in Eurasia

planned economies exhibit the phenomenon of 'soft budget constraints', meaning that chronically loss-making state-owned enterprises are not allowed to fail. Instead, they were consistently bailed out with financial subsidies or other instruments and could assume that they would survive even after chronic losses. This expectation, in turn, influenced their behaviour.[31]

Additionally, the technological knowledge and industrial capabilities of the Soviet Union were not sophisticated enough to meet the geographic and climatic challenges that the oil (and later the gas) industry had to cope with since the late 1950s. With the production regions moving farther away from the consumption centres, the pipeline system connecting them had to traverse greater distances. The increasing distances and a growing demand required larger pipe diameters to make construction viable. However, these large-diameter pipes and the necessary compressor stations could not be manufactured in the Soviet Union or the CMEA member states in sufficient numbers and, more important, in ways that delivered the necessary quality and specifications. The lack of sufficient equipment caused delays in several pipeline projects.[32]

As a solution to this problem, the Soviet leadership resorted to importing Western equipment,[33] especially steel pipes with diameters greater than one metre (or 40 inches) and powerful compressor stations (25 megawatts).[34]

> In the gas industry, Western equipment has consistently provided the leading edge of Soviet advances for over 20 years, supplying the most advanced generation of equipment, especially for transmission, as the Soviet gas industry has moved toward bigger pipe, higher pressure, and longer lines.[35]

However, this 'fast start' option required hard currency and tended to inhibit domestic innovation.[36] To mitigate its hard currency shortage, the Soviet Union increased its exports of natural resources to Western countries and resorted to barter transactions. For instance, oil and gas exports were used to finance the further development of the

---

31 Cf. Kornai, János: Resource-constrained versus demand-constrained systems, in: Econometrica, 1979 (vol. 47), no. 4, pp. 801–819; Kornai, János: Economics of shortage, Amsterdam: North-Holland, 1980; Kornai, János: The soft budget constraint, in: Kyklos, 1986 (vol. 39), no. 1, pp. 3–30.
32 Stent, Angela E.: From embargo to Ostpolitik: the political economy of West German-Soviet relations, 1955–1980, Cambridge: Cambridge University Press, 1981; Högselius, Per: Red gas: Russia and the origins of European energy dependence, New York: Palgrave Macmillan, 2013.
33 Cf. e.g., Holliday, George D.: Transfer of technology from West to East: a survey of sectoral case studies, in: OECD (ed.): East–West technology transfer, Paris: OECD, 1984, pp. 53–94; Crawford, Beverly: Western control of East–West trade finance: the role of U.S. power and the international regime, in: Bertsch, Gary (ed.): The control of East–West trade: power, politics and policy, Durham, NC: Duke University Press, 1988, pp. 280–312.
34 Stent, Angela E.: From embargo to Ostpolitik: the political economy of West German-Soviet relations, 1955–1980, Cambridge: Cambridge University Press, 1981, p. 101; Högselius, Per: Red gas: Russia and the origins of European energy dependence, New York: Palgrave Macmillan, 2013, pp. 23–26, 142, 189.
35 Gustafson, Thane: The Soviet gas campaign: politics and policy in Soviet decisonmaking, Santa Monica, CA: RAND, 1983, p. 84.
36 Ibid.

industry and the pipeline grid. This led to a priority of exports over domestic consumption, resulting in periodic supply shortages in the Soviet Union. Due to the political climate during the Cold War, exports of technology and equipment were often subject to restrictions. However, European companies (partly with the tacit support of their national governments) often attempted to circumvent the sanctions imposed by the US or international organisations.[37]

0.3.1. Oil

The origins of Soviet oil industry can be found in the late 19th century Russian Empire, where crude oil production was concentrated in the Caucasus (Baku, Grozny and Maikop) and along the north and east coasts of the Caspian Sea. Some minor production also occurred in the Fergana valley and on Sakhalin Island. However, the most important oil-producing province was Baku in Azerbaijan. Consequently, an oil pipeline from Baku to the Black Sea port of Batumi was constructed between 1897 and 1905, which enabled oil exports to the world market. It was the only export pipeline at the time; the remainder of the Russian Empire's crude oil was transported by rail and ship. However, only approximately 10% of the crude oil produced was exported.[38]

The Russian civil war of 1917–1921 and the subsequent upheavals considerably disrupted the oil industry and the economy as a whole: domestic demand for crude oil and oil products was very limited in the early days of the Soviet Union. The Soviet Union only began to systematically develop its oil industry and, as a result, substantially increase geological exploration following the implementation of the planned economy in 1927. Between 1930 and 1940, the focus of these explorations was the Volga-Urals region where the Soviet government hoped to discover a new oil-producing province on a par with Baku. In 1938, however, Baku remained by far the most important oil-producing region in the Soviet Union, accounting for more than two-thirds of total oil production.[39]

Due to the limited domestic demand for oil products, an increasing share of Soviet oil was exported after 1923. Thus, by the first five-year plan (1928–1932) a widespread domestic pipeline grid was envisioned: new pipelines would be constructed and measures to promote oil exports were implemented. These exports increased steadily until

---

37 Stent, Angela E.: From embargo to Ostpolitik: the political economy of West German-Soviet relations, 1955–1980, Cambridge: Cambridge University Press, 1981, pp. 93–126; Demidova, Ksenia: The deal of the century: the Reagan administration and the Soviet pipeline, in: Patel, Kiran Klaus/ Weisbrode, Kenneth (eds): European integration and the Atlantic community in the 1980s, Cambridge: Cambridge University Press, 2013, pp. 59–82; Högselius, Per: Red gas: Russia and the origins of European energy dependence, New York: Palgrave Macmillan, 2013, pp. 188–190.
38 Hassmann, Heinrich: Erdöl in der Sowjetunion. Geschichte, Gebiete, Probleme, Hamburg: Indstrieverlag von Hernhaussen KG, 1951, pp. 34–35, 42, 45.
39 Ibid., pp. 52, 62–63, 71, 105.

0. Introduction: Export Pipelines in Eurasia

1932; in the following years, however, exports declined rapidly, and on the eve of the Second World War, they nearly completely ceased because domestic demand had increased substantially during the second five-year plan.[40]

Spurred by intensified exploration during the Second World War, the centre of gravity in the Soviet oil industry gradually shifted from Baku and the North Caucasus to the Volga-Urals region.[41] Until 1950, the production of crude oil in the Volga-Urals region amounted to 28% of total Soviet oil production (or more than ten million tonnes per year), while Baku was still responsible for a production share of 45%.[42]

Overall, Soviet oil production increased rapidly in the 1950s (Table 0-1). However, Baku never recovered its pre-war production levels, and by the mid-1950s, the Volga-Urals region became the most important producing region in the Soviet Union, representing more than 50% of total oil production. By the mid-1960s, the Volga-Urals region accounted for approximately 72% of total Soviet oil production; however, its absolute production level peaked in 1975.[43]

Table 0-1: Soviet Oil and Gas Production (in Millions of Tonnes and Billions of Cubic Metres)

	Oil production	Oil exports	Gas production	Gas exports
1950	37.9	0.3	5.8	0.05
1955	70.8	2.9	9.0	0.16
1960	147.9	17.8	45.3	0.24
1965	242.9	43.4	127.7	0.39
1970	353.0	66.8	197.9	3.3
1975	489.3	93.1	289.3	19.3
1980	603	159.0	435	57.2
1985	595	164.9	643	71.0

Sources: Block, Herbert: Energy syndrome, Soviet version, in: Annual Review of Energy, 1977 (vol. 2), pp. 455–497, here pp. 482, 486; Bethkenhagen, Jochen: Soviet energy: rapid rise in output and exports, in: Economic Bulletin, 1988 (vol. 24), no. 12, pp. 7–13, here p. 8, 11.

With the shift in production regions and the substantial increase in oil output, the Soviet pipeline system evolved with the Volga-Urals region as its focus. By the 1950s, domestic pipeline construction was well underway. Between 1950 and 1955, the length

---

40 Ibid., pp. 62–63, 71, 72–73.
41 The so-called 'second Baku' comprises the regions of Bashkortostan (formerly the Bashkir ASSR), Samara (formerly Kuybyshev) and Saratov. Block, Herbert: Energy syndrome, Soviet version, in: Annual Review of Energy, 1977 (vol. 2), pp. 455–497, here p. 478; Dienes, Leslie/ Shabad, Theodore: The Soviet energy system: resource use and policies, Washington DC: Winston & Sons, 1979, p. 52.
42 Hassmann, Heinrich: Erdöl in der Sowjetunion. Geschichte, Gebiete, Probleme, Hamburg: Indstrieverlag von Hernhaussen KG, 1951, pp. 66, 108.
43 Hodgkins, Jordan A.: Soviet power: energy resources, production and potentials, Englewood Cliffs, NJ: Prentice-Hall, 1961, p. 122; Dienes, Leslie/ Shabad, Theodore: The Soviet energy system: resource use and policies, Washington DC: Winston & Sons, 1979, pp. 51–52.

of pipeline in the Soviet Union doubled to 10,400 kilometres (km); within a decade, the Soviet pipeline grid tripled in size at a length of 16,700km. By the end of the 1970s, 85% of all Soviet crude oil was transported by pipeline, 10% by rail and 5% by ship.[44]

0.3.1.1. The First and Only Large Export Pipeline: Druzhba (1962/1964)

The tremendous increase in output also enabled oil exports because hard currency earnings were needed to finance Soviet imports. However, no new export links were built in the 1950s; only the pipeline to the port of Batumi could potentially be used to export crude oil.[45] In 1956, the Soviet Union decided to build a dedicated marine crude oil export terminal at Tsemesskaya Bay, near Novorossiisk. However, the first berth was not completed until 1964; associated tank farms were completed in 1969.[46]

Nevertheless, by the end of the 1950s, crude oil from the Volga-Urals region was being exported to Czechoslovakia, Yugoslavia, Poland and other countries of the CMEA by rail and ship due to the development and expansion of petrochemical industries in these countries. These oil exports were vital to modernisation and economic growth in Eastern Europe.[47]

To replace the cost-intensive transportation of oil via rail and ship, in December 1958, the CMEA signed an agreement on the construction of a trunk pipeline system carrying Volga-Urals crude oil to new refineries in Płock (Poland), Schwedt (East Germany), Bratislava (Czechoslovakia) and Százhalombatta (Hungary). The construction of the first string of the 'Druzhba' (Friendship) pipeline commenced in 1960, and it was inaugurated in 1962 when the first oil reached Czechoslovakia. In September 1963, the oil deliveries were extended to Hungary, to Poland in November 1963, and to the German Democratic Republic (GDR) in December 1963. The entire pipeline was completed in October 1964. The total length of this pipeline including all of its branches exceeded 6,000km. It started in the Samara (formerly Kuybyshev) region and divided into two branches in Mozyr (Belarus). Its northern branch is 700km long and crosses

---

44 Hodgkins, Jordan A.: Soviet power: energy resources, production and potentials, Englewood Cliffs, NJ: Prentice-Hall, 1961, pp. 123, 127; Dienes, Leslie/ Shabad, Theodore: The Soviet energy system: resource use and policies, Washington DC: Winston & Sons, 1979, pp. 63–64.

45 Hodgkins, Jordan A.: Soviet power: energy resources, production and potentials, Englewood Cliffs, NJ: Prentice-Hall, 1961, pp. 123, 128; Dienes, Leslie/ Shabad, Theodore: The Soviet energy system: resource use and policies, Washington DC: Winston & Sons, 1979, pp. 33, 35.

46 United Nations Development Programme/ World Bank: Cross-border oil and gas pipelines: problems and prospects, Washington DC: Energy Sector Management Assistance Programme, 2003, p. 55, http://siteresources.worldbank.org/INTOGMC/Resources/crossborderoilandgaspipelines.pdf.

47 Hodgkins, Jordan A.: Soviet power: energy resources, production and potentials, Englewood Cliffs, NJ: Prentice-Hall, 1961, p. 109; Dienes, Leslie/ Shabad, Theodore: The Soviet energy system: resource use and policies, Washington DC: Winston & Sons, 1979, p. 35; Balmaceda, Margarita M.: Der Weg in die Abhängigkeit: Ostmitteleuropa am Energietropf der UdSSR, in: Osteuropa, 2004 (vol. 54), no. 9–10, pp. 162–179, here p. 163.

0. Introduction: Export Pipelines in Eurasia 25

Poland into the GDR; its 400km long southern branch runs across Czechoslovakia and into Hungary. The Druzhba pipeline reached its maximal throughput capacity of 50 million tonnes (mt) per year in 1975. In the mid-1970s, the pipeline system was further expanded with the construction of a second line.[48] The pipeline was designed to 'telescope' down in size and, therefore, throughput capacity as the system extended farther from the Russian border.[49]

Since the early 1960s, approximately 25–30% of Soviet oil production was shipped abroad. Thus, between 1960 and 1965, Soviet exports of oil and oil products more than doubled from 8.4mt to 19mt annually.[50]

Additionally, the Druzhba pipeline system also played an important role in serving domestic refineries. A branch line, beginning in Unecha, reached the Belorussian refinery at Novopolotsk in 1965 (with a capacity of 39.2mt annually) and in 1968 was extended to the Latvian oil-export terminal of Ventspils. Since 1980, the northern spur also supplied crude oil to a refinery at Mažeikiai in Lithuania. Another branch line of the Druzhba pipeline, beginning at Nikolskoye (with a capacity of 17mt annually) in Tambov oblast, reached the Ukrainian refinery of Kremenchug in 1974.[51]

The gradual decline of the Volga-Urals production region in the 1970s encouraged exploration in Western Siberia. In contrast to production regions situated in reasonable proximity to the main markets in the European part of the Soviet Union, Western Siberia was nearly unpopulated and had no infrastructure available. 'The onrush of West Siberian oil required further expansion of the pipeline network, with increasingly large pipe diameters'.[52] These new pipelines originating in Western Siberia were then connected to the Druzhba system.

---

48 Russell, Jeremy: Energy considerations in Comecon policies, in: The World Today, 1976 (vol. 32), no. 2, pp. 39–48, here p. 41; Dienes, Leslie/ Shabad, Theodore: The Soviet energy system: resource use and policies, Washington DC: Winston & Sons, 1979, p. 64; Balmaceda, Margarita M.: Der Weg in die Abhängigkeit: Ostmitteleuropa am Energietropf der UdSSR, in: Osteuropa, 2004 (vol. 54), no. 9–10, pp. 162–179, here p. 163.
49 United Nations Development Programme/ World Bank: Cross-border oil and gas pipelines: problems and prospects, Washington DC: Energy Sector Management Assistance Programme, 2003, p. 55, http://siteresources.worldbank.org/INTOGMC/Resources/crossborderoilandgaspipelines.pdf.
50 Dienes, Leslie/ Shabad, Theodore: The Soviet energy system: resource use and policies, Washington DC: Winston & Sons, 1979, p. 35; Balmaceda, Margarita M.: Der Weg in die Abhängigkeit: Ostmitteleuropa am Energietropf der UdSSR, in: Osteuropa, 2004 (vol. 54), no. 9–10, pp. 162–179, here p. 163.
51 Dienes, Leslie/ Shabad, Theodore: The Soviet energy system: resource use and policies, Washington DC: Winston & Sons, 1979, pp. 64–65; United Nations Development Programme/ World Bank: Cross-border oil and gas pipelines: problems and prospects, Washington DC: Energy Sector Management Assistance Programme, 2003, p. 67, http://siteresources.worldbank.org/INTOGMC/Resources/crossborderoilandgaspipelines.pdf.
52 Dienes, Leslie/ Shabad, Theodore: The Soviet energy system: resource use and policies, Washington DC: Winston & Sons, 1979, pp. 56–57, quote p. 65.

The smaller oil-production region in Kazakhstan was also connected to the Druzhba system via the Atyrau–Samara pipeline. Since 1970, a 691km-long pipeline with an annual capacity of 13mt, starting at the Uzen oil field, links Atyrau (formerly Gur'yev) with Samara.[53]

0.3.1.2. Smaller Export Pipelines

In addition to the Druzhba system, a smaller export pipeline from Samara to the Black Sea export terminal of Novorossiisk was completed in 1974. The tanker port had an annual capacity of 30–35mt of crude oil at the time.[54]

> In 1977 another direct pipeline, of 48-inch diameter and 676 miles long, was completed between Kuybyshev and the Ukrainian refinery at Lisichansk. It was then extended to the Kremenchug refinery, also in the Ukraine, and to the Odessa oil terminal on the Black Sea, which has been handling a growing volume of exports.[55]

Table 0-2: Soviet Oil Export Pipelines

Pipeline	Route	Year commissioned	Designed capacity	Pipeline operator
Atyrau–Samara*	(Uzen)–Atyrau–Samara	1970	13mt	Glavtransneft
Baku–Batumi	Baku–Tbilisi–Batumi	1905	N/A	Glavtransneft
Druzhba (northern branch)	Samara–Unecha–Gomel–Mozyr–Płoch–Schwedt	1962–64	50mt (1975), later extended to 81mt 43mt at the Belarussian border, 27mt at the German border	Glavtransneft
Druzhba (southern branch)	Samara–Unecha–Gomel–Mozyr–Brody–Drogobych–Uzhgorod–Slovakia or Hungary	1962–64	50mt (1975), later extended to 81mt 38mt at the Belarussian border, 27mt at the Slovak and Hungarian border	Glavtransneft
Nikolskoye–Kremenchug*	Nikolskoye–Kremenchug	1974	17mt	Glavtransneft

---

53 Ibid., pp. 54–55. In this chapter, the pipeline will be called the Atyrau–Samara pipeline because in Atyrau the production of several Kazakh oil fields (and not just from the Uzen field) was pooled and transported north to Samara.
54 Dienes, Leslie/ Shabad, Theodore: The Soviet energy system: resource use and policies, Washington DC: Winston & Sons, 1979, p. 67.
55 Ibid.

0. Introduction: Export Pipelines in Eurasia

Pipeline	Route	Year commissioned	Designed capacity	Pipeline operator
Samara–Novorossiisk	Samara–Volgograd–Tikhoretsk–Novorossiisk	1974	N/A	Glavtransneft
Samara–Odessa	Samara–Velikotsk–Lisishansk–Kremenchug–Odessa	1977	Samara–Lisichansk: 82mt Velikotsk–Odessa: 16mt	Glavtransneft
Unecha–Novopolotsk*	Unecha–Novopolotsk; 1968 extended to Ventspils	1965	39.2mt	Glavtransneft

*Denotes a pipeline that was not used for exports during the Soviet period but was used for this purpose after its dissolution.
Sources: see text and United Nations Development Programme/ World Bank: Cross-border oil and gas pipelines: problems and prospects, Washington DC: Energy Sector Management Assistance Programme, 2003, p. 67, http://siteresources.worldbank.org/INTOGMC/Resources/crossborderoilandgaspipelines.pdf.

Difficulties plagued the Soviet energy sector in the late 1970s: a declining reserve base led to dramatically slowing growth rates in oil production; simultaneously, problems occurred in the coal industry, which was intended to compensate for declines in oil production.[56] Thus, the Soviet government decided to prioritise natural gas production and increase gas output by 45% by 1985.[57] As a result, the natural gas industry unintentionally became the new star of the Soviet energy sector.

0.3.2. Gas

Until the Second World War, the natural gas industry in the Soviet Union was in its infancy. Gas was only extracted in the form of associated gas accruing during oil production; such production was centred in Saratov.[58] With the annexation of Galicia during the Second World War and its integration into the Ukrainian SSR, the Soviet Union came into possession of producing gas fields.[59]

---

56 Coal production failed to meet expectations (that is, plan targets) due to outdated technology and underinvestment. Gustafson, Thane: The Soviet gas campaign: politics and policy in Soviet decisonmaking, Santa Monica, CA: RAND, 1983, p. 30.
57 Ibid, p. 29; Green, Milford B./ Sagers, Matthew J.: Changes in Soviet natural gas flows: 1970–1985, in: Professional Geographer, 1985 (vol. 37), no. 3, pp. 310–319, here p. 310.
58 Hassmann, Heinrich: Erdöl in der Sowjetunion. Geschichte, Gebiete, Probleme, Hamburg: Indstrieverlag von Hernhaussen KG, 1951, p. 107; Hodgkins, Jordan A.: Soviet power: energy resources, production and potentials, Englewood Cliffs, NJ: Prentice-Hall, 1961, p. 135; Dienes, Leslie/ Shabad, Theodore: The Soviet energy system: resource use and policies, Washington DC: Winston & Sons, 1979, p. 69.
59 Högselius, Per: Red gas: Russia and the origins of European energy dependence, New York: Palgrave Macmillan, 2013, p. 13.

Natural gas only emerged as an important source of energy in the Soviet Union at the end of the 1950s, as thermoelectric stations became the largest consumers of gas. Future production plans accorded natural gas a prominent place in the energy balance of consumed fuels. The gasification of 160 cities was planned, and a network of 26,000km of pipelines was scheduled for installation between 1959 and 1965.[60]

By the end of the 1950s, approximately 87% of the known reserves of natural gas were located in the European part of the Soviet Union (chiefly in the Ukrainian SSR, the Volga region, the North Caucasus, the Azerbaijan SSR and the Komi ASSR), and the remainder was located in the recently discovered Bukhara-Khiva gas fields in Uzbekistan.[61]

However, prior to the late 1960s, the Soviets had no large-scale export plans; the country was still a net importer of gas.[62] Gas imports from Afghanistan and Iran freed Soviet gas supplies for export without the need to exploit Siberian gas fields.[63] 'Despite increased production levels, the country only became a net exporter of natural gas in 1974, when imports from Iran and Afghanistan were exceeded by exports to Europe'.[64]

0.3.2.1. Domestic Pipelines

Until the 1950s, this natural gas was only consumed locally due to a lack of transport infrastructure. Only two major pipelines from Saratov to Moscow (inaugurated in 1946) and from the Dashava gas field in Galicia to Kiev (inaugurated in 1948 and subsequently extended eastward, reaching Moscow in 1951) existed for the supply of distant centres; interregional gas pipelines were few in number. The Samara–Moscow and Dashava–Kiev–Moscow pipelines were a response to necessities rather than products of long-term planning.[65]

---

60   Hodgkins, Jordan A.: Soviet power: energy resources, production and potentials, Englewood Cliffs, NJ: Prentice-Hall, 1961, p. 145.
61   Ibid., p. 138.
62   Ibid., p. 145.
63   Block, Herbert: Energy syndrome, Soviet version, in: Annual Review of Energy, 1977 (vol. 2), pp. 455–497, here p. 490; Green, Milford B./ Sagers, Matthew J.: Changes in Soviet natural gas flows: 1970–1985, in: Professional Geographer, 1985 (vol. 37), no. 3, pp. 310–319, here p. 312; Estrada, Javier/ Bergesen, Helge Ole/ Moe, Arild/ Sydnes, Anne Kristin: Natural gas in Europe: markets, organisation and politics, London: Pinter Publisher, 1988, p. 174; Högselius, Per: Red gas: Russia and the origins of European energy dependence, New York: Palgrave Macmillan, 2013, p. 40.
64   Stern, Jonathan P.: Soviet energy prospects in the 1980s, in: The World Today, 1980 (vol. 36), no. 5, pp. 188–195, here p. 191.
65   Hassmann, Heinrich: Erdöl in der Sowjetunion. Geschichte, Gebiete, Probleme, Hamburg: Indstrieverlag von Hernhaussen KG, 1951, pp. 84–85; Hodgkins, Jordan A.: Soviet power: energy resources, production and potentials, Englewood Cliffs, NJ: Prentice-Hall, 1961, p. 142; Dienes, Leslie/ Shabad, Theodore: The Soviet energy system: resource use and policies, Washington DC: Winston & Sons, 1979, p. 76; Högselius, Per: Red gas: Russia and the origins of European energy dependence, New York: Palgrave Macmillan, 2013, pp. 13–15.

0. Introduction: Export Pipelines in Eurasia

The Soviet government only began to devote greater attention to natural gas in the mid-1950s, and the surge in gas exploration and production was accompanied by the construction of numerous long-distance pipelines.[66] 'Rather than connecting a certain gas field with a specific user region, the aim was from now on to build pipelines with an eye toward a coherent whole'.[67] Such an integrated system of pipelines would provide consumption centres with access to gas from different sources; it would also ensure the political and economic integration of newly annexed territories.[68]

In 1956, the decision was made to connect the Dashava gas field in western Ukraine with Minsk in Belarus and the new Baltic Soviet republics of Lithuania and Latvia. The supply to Minsk began in 1960, to Lithuania in 1961 and to Latvia in 1962. In 1972, an additional pipeline from Valdai (near Torzhok) to Riga was commissioned to supply Latvia with gas from Komi and Siberia. However, the diversification was of limited practical value for Latvia because of problems associated with supplying sufficient volumes of gas from Komi.[69]

In addition to gas fields in the Ukraine, natural gas was shipped from the oil-producing regions in Azerbaijan, the North Caucasus and the Volga-Urals region to the consumption centres in Moscow and St. Petersburg (formerly Leningrad), and to the Black Sea coast, in the 1950s and 1960s. While one major pipeline from Saratov to Moscow had come online in 1946, the Central Pipeline System from the gas fields in the Stavropol region of the North Caucasus to Moscow was completed in late 1956 and was extended to St. Petersburg in 1959.[70] The few existing long-distance pipelines served the purpose of supplying fuel to major Soviet cities; industrial centres were generally supplied locally with associated gas.[71]

Moreover, the Soviets also commissioned an interconnecting pipeline grid between the North and South Caucasus: in 1959, a pipeline from Stavropol to Grozny was constructed, with an extension to Tbilisi in 1963. In 1959, the pipeline from Baku to Tbilisi became operational and was extended to Yerevan in 1960. In 1971, the pipeline between Baku and Tbilisi was extended to accommodate Iranian gas imports.[72]

---

66 Högselius, Per: Red gas: Russia and the origins of European energy dependence, New York: Palgrave Macmillan, 2013, pp. 15, 20.
67 Ibid., p. 21.
68 Ibid.
69 Ibid., pp. 21, 146–147.
70 Dienes, Leslie/ Shabad, Theodore: The Soviet energy system: resource use and policies, Washington DC: Winston & Sons, 1979, p. 75.
71 Högselius, Per: Red gas: Russia and the origins of European energy dependence, New York: Palgrave Macmillan, 2013, p. 20.
72 Dienes, Leslie/ Shabad, Theodore: The Soviet energy system: resource use and policies, Washington DC: Winston & Sons, 1979, pp. 74–75; Högselius, Per: Red gas: Russia and the origins of European energy dependence, New York: Palgrave Macmillan, 2013, p. 22.

0.3.2.1.1. Bukhara–Urals Pipeline

'The importance of the Central Asian region as a potential producer of natural gas emerged in the middle 1950s, but began to be translated into actual production for the Central Russian market only a decade later'.[73] The first Central Asian regional pipeline connected the Dzharkak and Mubarek gas fields in Uzbekistan with Kyrgyzstan and southern Kazakhstan; the first section was completed in 1960, and the second reached Tashkent in 1968 and was extended to Bishkek (formerly Frunze) in 1970 and Almaty (formerly Alma-Ata) in 1971.[74]

The discovery of substantial gas reserves in Uzbekistan (from the Gazli field near Bukhara) that neither Uzbekistan nor the other Central Asian republics could absorb initiated the construction of a twin pipeline with an annual capacity of 21 bcm to the industrial cities of Chelyabinsk and Yekaterinburg (formerly Sverdlovsk) in the Urals. Construction began in 1960, and the first gas reached Chelyabinsk in 1963 and Yekaterinburg in 1965.[75]

After the dissolution of the Soviet Union, the Bukhara–Urals pipeline was retired by Uzbekistan. In September 1997, Kazakhstan reversed the flow of its section of the pipeline to import Russian gas to replace some of the Turkmen gas that formerly arrived via the Central Asia–Centre pipeline (see below).[76] In 1999, Turkmen gas began to flow through the Central Asia–Centre pipeline again and the Bukhara–Urals pipeline was retired (see the chapter by Barkanov). In 2001, the lack of maintenance on the Central Asia–Centre pipeline caused Uzbekistan to re-open the Bukhara–Urals pipeline to transit increasing volumes of Turkmen gas. The pipeline operates at a capacity of only approximately 5.0bcm annually and needs refurbishing.[77]

---

73   Dienes, Leslie/ Shabad, Theodore: The Soviet energy system: resource use and policies, Washington DC: Winston & Sons, 1979, p. 79.
74   Ibid. Occasionally, this intraregional gas pipeline is incorrectly considered a part of the Central Asia–Centre pipeline. Instead, it supplies consumption centres in Kyrgyzstan and southern Kazakhstan with gas from Uzbekistan.
75   Yenikeyeff, Shamil M.: Kazakhstan's gas: export markets and export routes, Oxford: Oxford Institute for Energy Studies, 2008, p. 36, http://www.oxfordenergy.org/wpcms/wp-content/uploads/2010/11/NG25-KazakhstansgasExportMarketsandExportRoutes-ShamilYenikeyeff-2008.pdf; Grigoriev, Leonid: Russia, Gazprom and the CAC, in: Dellecker, Adrian/ Gomart, Thomas (eds): Russian energy security and foreign policy, London: Routledge, 2011, pp. 147–169, here p. 150; Högselius, Per: Red gas: Russia and the origins of European energy dependence, New York: Palgrave Macmillan, 2013, pp. 22–23.
76   International Energy Agency: Caspian oil and gas: the supply potential of Central Asia and Transcaucasus, Paris: IEA, 1998, p. 225; Yenikeyeff, Shamil M.: Kazakhstan's gas: export markets and export routes, Oxford: Oxford Institute for Energy Studies, 2008, p. 36, http://www.oxfordenergy.org/wpcms/wp-content/uploads/2010/11/NG25-KazakhstansgasExportMarketsandExportRoutes-ShamilYenikeyeff-2008.pdf.
77   EIA Country Analysis Briefs, Uzbekistan, 19 January 2012, http://www.eia.gov/countries/analysisbriefs/cabs/Uzbekistan/pdf.pdf.

0. Introduction: Export Pipelines in Eurasia

0.3.2.1.2. Central Asia–Centre Pipeline (CAC)

'The next step was to connect the emerging European and non-European pipeline systems with each other.' In this regard, connecting Central Asia to Moscow became a primary task.[78]

In 1965, construction began on the first pipeline delivering Central Asian gas to the European part of the Soviet Union, the Central Asia–Central Russia or Central Asia–Centre (CAC) transmission system; it was inaugurated in 1967 with the opening of the first line from the Gazli field in Uzbekistan to Moscow, a distance of more than 2,690km. The CAC was expanded in five stages. The second stage of the CAC pipeline was completed in 1970; the combined capacity of the first two lines was 25bcm annually.[79]

The Shatlyk field in Turkmenistan, discovered in 1968, is the point of origin of the third stage of the CAC pipeline. Production at the Shatlyk field began in 1973, when the first of the two branch lines was laid to join the CAC transmission system; the second string was completed in 1975, increasing the total capacity of the CAC's third stage to 40bcm annually.[80]

The western branch on the CAC pipeline runs along the eastern shore of the Caspian Sea. It originates at Ekarem, near the Turkmen–Iranian border, and runs north. Construction began in 1970 and was finalised in 1976. The pipeline, however, was slow to achieve its designed capacity of 15bcm because of delays in the installation of compressor stations along the route. The pipeline continues via Uzen in Kazakhstan to the Beineu compressor station, where it meets the eastern branch of the CAC.[81]

In the fifth stage, on the eastern branch of the CAC, a pipeline from the Dauletabad gas field (formerly Sovietabad) in Turkmenistan to Khiva in Uzbekistan was commissioned in 1985 and completed between 1986 and 1988 with a capacity of 10bcm annually.[82]

At the end of the 1980s, the CAC pipeline network had a total annual capacity of approximately 90bcm.[83] The five CAC feeder pipelines converge in Beineu in northwest

---

78 Högselius, Per: Red gas: Russia and the origins of European energy dependence, New York: Palgrave Macmillan, 2013, p. 23.
79 Dienes, Leslie/ Shabad, Theodore: The Soviet energy system: resource use and policies, Washington DC: Winston & Sons, 1979, pp. 81–82.
80 Ibid., p. 82.
81 Ibid., p. 84.
82 Götz, Roland: Mythos Diversifizierung. Europa und das Erdgas des Kaspiraums, in: Osteuropa, 2007 (vol. 57), no. 8–9, pp. 449–462, here p. 459, Table 4.
83 Olcott, Martha B.: International gas trade in Central Asia: Turkmenistan, Iran, Russia and Afghanistan, Stanford, CA: Stanford Institute for International Studies, 2004, p. 24, footnote 113, (Working Paper 28), http://iis-db.stanford.edu/pubs/20605/Turkmenistan_final.pdf; Götz, Roland: Mythos Diversifizierung. Europa und das Erdgas des Kaspiraums, in: Osteuropa, 2007 (vol. 57), no. 8–9, pp. 449–462, here p. 457; Yenikeyeff, Shamil M.: Kazakhstan's gas: export markets and export routes, Oxford: Oxford Institute for Energy Studies, 2008, p. 34, http://www.oxfordenergy.org/wpcms/wp-content/uploads/2010/11/NG25-KazakhstansgasExportMarketsandExportRoutes-ShamilYenikeyeff-2008.pdf.

Kazakhstan and run 748km to Aleksandrov Gai, in Saratov oblast, where the CAC pipeline network meets the Soyuz (Orenburg–Uzhgorod) pipeline.[84] At Aleksandrov Gai, the system divides into two arms, one going northwest in the direction of Moscow and the other turning westward to join the North Caucasus–Moscow transmission system at Ostrogozhsk to replenish the southern gas supply after the depletion of the North Caucasus fields.[85]

After the dissolution of the Soviet Union, the CAC carried an average of approximately 35–40bcm of gas per year; in 2008, after prolonged refurnishing, its capacity was estimated at not more than 47bcm.[86] The western branch of the CAC system (the Ekarem–Beineu pipeline), particularly its northernmost segment, is severely underutilised due to declining gas production. It currently only has an annual capacity of 1bcm.[87]

0.3.2.2. Gas Exports[88]

By 1944, a small export pipeline that had been constructed by Nazi Germany the year before was transporting gas from the Dashava gas field in Galicia to Poland.[89] However, far-reaching export plans were not developed before the discovery of the Shebelinka gas field.

0.3.2.2.1. Bratstvo Pipeline

The Shebelinka gas field, which became operational in 1956 and contained explored reserves of 530 billion cubic metres (bcm), made the Ukraine the Soviet Union's principle

---

84 At Makat, some 150km from Atyrau on the Kazakh Caspian Sea coast, a 28bcm per year line branches off the CAC (Ekarem–Beineu) system, heads southeastward to Atyrau and continues along the north coast of the Caspian Sea via Astrakhan, Kalmykia and Dagestan in Russia to link up with the North Caucasian pipeline network. International Energy Agency: Caspian oil and gas: the supply potential of Central Asia and Transcaucasus, Paris: IEA, 1998, p. 255.
85 Dienes, Leslie/ Shabad, Theodore: The Soviet energy system: resource use and policies, Washington DC: Winston & Sons, 1979, p. 84; Götz, Roland: Mythos Diversifizierung. Europa und das Erdgas des Kaspiraums, in: Osteuropa, 2007 (vol. 57), no. 8–9, pp. 449–462, here p. 457.
86 Götz, Roland: Mythos Diversifizierung. Europa und das Erdgas des Kaspiraums, in: Osteuropa, 2007 (vol. 57), no. 8–9, pp. 449–462, here p. 457; Yenikeyeff, Shamil M.: Kazakhstan's gas: export markets and export routes, Oxford: Oxford Institute for Energy Studies, 2008, p. 35, http://www.oxfordenergy.org/wpcms/wp-content/uploads/2010/11/NG25-KazakhstansgasExportMarkets andExportRoutes-ShamilYenikeyeff-2008.pdf; Grigoriev, Leonid: Russia, Gazprom and the CAC, in: Dellecker, Adrian/ Gomart, Thomas (eds): Russian energy security and foreign policy, London: Routledge, 2011, pp. 147–169, here p. 150.
87 International Energy Agency: Caspian oil and gas: the supply potential of Central Asia and Transcaucasus, Paris: IEA, 1998, p. 254; Götz, Roland: Mythos Diversifizierung. Europa und das Erdgas des Kaspiraums, in: Osteuropa, 2007 (vol. 57), no. 8–9, pp. 449–462, here p. 459, Table 4.
88 On Soviet gas exports to Western Europe, see Heinrich, Andreas: Internationalisation, market structures and enterprise behavior: the Janus-faced Russian gas monopoly Gazprom, in: Liuhto, Kari (ed.): East goes West: the internationalization of Eastern enterprises, Lappeenranta: University of Technology, 2001, pp. 51–87, especially pp. 64–68.
89 Högselius, Per: Red gas: Russia and the origins of European energy dependence, New York: Palgrave Macmillan, 2013, p. 14.

0. Introduction: Export Pipelines in Eurasia 33

gas-producing region in the 1960s and early 1970s. This development gave rise to a high-capacity Ukrainian gas transmission system and generated the first significant exports of natural gas, first to Eastern and later Western Europe.[90]

The development of the Ukraine as the main gas province in the Soviet Union led to an agreement between the Soviet Union and Czechoslovakia in 1964 on the construction of a gas pipeline for exports of Soviet gas.[91] For this purpose, the 'Bratstvo' (Brotherhood) export pipeline system was constructed. Its first section opened in 1967, carrying gas from Ukrainian gas fields to Czechoslovakia and Austria in 1968; a separate line also delivered gas to Poland. Bratstvo was extended in 1973 to transport gas to the GDR, as well as West Germany (1973) and Italy in 1974. Deliveries to Bulgaria began in 1974, to Hungary in 1976, to Yugoslavia in 1979 and to Romania in 1980.[92]

The decision to export gas to the West was made in 1966. Exports to Austria in 1968 were possible because of the existing pipeline infrastructure in Czechoslovakia; the section of the Bratstvo pipeline had been completed a year earlier, reaching Bratislava in June 1967.[93]

The gas fields in western Ukraine had been charged with the task of supplying Czechoslovakia until Siberian gas became available. However, the Ukrainian gas industry was already declining in the late 1970s. Supplies of Siberian gas, which were intended to replace Ukrainian gas for export, were delayed.[94] As a result, the Soviet Union was unable to fulfil its export obligations due to a lack of long-distance pipelines that could bring abundant gas from Central Asia or Siberia to Ukraine and the Bratstvo pipeline; before 1970, this resulted in repeated delivery failures. Thus, Galician gas fields had been overexploited, resulting in rapidly decreasing production volumes. Simultaneously, the demand in Belarus, Lithuania and Latvia was growing.[95]

To mitigate these supply problems, a plan for a new pipeline from the eastern Ukrainian Shebelinka gas field to Kiev and then on to Galicia was developed in 1967;

---

90   Dienes, Leslie/ Shabad, Theodore: The Soviet energy system: resource use and policies, Washington DC: Winston & Sons, 1979, pp. 75–76.
91   Högselius, Per: Red gas: Russia and the origins of European energy dependence, New York: Palgrave Macmillan, 2013, p. 35.
92   Dienes, Leslie/ Shabad, Theodore: The Soviet energy system: resource use and policies, Washington DC: Winston & Sons, 1979, p. 76; Balmaceda, Margarita M.: Der Weg in die Abhängigkeit: Ostmitteleuropa am Energietropf der UdSSR, in: Osteuropa, 2004 (vol. 54), no. 9–10, pp. 162–179, here p. 163.
93   Block, Herbert: Energy syndrome, Soviet version, in: Annual Review of Energy, 1977 (vol. 2), pp. 455–497, here p. 491; Stern, Jonathan P.: Specters and pipe dreams, in: Foreign Policy, 1982, no. 48, pp. 21–36, here p. 22; Högselius, Per: Red gas: Russia and the origins of European energy dependence, New York: Palgrave Macmillan, 2013, pp. 9, 90.
94   Dienes, Leslie/ Shabad, Theodore: The Soviet energy system: resource use and policies, Washington DC: Winston & Sons, 1979, pp. 75–76; Högselius, Per: Red gas: Russia and the origins of European energy dependence, New York: Palgrave Macmillan, 2013, p. 90.
95   Högselius, Per: Red gas: Russia and the origins of European energy dependence, New York: Palgrave Macmillan, 2013, pp. 93–95.

the pipeline reached Kiev in the winter of 1969 and Galicia in the summer of 1970. However, the pipeline proved too small to compensate for the steep decline in Galician gas production. In reaction, Moscow chose to prioritise exports over domestic supplies; as a result, Belarus suffered a supply shortfall.[96]

Additionally, a trunk pipeline system from the Shebelinka gas field to the industrial centres of Dnepropetovsk and Kharkov and Odessa was opened in 1966 and from there to the Moldovan SSR; it was extended to Bulgaria and Romania between 1974 and 1978.[97]

The Shebelinka gas field had to play a larger role in Soviet exports than originally planned due to delays in the exploration of Siberian gas fields. Shebelinka gas was supposed to be exported to Czechoslovakia and delivered to Belarus and the Baltic republics. As the field already showed signs of decline, arrangements were made for the export of Central Asian gas.[98]

'The Bratstvo pipeline, which supplied Czechoslovakia and Austria, was no longer sufficient, but had to be complemented by additional lines'.[99] Thus, to enable exports to Austria, Italy, France and East and West Germany, a double pipeline was laid parallel to the existing Bratstvo pipeline from Shebelinka to Uzhgorod. Construction began in 1971.[100] Due to construction delays in 1973, exports to West Germany were initially transported through the original Bratstvo pipeline. The use of the Bratstvo pipeline for exports to West Germany in 1973, however, left the Ukrainian SSR undersupplied with gas; Belarus also suffered supply problems. In early 1974, the last section of the troubled Shebelinka–Uzhgorod export pipeline became operational.[101]

Natural gas from the Orenburg field in the Volga-Urals region was used to repay CMEA countries for their contributions to the construction of the Bratstvo pipeline; the gas was delivered via the Orenburg–Uzhgorod pipeline (this so-called Soyuz pipeline is part of the Bratstvo system), the construction of which began in 1974, was completed in 1978 and reached full capacity in 1980.[102] The Soyuz pipeline helped double Soviet gas exports to CMEA countries.[103]

---

96 Ibid., pp. 100–102.
97 Dienes, Leslie/ Shabad, Theodore: The Soviet energy system: resource use and policies, Washington DC: Winston & Sons, 1979, p. 76; Högselius, Per: Red gas: Russia and the origins of European energy dependence, New York: Palgrave Macmillan, 2013, pp. 21, 144.
98 Högselius, Per: Red gas: Russia and the origins of European energy dependence, New York: Palgrave Macmillan, 2013, p. 144.
99 Ibid., p. 135.
100 Ibid., pp. 151, 154.
101 Ibid., pp. 159, 160, 163.
102 Dienes, Leslie/ Shabad, Theodore: The Soviet energy system: resource use and policies, Washington DC: Winston & Sons, 1979, pp. 76, 77–78; Högselius, Per: Red gas: Russia and the origins of European energy dependence, New York: Palgrave Macmillan, 2013, p. 171.
103 Balmaceda, Margarita M.: Der Weg in die Abhängigkeit: Ostmitteleuropa am Energietropf der UdSSR, in: Osteuropa, 2004 (vol. 54), no. 9–10, pp. 162–179, here p. 171.

0. Introduction: Export Pipelines in Eurasia   35

0.3.2.2.2. Northern Lights Pipeline
Before the discovery of the giant Vuktyl' gas field near Ukhta in 1964, the Komi ASSR had been a minor, but significant, element in the Soviet gas industry. Production at Vuktyl' began in 1968, reaching a volume of 17.8bcm in 1975. The development of the Vuktyl' field initiated the construction of another major gas transmission system, 'Northern Lights' (Siyanie Severa), which was designed to transport natural gas from the Komi region across the European part of the Soviet Union through Torzhok (at the intersection with the Moscow–St. Petersburg pipeline) and Minsk to the export point of Uzhgorod on the Czechoslovak border. Construction began in 1967 with the possibility of integrating the pipeline into a Siberian network. It became operational in 1969 when the line from Vuktyl' reached Torzhok; it was extended to Minsk by 1974. A second line to Torzhok was completed in 1973, and the first branch from West Siberia began feeding into the Northern Lights system in 1976. The Northern Lights system had a total annual capacity of 55bcm, with 20bcm coming from Vuktyl' and more than 30bcm from West Siberia.[104]

0.3.2.2.3. Finland Connector
In 1971, Finland and the Soviet Union signed the first delivery contract for 1.4bcm of gas per year beginning in January 1974; however, the earmarked Siberian gas was not yet available.[105] 'As a result, Finland's gas-consuming industries competed with gas users in St. Petersburg (formerly Leningrad) and the Baltics for scarce Komi and Caucasian gas'.[106] Although the pipeline was expanded to a capacity of 20bcm per year in 1999, it remains underutilised, as gas demand in Finland has not increased as expected and a pipeline extension to Sweden has not materialised.[107]

0.3.2.2.4. Yamal (Urengoi–Uzhgorod) Pipeline
In the mid-1960s, a series of deposits were discovered in Siberia (Gubkinsk and Zapolyarny in 1965, Komsomolsk and Urengoi in 1966, Medvezhie in 1967, and Yamburg

---

104 Dienes, Leslie/ Shabad, Theodore: The Soviet energy system: resource use and policies, Washington DC: Winston & Sons, 1979, p. 86; Högselius, Per: Red gas: Russia and the origins of European energy dependence, New York: Palgrave Macmillan, 2013, p. 138.
105 Högselius, Per: Red gas: Russia and the origins of European energy dependence, New York: Palgrave Macmillan, 2013, p. 163; see also Estrada, Javier/ Bergesen, Helge Ole/ Moe, Arild/ Sydnes, Anne Kristin: Natural gas in Europe: markets, organisation and politics, London: Pinter Publisher, 1988, pp. 173–174; Stern, Jonathan P.: Soviet oil and gas exports to the West: commercial transaction or security threat?, Aldershot: Gower, 1989, p. 31.
106 Gazprom: Helsinki hosts celebrations dedicated to 30-year anniversary of Russian gas deliveries to Finland, press relaese, 24 September 2004, http://www.gazprom.com/press/news/2004/september/article62878/; Högselius, Per: Red gas: Russia and the origins of European energy dependence, New York: Palgrave Macmillan, 2013, pp. 147, 163.
107 Heinrich, Andreas: Russlands Exportpipelines: Diversifizierung oder Bestandssicherung?, Russland-Analysen, no. 217, 25 March 2011, pp. 18–23, here, p. 23, http://www.laender-analysen.de/russland/pdf/Russlandanalysen217.pdf.

in 1969) of a size unlike any previously identified in the Soviet Union. This convinced the authorities that this gas has to be piped westward.[108] Consequently, as gas production in many fields of the European USSR had been peaking since the early 1970s, the centre of the industry increasingly shifted to Western Siberia, where the first giant gas field began operations in 1972. Production from these fields began slowly; nevertheless, it marked the beginning of the 'Siberian period'.[109]

These new Siberian supplies were injected into the existing pipeline network.[110]

> The long-term strategy for the transmission of Urengoy gas to the European USSR and for export appeared to envisage three basic pipeline routes: (1) the Northern Lights system diagonally across the European USSR to the Czechoslovak border for export; (2) the existing transmission system for Medvezh'ye gas southwest to the Urals and on to Perm' and Kazan' with a projected extension to Yelets, joining the old North Caucasus–Moscow system and on to Kiev; (3) the new southerly route through Surgut and Chelyabinsk with an extension to Kuybyshev. This system may ultimately also be extended to the Czechoslovak border to provide additional export capacity.[111]

In early 1978, the Soviet Union expressed an interest in new export agreements with Western Europe.

> Following up on the principle deployed in the Soyuz project—in which gas from Orenburg in central Russia would be exported to communist Central Europe—it was proposed that a dedicated export pipeline be constructed that would not be used for any other purpose than shipments to the West. This meant that export flows and domestic flows would be separated from each other to a much greater extent than in previous East–West projects.[112]

The new project was intended to exploit the large Yamburg gas field in the Yamal-Nenets Autonomous Region. Thus, the pipeline was termed the Yamal pipeline. The Yamal pipeline to European Russia was 1,860km long and inaugurated in late 1983; it had a designed annual throughput capacity of 40bcm. Because the Yamburg gas field

---

108 Dienes, Leslie/ Shabad, Theodore: The Soviet energy system: resource use and policies, Washington DC: Winston & Sons, 1979, p. 87; Balmaceda, Margarita M.: Der Weg in die Abhängigkeit: Ostmitteleuropa am Energietropf der UdSSR, in: Osteuropa, 2004 (vol. 54), no. 9–10, pp. 162–179, here p. 169; Högselius, Per: Red gas: Russia and the origins of European energy dependence, New York: Palgrave Macmillan, 2013, p. 40.
109 Dienes, Leslie/ Shabad, Theodore: The Soviet energy system: resource use and policies, Washington DC: Winston & Sons, 1979, p. 68; Victor, Nadejda M./ Victor, David G.: Bypassing Ukraine: exporting Russian gas to Poland and Germany, in: Victor, David G./ Jaffe, Amy M./ Hayes, Mark H. (eds): Natural gas and geopolitics: from 1970 to 2040, Cambridge: Cambridge University Press, 2006, pp. 122–168, here pp. 128–129.
110 Victor, Nadejda M./ Victor, David G.: Bypassing Ukraine: exporting Russian gas to Poland and Germany, in: Victor, David G./ Jaffe, Amy M./ Hayes, Mark H. (eds): Natural gas and geopolitics: from 1970 to 2040, Cambridge: Cambridge University Press, 2006, pp. 122–168, here pp. 129–130.
111 Dienes, Leslie/ Shabad, Theodore: The Soviet energy system: resource use and policies, Washington DC: Winston & Sons, 1979, p. 92.
112 Högselius, Per: Red gas: Russia and the origins of European energy dependence, New York: Palgrave Macmillan, 2013, pp. 179–180, quote p. 180.

0. Introduction: Export Pipelines in Eurasia 37

could not be brought online in time, gas from the Urengoi field was used to fulfil supply contracts. Thus, the pipeline was renamed the Urengoi–Uzhgorod export pipeline.[113]

Delivering gas to Europe through a northern route via Belarus and Poland had been under discussion in the Soviet Union as early as 1978 and was driven by a keen interest in tapping the large Yamal fields. The Yamal project was delayed again in 1981 in favour of focusing on the Urengoi field, already under development and at the time the largest producing gas field in the world. Once Urengoi was in production, next in line logically was the Yamburg field just to the northwest."[114]

Table 0-3: Soviet Gas Export Pipelines

Pipeline	Route	Year commissioned	Designed capacity	Operator
Bratstvo (northern branch, including Soyuz)	Shebelinka–Kharkov–Uzhgorod–Slovakia	1967	100bcm	Mingazprom
Bratstvo (southern branch)	Shebelinka–Dnepropetrovsk–Izmail–Moldova–Romania–Bulgaria	1966, 1974–78	28bcm	Mingazprom
Bukhara–Urals*	Gazli (Bukhara)–Taldyk–Chelyabinsk–Yekaterinburg	1963–65	21bcm	Mingazprom
Central Asia–Centre*	Gazli/ Shatlyk/ Ekarem/ Dauletabad–Aleksandrov Gai–Petrovsk–Algasovo–Moscow	1967–85	90bcm	Mingazprom
Finland Connector	St. Petersburg–Helsinki	1974	20bcm	Mingazprom
Northern Lights	Vuktyl'–Ukhta–Torzhok–Smolensk–Minsk–Uzhgorod	1969	55bcm	Mingazprom
Progress	Yamburg–Uzhgorod	1989	30bcm	Mingazprom
Yamal	Urengoi–Uzhgorod	1983	40bcm	Mingazprom

*Denotes a pipeline that was not used for exports during the Soviet period but was used for this purpose after its dissolution.
Sources: see text and Götz, Roland: Mythos Diversifizierung. Europa und das Erdgas des Kaspiraums, in: Osteuropa, 2007 (vol. 57), no. 8–9, pp. 449–462, here p. 460, Table 5.

In 1985, a multilateral agreement with the CMEA countries was signed for the construction of the 'Progress' pipeline from the giant Yamburg gas field in Western Siberia to Uzhgorod; the pipeline has an annual capacity of 30bcm and runs parallel to the Yamal (Urengoi–Uzhgorod) pipeline. For their financial and technical support, the

---
113 Ibid., pp. 197–198.
114 Victor, Nadejda M./ Victor, David G.: Bypassing Ukraine: exporting Russian gas to Poland and Germany, in: Victor, David G./ Jaffe, Amy M./ Hayes, Mark H. (eds): Natural gas and geopolitics: from 1970 to 2040, Cambridge: Cambridge University Press, 2006, pp. 122–168, here p. 146.

CMEA countries received natural gas from the Yamburg field from 1989 onward for a period of ten years.[115]

0.4. Post-Soviet Eurasia

With the dissolution of the Soviet Union, the geopolitical situation of energy exports from Eurasia became vastly more complex. First, while Russia remained the largest single producer and reserve holder, a substantial amount of natural gas and crude oil production and reserves passed to the control of the newly independent states of Central Asia and the Caucasus (especially Azerbaijan, Kazakhstan and Turkmenistan). Second, the unified oil and gas pipeline systems were broken up, and each new state of the former Soviet Union (FSU) gained control of the pipelines on its territory, as did the former CMEA members in Central and Eastern Europe, because the export infrastructure passed through their territory. This gave those states control over export pipelines to Europe, making them transit states for the export of both oil and gas from the FSU. However, Russia retained nearly total control over the ability of Central Asian energy producers to export to the West, as all of their export pipelines had to pass through Russia before entering the transit states.[116] However, Russia was reluctant to relinquish market share to the new competitors, especially for natural gas; it significantly limited access to the Russian pipeline system.[117]

The lack of adequate export infrastructure was one of the most pressing problems facing the energy sectors of the FSU countries. This situation hampered the inflow of needed foreign direct investments into their energy sectors; thus, the construction of new export pipelines became a priority in these countries.[118] 'However, most routing options are fraught with technical, financial, legal and/or political difficulties'.[119]

---

115 Bethkenhagen, Jochen: Soviet energy: rapid rise in output and exports, in: Economic Bulletin, 1988 (vol. 24), pp. 7–13, here p. 7; Balmaceda, Margarita M.: Der Weg in die Abhängigkeit: Ostmitteleuropa am Energietropf der UdSSR, in: Osteuropa, 2004 (vol. 54), no. 9–10, pp. 162–179, here p. 171; Victor, Nadejda M./ Victor, David G.: Bypassing Ukraine: exporting Russian gas to Poland and Germany, in: Victor, David G./ Jaffe, Amy M./ Hayes, Mark H. (eds): Natural gas and geopolitics: from 1970 to 2040, Cambridge: Cambridge University Press, 2006, pp. 122–168, here p. 143.
116 Cf. e.g., International Energy Agency: Caspian oil and gas: the supply potential of Central Asia and Transcaucasus, Paris: IEA, 1998, p. 36; Laurila, Juhani: Transit transport between the European Union and Russia in light of Russian geopolitics and economics, in: Emerging Markets Finance and Trade, 2003 (vol. 39), no. 5, pp. 27–57; Ericson, Richard E.: Eurasian natural gas: significance and recent developments, in: Eurasian Geography and Economics, 2012 (vol. 53), no. 5, pp. 615–648.
117 International Energy Agency: Caspian oil and gas: the supply potential of Central Asia and Transcaucasus, Paris: IEA, 1998, p. 36.
118 Ibid.
119 Ibid.

0. Introduction: Export Pipelines in Eurasia

In addition to the political problems caused by the dissolution of the Soviet Union and the unified pipeline network, pipeline construction in the FSU had to cope with economic considerations that had been largely neglected under the planned economy. With the transition to a market-based economy, pipeline projects could no longer be realised through political will alone; there had to be an underlying economic and business rationale.[120] With economic considerations coming to the fore, the phenomenon of 'soft budget constraints' had to be overcome. Economies in transition are especially affected by this phenomenon, as they intend to reform and are thus more heavily reliant on market mechanisms for which the economy is not yet prepared. This often resulted in a near economic collapse.[121]

To develop the countries' energy sectors and construct the necessary export infrastructures, foreign partners were needed for technology and knowledge transfers (e.g., drilling technology, pipes, compressors) and project finance (e.g., through foreign direct investments and guarantees of commercial loans). These circumstances thus stress the need for cost efficiency and a solid business model for the construction of export pipelines to attract such foreign partners.

0.4.1. Russia

Of the FSU oil and gas producers, the Russian Federation was the least affected by the upheaval of the dissolution of the Soviet Union. Because the oil and gas production in Soviet Union was concentrated in the Russian SSR, its production regions were sufficiently developed to ensure production to satisfy domestic demand and exports. As the largest republic in the Soviet Union, Russia also in possessed most of the Soviet pipeline infrastructure.

Nevertheless, conditions for the transport of natural gas and oil from Russia to Western Europe also changed radically. The newly independent states introduced transit fees that made Russian oil and gas exports more expensive. Ukraine in particular has attempted to exploit its near monopoly position on Russian energy transit to offset its weak position as a customer for Russian gas and a debtor. Because of long-lasting quarrels with Ukraine, Russia developed plans for alternative transit routes to break Ukraine's transit monopoly and reduce transit across FSU countries to the greatest extent possible.

---

120 Cf. Guillet, Jérôme: How to get a pipeline built: myth and reality, in: Dellecker, Adrian/ Gomart, Thomas (eds.): Russian energy security and foreign policy, London: Routledge, 2011, pp. 58–73.
121 Cf. e.g., Frydman, Roman/ Gray, Cheryl/ Hessel, Marek/ Rapaczynski, Andrzej: The limits of discipline: ownership and hard budget constraints in the transition economies, in: Economics of Transition, 2000 (vol. 8), no. 3, pp. 577–601; Kornai, János: Hardening the budget constraint: the experience of the post-socialist countries, in: European Economic Review, 2001 (vol. 45), no. 9, pp. 1573–1599; Maskin, Eric/ Xu, Chenggang: Soft budget constraint theories: from centralization to the market, in: Economics of Transition, 2001 (vol. 9), no. 1, pp. 1–27.

0.4.1.1. Oil

The Soviet-era Druzhba pipeline continues to represent the backbone of Russian oil exports. However, Russia sought to expand its energy exports, bypass transit countries and enter new consumer markets. The Baltic Pipeline System (BPS) is a case in point. Until 2006, the Lithuanian refinery Mažeikių Nafta and the Latvian oil terminal Ventspils on the Baltic coast were supplied through a Druzhba branch line from Unecha on the Belarusian border. Since then, however, Russian oil has been exported through a Russian port on the Baltic Sea using the new BPS pipeline.

0.4.1.1.1. Baltic Pipeline System (BPS)
The Baltic Pipeline System consists of two different pipeline routes: BPS I runs from the Druzhba pipeline in Yaroslavl in central Russia to the oil terminal of Primorsk north of St. Petersburg, and BPS II runs from the Druzhba pipeline in Unecha on the Russian–Belarusian border to the oil terminal of Ust-Luga west of St. Petersburg.[122]

In an effort to create a second major oil export facility on Russian territory after Novorossiisk,

> the Russian state promoted the Baltic pipeline extension against the protestations of private industry, which felt that the extension would not be commercially optimal compared with other alternatives. These other options, however, would have involved other states downstream of the Russian border.[123]

Under discussion since 1995, construction of BPS I began in May 2000. In December 2001, the new export terminal opened and the pipeline became operational, while work on a second line began in September 2002. The first pipeline has an annual capacity of 12mt; the second line increased this capacity to a total of 30mt per year. BPS I has been extended since; its overall annual throughput capacity is now 76.5mt of oil.[124]

The construction of the BPS II oil pipeline, designed for a total capacity of 50mt annually, began in 2009. The new pipeline will be constructed in two stages. The first branch, which has an annual capacity of 30mt of oil, became operational in the first

---

[122] United Nations Development Programme/ World Bank: Cross-border oil and gas pipelines: problems and prospects, Washington DC: Energy Sector Management Assistance Programme, 2003, p. 119, http://siteresources.worldbank.org/INTOGMC/Resources/crossborderoilandgas pipelines.pdf.

[123] United Nations Development Programme/ World Bank: Cross-border oil and gas pipelines: problems and prospects, Washington DC: Energy Sector Management Assistance Programme, 2003, p. 31, http://siteresources.worldbank.org/INTOGMC/Resources/crossborderoilandgaspipelines.pdf.

[124] United Nations Development Programme/ World Bank: Cross-border oil and gas pipelines: problems and prospects, Washington DC: Energy Sector Management Assistance Programme, 2003, pp. 123–124, http://siteresources.worldbank.org/INTOGMC/Resources/crossborderoilandgas pipelines.pdf; Laurila, Juhani: Transit transport between the European Union and Russia in light of Russian geopolitics and economics, in: Emerging Markets Finance and Trade, 2003 (vol. 39), no. 5, pp. 27–57, here p. 46; Interfax, 22 March 2012.

0. Introduction: Export Pipelines in Eurasia 41

quarter of 2012. The second branch will expand the pipeline's capacity by 8mt in 2014, resulting in a total export capacity of 38mt; capacity for another 12mt per year are set aside for delivery to the large Kirishi refinery in Leningrad oblast.[125]

0.4.1.1.2. East Siberia–Pacific Ocean (ESPO)
While Russian oil exports via pipeline have thus far only been shipped westward, a pipeline running from eastern Siberia to the Russian Pacific coast is not only intended to connect eastern Siberian oil fields to the national pipeline grid but also to gain access to new consumer markets in the Asia-Pacific region. The first section of the East Siberia–Pacific Ocean (ESPO I) pipeline from Taishet to Skovorodino was finished at the end of 2009; this section is 2,700km long and has an annual capacity of 50mt (after its expansion from 30mt).[126]

The second section (ESPO II) from Skovorodino to the oil terminal at Kozmino Bay near Nakhodka was completed in December 2012 and is 2,100km long and has a total annual capacity of 50mt. Prior to its completion, oil for export was transported by rail from Skovorodino to the Kozmino oil terminal.[127]

Table 0-4: Russia's Oil Export Pipelines

Pipeline	Route	Year commissioned	Designed capacity	Operator
BPS I	Yaroslavl–Primorsk	2001	76.5mt	Transneft
BPS II	Unecha–Ust-Luga	2012/ 2014	38mt (excluding 12mt for the Kirishi refinery)	Transneft
ESPO I	Taishet–Skovorodino	2009	50mt	Transneft
ESPO II	Skovorodino–Kozmino Bay (near Nakhodka)	2012	50mt	Transneft
ESPO-China	Skovorodino–Chinese border	2010	15mt	Transneft

Sources: see text.

The completion of the first section of the ESPO pipeline enabled the construction of an export pipeline to China. In February 2009, the China Development Bank granted the Russian state oil company Rosneft and the state-owned Russian oil pipeline operator Transneft a loan of US$25 billion for the delivery of 15mt of oil per year to China over a period of 20 years. The loan enabled the Russian companies to build a branch line

---

125 Socor, Vladimir: Russia completing Baltic Pipeline System construction, reducing Druzhba pipeline flow, in: Eurasia Daily Monitor, 24 February 2012 (vol. 9), no. 39, http://www.jamestown.org/single/?no_cache=1&tx_ttnews[tt_news]=39055&tx_ttnews[backPid]=7&cHash=7a5cfb4f5d3ce3909b4d0a2ca2e5a267.
126 Cf. http://www.transneft.ru/projects/118/10020/; http://www.transneft.ru/projects/118/10709/ (accessed 6 September 2013).
127 Cf. http://www.transneft.ru/projects/118/10709/ (accessed 6 September 2013).

from the ESPO pipeline to China and develop and exploit the eastern Siberian oil fields that are intended to feed this pipeline. The 64km branch line from Skovorodino to China was completed in December 2010 and has an initial annual capacity of 15mt, which will be increased to 30mt of oil per year. This branch line connects to a 960km-long pipeline to the Chinese refinery in Daquing.[128]

0.4.1.2. Gas

As conflicts with transit countries resulting in supply interruptions have been the most damaging for Russia and its gas monopoly Gazprom, the construction of pipelines that would bypass the most troublesome transit countries (i.e., Ukraine) or avoid transit countries altogether became a priority.

0.4.1.2.1. Yamal–Europe Gas Pipeline
The Yamal–Europe gas pipeline begins at in Torzhok (Tver oblast), where it receives gas from Western Siberia, and runs across Belarus to Poland and Germany. The current overall length of this gas pipeline exceeds 2,000km. The original intention was to construct two lines with a total capacity of 66bcm; however, capacity was adjusted to meet substantially lower actual market demand (especially in Poland).

The construction of the Yamal–Europe gas pipeline began in 1994 close to the German and Polish borders. The pipeline was constructed in a stepwise process from west to east, partly using existing pipelines. The first sections of the pipeline between Poland and Germany were brought online as early as 1996. However, Gazprom's payment problems, impeding the delivery of pipes and causing the withdrawal of international credit after the financial crisis in 1998, delayed construction. The pipeline became operational in 2000, and it took until 2006 to reach its full capacity of 33bcm per year after the last compressor station had been installed.[129]

While Gazprom is the sole owner of the Belarusian gas pipeline section, the Polish section was built by EuRoPol GAZ, a joint venture of Gazprom and Polish PGNiG. The German end of the project was overseen by Wingas, a joint venture between Gazprom and Wintershall.[130]

128　Cf. http://www.transneft.ru/projects/118/10019/ (accessed 6 September 2013); Energy Charter Secretariat: Bringing oil to the market: transport tariffs and underlying methodologies for cross-border crude oil and products pipelines, Brussels: Energy Charter Secretariat, 2012, p. 72.
129　Petroleum Economist (1998), no. 10, p. 36; Victor, Nadejda M./ Victor, David G.: Bypassing Ukraine: exporting Russian gas to Poland and Germany, in: Victor, David G./ Jaffe, Amy M./ Hayes, Mark H. (eds): Natural gas and geopolitics: from 1970 to 2040, Cambridge: Cambridge University Press, 2006, pp. 122–168, here pp. 152–154.
130　Cf. Gazprom, http://www.gazprom.com/about/production/projects/pipelines/; http://www.gazpromquestions.ru/?id=2#c536 (accessed 6 September 2013).

0. Introduction: Export Pipelines in Eurasia   43

The construction of the second string of the gas pipeline (a project called Yamal–Europe II) was discussed in one form or another for several years but shelved by Gazprom in the mid-2000s.[131] However, in April 2013, Russian President Vladimir Putin tasked Gazprom with exploring the possibility of implementing Yamal–Europe II, envisaging the construction of a gas pipeline from the Belarusian border via Poland to Slovakia. The throughput is forecasted to be no less than 15bcm per year.[132]

0.4.1.2.2. Blue Stream Gas Pipeline[133]
While the Yamal-Europe pipeline reduced the number of transit countries and excluded troublesome Ukraine, Russia soon began to plan pipelines that would avoid transit countries altogether. In December 1997, Russia and Turkey signed an intergovernmental agreement to supply of 365bcm of gas to Turkey over 25 years. For this purpose, the Blue Stream (Goluboi Potok) gas pipeline was constructed to directly supply Russian gas to Turkey and bypass transit countries. The 1,213km long gas pipeline consists of an overland and a subsea section. It begins in the vicinity of Dzhugba (Stavropol krai) and ends at the Durusu terminal in Turkey. The construction of the offshore section began in September 2001. The submerged section of the pipeline is 393km long at depths of up to 2,150 metres. The construction of the pipeline was completed in December 2002. However, although it has been fully operational since February 2003, the official inauguration ceremony was delayed until November 2005 because Turkey demanded a price revision and lower supply volumes. Even before this dispute, Turkey had negotiated with Gazprom to reduce by half the gas flow contracted to start in 2003.[134]

The gas pipeline's design capacity is 16bcm of gas per year. Thus far, Blue Stream has been underutilised. Nevertheless, in 2010, Gazprom attempted to persuade Ankara to endorse a plan to lay a second line under the Black Sea to double the annual capacity of the original Blue Stream pipeline to 30bcm. After this attempt failed, Moscow unveiled the alternative 'South Stream' gas pipeline project.[135]

---

131 Cf. e.g., Heinrich, Andreas: Poland as a transit country for Russian natural gas: potential for conflict, Koszalin: KICES, 2007.
132 Cf. Gazprom, http://www.gazprom.com/about/production/projects/pipelines/; http://www.gazpromquestions.ru/?id=2#c536 (accessed 6 September 2013).
133 Cf. Gazprom, http://www.gazprom.com/about/production/projects/pipelines/; http://www.gazpromquestions.ru/?id=2#c536 (accessed 6 September 2013).
134 Victor, Nadejda M./ Victor, David G.: Bypassing Ukraine: exporting Russian gas to Poland and Germany, in: Victor, David G./ Jaffe, Amy M./ Hayes, Mark H. (eds): Natural gas and geopolitics: from 1970 to 2040, Cambridge: Cambridge University Press, 2006, pp. 122–168, here pp. 143–144; Krauer-Pacheco, Ksenia: Turkey as a transit country and energy hub: the link to its foreign policy aims, Bremen: Forschungsstelle Osteuropa, 2011, p. 34, http://www.forschungsstelle.uni-bremen.de/UserFiles/file/fsoAP118.pdf.
135 Krauer-Pacheco, Ksenia: Turkey as a transit country and energy hub: the link to its foreign policy aims, Bremen: Forschungsstelle Osteuropa, 2011, p. 34, http://www.forschungsstelle.uni-bremen.de/UserFiles/file/fsoAP118.pdf.

0.4.1.2.3. Nord Stream Gas Pipeline[136]
Building on the (at least technical) success of the Blue Stream project, the Nord Stream gas pipeline constitutes another new export route for Russian gas that bypasses all transit countries. Nord Stream runs 1,224km across the Baltic Sea from Portovaya Bay (near Vyborg) to the German coast at Greifswald.

Construction of the first line with a throughput capacity of 27.5bcm per year was completed in September 2011. In November 2011, the gas pipeline's first line was commissioned, and commercial gas supplies to European consumers began. The work on Nord Stream's second line commenced in May 2011 and was completed in April 2012; it was commissioned in October 2012. The second line increased the gas pipeline's annual total capacity from 27.5 to 55bcm.

The Nord Stream project is administered by Nord Stream AG, a joint venture among Gazprom (51%), BASF/Wintershall and E.ON with a shareholding of 15.5% each, and Gasunie and GDF Suez with 9% each. The Nord Stream project participants are contemplating the possibility of constructing the third and fourth lines, which they consider economically and technically viable.

0.4.1.2.4. South Stream Gas Pipeline
In similar vein, the South Stream pipeline project (Yuzhn Potok), led by Gazprom and Italy's Eni, was initiated in June 2007 and will transport Russian gas directly to, among others, Bulgaria, Austria, Hungary, Greece and Italy. The South Stream pipeline begins on the Russian Black Sea coast near Anapa (Krasnodar krai) and runs beneath the Black Sea to the Pasha Dere terminal near Varna, Bulgaria. Its offshore section will be 925km long, reaching a maximum depth of 2,250 metres. South Stream is proposed to have an annual capacity of 63bcm of gas; it is expected to cost between 15–25 billion Euros and be operational by 2015, reaching full capacity in 2019.[137]

Russia argues that this pipeline project is designed to avoid the recurring gas disputes with Ukraine and, together with the Nord Stream pipeline, is part of diversification strategy to avoid transit bottlenecks. However, South Stream must also be regarded

---

136 Cf. Gazprom, http://www.gazpromquestions.ru/?id=2#c536; see also http://www.nord-stream.com/?r=1 (accessed 6 September 2013).
137 Kusznir, Julia: The Nabucco gas pipeline project and its impact on EU energy policy in the South Caucasus, in: Caucasus Analytical Digest, 12 December 2011, no. 33, pp. 9–13, here p. 11, http://www.isn.ethz.ch/Digital-Library/Publications/Detail/?ots591=0c54e3b3-1e9c-be1e-2c24-a6a8c7060233&lng=en&id=135318; Krauer-Pacheco, Ksenia: Turkey as a transit country and energy hub: the link to its foreign policy aims, Bremen: Forschungsstelle Osteuropa, 2011, pp. 34–35, http://www.forschungsstelle.uni-bremen.de/UserFiles/file/fsoAP118.pdf; Sidar, Cenk/Winrow, Gareth: Turkey and South Stream: Turco-Russian rapprochement and the future of the Southern Corridor, in: Turkish Policy Quarterly, 2011 (vol. 10), no. 2, pp. 51–61.

0. Introduction: Export Pipelines in Eurasia

as competitor of the Nabucco project for the transport of gas from the Caspian Basin to Europe.[138]

In early 2012, Russian Prime Minister Vladimir Putin secured approval to hasten the South Stream project and begin construction in December 2012, one year earlier than planned. Although the event was symbolic, as construction began in earnest only in October 2013, Russia has clearly demonstrated its determination to expand its market share and discourage competing pipeline projects.[139]

0.4.1.2.5. Altai–China Gas Pipeline

While all of these new export pipelines run westward to European customers, Russia has also attempted to obtain new markets for its gas in Asia. In March 2006, Gazprom and China National Petroleum Corporation (CNPC) signed a memorandum of understanding on natural gas supplies from Russia to China. However, the construction of the planned export pipeline from Western Siberia via Novosibirsk and the Republic of Altai to China has made no progress; the first gas deliveries had to be postponed from 2011 to 2015 because the two parties could not agree on the gas-pricing scheme. The Altai–China pipeline is planned to have an annual capacity of 30bcm with estimated costs of US$12 billion.[140]

Table 0-5: Russia's Gas Export Pipelines

Pipeline	Route	Year of commissioning	Designed capacity	Operator
Altai–China	Purpe (south of Novy Urengoi)–Argan–Volodino–Novosibirsk–Chuisk–China	Under consideration	30bcm	Gazprom
Blue Stream	Dzhugba (Staropol krai)–Durusu terminal (near Samsun)	2005	16bcm	Gazprom
Nord Stream	Vyborg–Greifswald	2011	55bcm	Gazprom
South Stream	Anapa (Krasnodar krai)–Pasha Dere terminal (near Varna)	2015 (plan)	63bcm	Gazprom
Yamal–Europe	Torzhok–Minsk–Wlocawek–Mallnow	2000	33bcm	Gazprom, EuRoPol GAZ, Wingas

Sources: see text.

---

138 Cf. Locatelli, Catherine: Russian and Caspian hydrocarbons: energy supply stakes for the European Union, in: Europe-Asia Studies, 2010 (vol. 62), no. 6, pp. 959–971, here p. 967.
139 Kusznir, Julia: TAP, Nabucco West, and South Stream: the pipeline dilemma in the Caspian Sea basin and its consequences for the development of the Southern Gas Corridor, in: Caucasus Analytical Digest, 18 February 2013, no. 47, pp. 2–8, here p. 2, http://www.isn.ethz.ch/Digital-Library/Publications/Detail/?ots591=0c54e3b3-1e9c-be1e-2c24-a6a8c7060233&lng=en&id=160678.
140 Cf. Gazprom, http://www.gazprom.com/about/production/projects/pipelines/ (accessed 7 September 2013).

0.4.2. Kazakhstan

Unlike Russia, Kazakhstan's oil and gas industry had largely been neglected during the Soviet period, as the Soviet government focused on the Siberian production regions. Thus, the country suffered from a lack of developed deposits and technological knowledge and financing to exploit its reserves. Additionally, the country was completely dependent on Russia to reach any consumer market for its energy resources. As a consequence, the Kazakh government opened its energy sector to foreign investors to modernise its oil and gas industry and construct the necessary export pipelines.

0.4.2.1. Oil

Many Kazakh oil production areas were not connected to refining and consumption centres within the country but were connected to those in neighbouring countries, and vice versa. As a result, much of Kazakhstan's oil was exported to Russian refineries in Samara, Ufa, and Orsk in government-to-government swaps for the Siberian crude that is refined in Kazakhstan's Pavlodar refinery.[141] The domestic pipeline grid remains rudimentary; newly built export pipelines have been utilised to connect oil fields to domestic consumption centres (this situation also applies to the gas sector, see below).

0.4.2.1.1. Atyrau–Samara Pipeline

Until 2001, all oil exported by pipeline had to pass through Russia's pipeline grid. Kazakhstan's main export outlet had been the Soviet-built Atyrau–Samara pipeline to Russia and its pipeline system. From there, Transneft transported Kazakhstan's oil to the ports of Primorsk and Novorossiisk and through the Druzhba pipeline system. Originally designed with a capacity of 13mt, the pipeline's throughput had been reduced to approximately 10mt per year during the mid-1990s. Following an expansion between 1999 and 2001, the annual capacity of the Atyrau–Samara pipeline increased to 15.4mt. There are further plans to expand the pipeline's capacity to 25mt, but formal agreements have yet to be reached.[142]

0.4.2.1.2. Caspian Pipeline Consortium (CPC)

As of mid-1997, the export capacity of the Kazakh pipeline system via Russia was a mere 16mt per year. The creation of an independent export route was therefore a priority for Kazakhstan to transport its projected oil output. The first major project was the

---

141 International Energy Agency: Caspian oil and gas: the supply potential of Central Asia and Transcaucasus, Paris: IEA, 1998, pp. 209–210.
142 Ibid., p. 210; Babali, Tuncay: Prospects of export routes for Kashagan oil, in: Energy Policy, 2009 (vol. 37), no. 4, pp. 1298–1308, here p. 1300; KazMunaiGaz, http://www.kmg.kz/en/manufacturing/oil/atyrau_samara/ and http://www.kmgep.kz/eng/the_company/our_business/transportation_and_sales/ (accessed 25 June 2013).

0. Introduction: Export Pipelines in Eurasia 47

construction of a pipeline to the Russian Black Sea port of Novorossiisk. Although this pipeline passes through Russian territory, it is not under the control of Russia's pipeline operator, Transneft.[143]

The Caspian Pipeline Consortium (CPC) was formed in 1992 by the governments of Kazakhstan and Oman, which were subsequently joined by the Russian government and several oil companies, to construct an export pipeline from Kazakhstan's Tengiz oil field to a new loading facility near Novorossiisk. Protracted negotiations, especially regarding the partners' actual capital investments, delayed the signing of the so-called Shareholders' Agreement until December 1996.[144] An agreement on the technical aspects of the project was signed in May 1997. However, by the beginning of 1998, the project had encountered difficulties in securing rights of way with several regional governments in Russia, as the pipeline was planned to pass through their territory.[145]

Construction of the CPC began in May 1999. The pipeline runs 1,440km around the north coast of the Caspian Sea via Atyrau, Komsomolskaya and Kropotkin to the new Yuzhnaya Ozerevka terminal, 15km north of existing port and storage facilities at Novorossiisk. The CPC employs existing pipeline infrastructure surrounding the Caspian Sea, notably between Tengiz and the Russian city of Astrakhan; only 50% of the pipeline, 750km from Komsomolsk to Novorossiisk (via Tikhoretsk to avoid Chechnya), had to be newly constructed.[146]

Construction work was completed in November 2000; the pipeline became operational in December 2001 with an initial annual capacity of 28mt, to be expanded to 67mt.[147] Moreover, two pipeline spurs from the Kenkiyak and Karachaganak oil fields

---

143 International Energy Agency: Caspian oil and gas: the supply potential of Central Asia and Transcaucasus, Paris: IEA, 1998, p. 211.
144 The new agreement gave Russia a 24% ownership stake; other shares are owned by Kazakhstan (19%) and Oman (7%), with the remaining 50% divided among the oil firms Chevron (15%), LukARCO (12.5%), Rosneft-Shell and Mobil (7.5% each), British Gas and Agip (2% each), Oryx Energy Corporation and Kaz Pipeline Ventures (1.75% each). In November 2008, Transneft bought Oman's shares in the CPC, which are held by CPC Company. Thus, Russia's share increased to 31%. International Energy Agency: Caspian oil and gas: the supply potential of Central Asia and Transcaucasus, Paris: IEA, 1998, p. 214; Babali, Tuncay: Prospects of export routes for Kashagan oil, in: Energy Policy, 2009 (vol. 37), no. 4, pp. 1298–1308, here p. 1301.
145 The pipeline passes through the Astrakhan region, Kalmykia, Krasnodar territory, Stavropol and Novorossiisk. Right-of-way agreements had to be negotiated with the local governments in each of these Russian regions. International Energy Agency: Caspian oil and gas: the supply potential of Central Asia and Transcaucasus, Paris: IEA, 1998, pp. 213–215.
146 Ibid., p. 214; United Nations Development Programme/ World Bank: Cross-border oil and gas pipelines: problems and prospects, Washington DC: Energy Sector Management Assistance Programme, 2003, p. 95, http://siteresources.worldbank.org/INTOGMC/Resources/crossborderoi landgaspipelines.pdf; Babali, Tuncay: Prospects of export routes for Kashagan oil, in: Energy Policy, 2009 (vol. 37), no. 4, pp. 1298–1308, here p. 1301.
147 Capacity will gradually be increased in several phases, primarily through the addition of new pumping stations, to reach 67mt per year by 2014. In 2012, capacity increased from 28mt to 35mt per year. The second phase designed to increase capacity to 48mt is supposed to be

joined the CPC at Atyrau and provide additional supplies.[148] The Russian pipeline monopoly Transneft was designated the operator of CPC and given the responsibility to maintain all facilities, including those on Kazakh territory. However, its role is exclusively technical and does not include the organisation of throughput schedules and allocations among shippers (for additional information on the CPC and its internal organisation, see the chapter by Sorbello).[149]

0.4.2.1.3. Kazakhstan–China Oil Pipeline
While oil production from the Tengiz field had found an outlet with the CPC, other production regions in Kazakhstan had only limited access to the export infrastructure. Unspecified plans to export Kazakh oil to China were developed shortly after the Kazakhstan's independence. However, oil export routes to China only became more likely after CNPC acquired the Kazakh oil producer Aktobemunai and obtained rights to exploit the Uzen oil field in 1997.

As part of CNPC's bid for Uzen, the company promised to submit detailed proposals for the construction of a pipeline to China. The preliminary proposal called for a pipeline stretching some 3,000km from CNPC's Kenkiyak and Zhanazhol fields in the Aktyubinsk region of western Kazakhstan to Xinjiang Province in western China. The first stage was a planned line between Kenkiyak and Kumkol, followed by a continuation to the Chinese border. The initial capacity of this pipeline was envisioned at approximately 20mt per year, eventually increasing to 40mt.[150]

However, critics soon argued that a pipeline from western Kazakhstan to China would need to transport as much as 50mt annually to be economically feasible. To simply fill the first stage of the pipeline, CNPC would have to find additional sources to supplement the 10–13mt per year it would be able to supply from its Aktobemunai and Uzen fields. Consequently, the first pipeline proposal to China was shelved in 1999 due to insufficient oil output.[151]

---

complete in 2013, while the final phase will increase capacity to the designed 67mt by 2014 (cf. http://www.cpc.ru/en/expansion/Pages/default.aspx, accessed 17 September 2013).
148 Babali, Tuncay: Prospects of export routes for Kashagan oil, in: Energy Policy, 2009 (vol. 37), no. 4, pp. 1298–1308, here p. 1301.
149 International Energy Agency: Caspian oil and gas: the supply potential of Central Asia and Transcaucasus, Paris: IEA, 1998, pp. 39, 214, 218; United Nations Development Programme/ World Bank: Cross-border oil and gas pipelines: problems and prospects, Washington DC: Energy Sector Management Assistance Programme, 2003, pp. 94, 101–102, http://siteresources.world bank.org/INTOGMC/Resources/crossborderoilandgaspipelines.pdf.
150 International Energy Agency: Caspian oil and gas: the supply potential of Central Asia and Transcaucasus, Paris: IEA, 1998, p. 218.
151 Ibid.; Peyrouse, Sébastien: Economic aspects of the Chinese-Central Asia rapprochement, Stockholm/ Washington, DC: Central Asia-Caucasus Institute & Silk Road Studies Program, 2007, p. 57, http://www.silkroadstudies.org/new/docs/Silkroadpapers/2007/0709China-Central_Asia. pdf.

0. Introduction: Export Pipelines in Eurasia 49

After the discovery of the giant offshore Kashagan deposit in 2000, Kazakhstan revived the project of a pipeline connecting the Caspian Sea to Xinjiang in 2002. This pipeline from Atyrau to Alashankou is divided into three sections. The first, located in the west of the country, connects the Kenkiyak field to Atyrau over a length of 448km. Operational since 2003, the Kenkiyak–Atyrau section had the capacity to transport 6mt annually but it was extended to 14mt in 2006. It allowed the oil extracted from the Chinese-owned fields in Kenkiyak and Zhanazhol to be transported westward and join the Atyrau–Samara and CPC pipelines for export to European markets, until the direction of flow was reversed after all stages of the Kazakhstan–China pipeline were completed in 2011.[152]

The second section in the east of the country connects Atasu (in Karaganda Region) to Alashankou on the Sino–Kazakh border. Atasu has the advantage of crossing the Omsk–Pavlodar–Chimkent pipeline, thereby allowing for the transport of Russian oil. The Atasu–Alashankou section, 988km in length, was inaugurated in December 2005 and has been operational since May 2006. The third section connects Kenkiyak to the Kumkol fields via the town of Aralsk over a length of 750km and was completed in 2011. The last part of the pipeline, between Kumkol and Atasu, has existed since the Soviet period.[153]

During this first phase, the export capacity of the 2,818km long pipeline is approximately 10mt, but this will increase to 20mt by 2014. However, to be profitable, the Kazakhstan–China pipeline must find additional supplies of oil from Russia (for additional information on this pipeline, see the chapter by Sorbello).[154]

0.4.2.1.4. Kazakhstan–Iran Oil Pipeline via Turkmenistan
The framework agreement signed between CNPC and the Kazakh government in September 1997 also calls for the Chinese company to submit proposals to build an oil pipeline to Iran. The initial stage would be a 200km pipeline from Kazakhstan to the border with Turkmenistan, to be extended later. With a planned capacity of 50mt per year, a length of 1,600km and projected costs of US$1.2 billion, this would be the

---

152 Peyrouse, Sébastien: Economic aspects of the Chinese-Central Asia rapprochement, Stockholm/Washington, DC: Central Asia-Caucasus Institute & Silk Road Studies Program, 2007, p. 57, http://www.silkroadstudies.org/new/docs/Silkroadpapers/2007/0709China-Central_Asia.pdf; Babali, Tuncay: Prospects of export routes for Kashagan oil, in: Energy Policy, 2009 (vol. 37), no. 4, pp. 1298–1308, here p. 1302.
153 Peyrouse, Sébastien: Economic aspects of the Chinese-Central Asia rapprochement, Stockholm/Washington, DC: Central Asia-Caucasus Institute & Silk Road Studies Program, 2007, pp. 57–58, http://www.silkroadstudies.org/new/docs/Silkroadpapers/2007/0709China-Central_Asia.pdf.
154 Ibid., pp. 58, 60.

shortest and least expensive route, but US sanctions against Iran have prevented its construction.[155]

0.4.2.1.5. Trans-Caspian Oil Pipeline from Kazakhstan

As an alternative to pipelines sending Central Asian oil westward via either Russia or Iran, several parties have proposed constructing an oil pipeline beneath the Caspian Sea from Kazakhstan to Azerbaijan. From there, it could be linked to the BTC oil pipeline and thereby reach Western markets. Such a pipeline would likely use some of an existing line near the Tengiz field and cross the Caspian near Kazakhstan's border with Turkmenistan. However, the construction of trans-Caspian pipelines is complicated by environmental concerns and the uncertainty of territorial boundaries in and the legal status of the Caspian Sea.[156]

Table 0-6: Kazakhstan's Oil Export Pipelines

Pipeline	Route	Year commissioned	Designed capacity	Operator
Atyrau–Samara	Atyrau–Samara	1970	13bcm (currently 15bcm)	Kaztransoil, Transneft
CPC	Tengiz–Atyrau–Astrakhan–Komsomolskaya–Kropotkin–Yuzhnaya Ozerevka (near Novorossiisk)	2001	35mt (to be expanded to 67mt by 2014)	Transneft (technical operator; managerial decisions are made by CPC)
Kazakhstan–China	Atyrau–Kenkiyak–Aralsk–Kumkol–Atasu–Alashankou	2011	10mt	Kaztransoil, China National Oil Development Corporation
Kazakhstan–Iran	Kazakhstan–Turkmenistan–Iran	Under consideration	50mt	N/A
Trans-Caspian oil pipeline	Kazakhstan–Azerbaijan	Under consideration	N/A	N/A

Sources: see text and Energy Charter Secretariat: Bringing oil to the market: transport tariffs and underlying methodologies for cross-border crude oil and products pipelines, Brussels: Energy Charter Secretariat, 2012, p. 74.

---

155 International Energy Agency: Caspian oil and gas: the supply potential of Central Asia and Transcaucasus, Paris: IEA, 1998, p. 219; Babali, Tuncay: Prospects of export routes for Kashagan oil, in: Energy Policy, 2009 (vol. 37), no. 4, pp. 1298–1308, here p. 1302.

156 International Energy Agency: Caspian oil and gas: the supply potential of Central Asia and Transcaucasus, Paris: IEA, 1998, pp. 39, 219.

0.4.2.2. Gas

For years, Kazakhstan has examined various routes to deliver gas to international markets, including via Afghanistan to Pakistan and India, beneath the Caspian Sea to Azerbaijan and on to Turkey, via Turkmenistan to Iran and Turkey, and to China. Most of these proposals essentially consist of connection lines to proposed pipelines anchored in Turkmenistan, which is expected to supply the bulk of gas exports from the region.[157] However, only the pipeline to China has been constructed.

0.4.2.2.1. Kazakhstan–China Gas Pipeline

In August 2007, the governments of Kazakhstan and China signed an agreement on the construction and operation of a Kazakhstan–China gas pipeline as one leg of the transcontinental Central Asia–China gas pipeline originating in Turkmenistan.[158] In November 2008, CNPC and the Kazakh state-owned company KazMunaiGaz signed a preliminary agreement on its construction. The Kazakhstan–China gas pipeline begins in Beineu and runs over 1,480km via Bozoi and Kyzylorda to Chimkent with an annual capacity of 10bcm; it will supply 5bcm of gas to China and approximately 5bcm to southern Kazakhstan. Initial Kazakh gas exports to China will rely on gas produced by CNPC at the Zhanazhol field.[159]

This pipeline project was initiated in 2005 and has undergone a number of changes, including the geographical route of the pipeline and its annual capacity. Initially, Kazakhstan and China planned to construct the pipeline with an annual capacity of 30bcm, which was projected to reach up 40bcm by 2015. Since then, the capacity of the pipeline has been decreased to 10bcm per year for economic reasons (i.e., lower than expected gas output, but this could be expanded to 15bcm, depending on demand and supply conditions, according to CNPC).[160] Additionally, a change in the pipeline's route enabled access to a larger number of gas resources and helped to mitigate the gasification problem in southern Kazakhstan.[161]

---

157 Ibid., p. 227.
158 Intergas: Intergas Central Asia celebrates the 50th anniversary of main gas pipelines in the Republic of Kazakhstan, press release, 8 September 2011, http://www.intergas.kz/eng/press/novosti/?cid=0&rid=127.
159 Yenikeyeff, Shamil M.: Kazakhstan's gas: export markets and export routes, Oxford: Oxford Institute for Energy Studies, 2008, p. 68, http://www.oxfordenergy.org/wpcms/wp-content/uploads/2010/11/NG25-KazakhstansgasExportMarketsandExportRoutes-ShamilYenikeyeff-2008.pdf.
160 Ling, Song Yen: China, Kazakhstan complete first stage of new gas pipeline, in: Platts, 9 September 2013, http://www.platts.com/latest-news/natural-gas/singapore/china-kazakhstan-complete-first-stage-of-new-27386568.
161 Apart from potential gas exports to China, the pipeline will help solve the gasification problem in the central, eastern and southern parts of Kazakhstan by linking the main gas routes of central Kazakhstan to a unified system that will provide Kazakhstan with considerable flexibility in supplying gas to different markets. Kazakhstan considers this pipeline to be an energy

Construction began in July 2012, and in September 2013, the first section of the Kazakhstan–China pipeline from Bozoi to Chimkent was completed. In Chimkent, it connects to the Kazakh section of the Central Asia–China pipeline network that has operated since 2009.[162] Completion of the 311km second section from Beineu to Bozoi is expected in 2015.[163]

Table 0-7: Kazakhstan's Gas Export Pipeline

Pipeline	Route	Year of commissioning	Designed capacity	Operator
Kazkhstan–China	Beineu–Bozoi–Kyzylorda–Chimkent–Central Asia–China pipeline	2015 (plan)	10bcm	Intergaz

Sources: see text.

0.4.3. Turkmenistan

Turkmenistan's situation after the dissolution of the Soviet Union was very similar to that of Kazakhstan, with the only exception being that the country had not opened its energy sector to foreign investors. This has hampered the development of its energy sector and the construction of new export pipelines; it even led to the increasing decay of existing infrastructure.

0.4.3.1. Oil

Turkmenistan has little oil production, and its exports are therefore very limited. The country only possesses a small network of oil pipelines, primarily in the western part of the country. Refined products are transported by rail from the two refineries to Ashgabat and other consumption centres in the south of the country. Turkmenistan also ships small amounts of refined products to other FSU states and somewhat larger volumes of crude and products by ship to Iran.[164]

---

security project that will ensure the country's domestic gas demand without the need to rely on imports. Yenikeyeff, Shamil M.: Kazakhstan's gas: export markets and export routes, Oxford: Oxford Institute for Energy Studies, 2008, pp. 66–68, http://www.oxfordenergy.org/wpcms/wp-content/uploads/2010/11/NG25-KazakhstansgasExportMarketsandExportRoutes-ShamilYenikeyeff-2008.pdf.

162 Ling, Song Yen: China, Kazakhstan complete first stage of new gas pipeline, in: Platts, 9 September 2013, http://www.platts.com/latest-news/natural-gas/singapore/china-kazakhstan-complete-first-stage-of-new-27386568; 'Leaders commission section of Kazakhstan-China Gas Pipeline', in: Oil & Gas Journal, 10 September 2013, http://www.ogj.com/articles/2013/09/leaders-com mission-section-of-kazakhstan-china-gas-pipeline.html.

163 'Leaders commission section of Kazakhstan–China Gas Pipeline', in: Oil & Gas Journal, 10 September 2013, http://www.ogj.com/articles/2013/09/leaders-commission-section-of-kazakhstan-china-gas-pipeline.html.

164 International Energy Agency: Caspian oil and gas: the supply potential of Central Asia and Transcaucasus, Paris: IEA, 1998, pp. 250–251.

0.4.3.2. Gas

Turkmenistan operates two separate gas pipeline systems, one in the east and the other in the west of the country. It is primarily based on the Soviet-era Central Asia–Centre pipeline system for which the Turkmen authorities claim an export pipeline capacity of 60–80bcm. However, Turkmenistan's CAC export pipelines are in a precarious state. Since the dissolution of the Soviet Union, these pipelines have lacked maintenance; long stretches are badly corroded and risk cracking if operated at design pressure. Thus, they have been utilised at less than half their designed capacities in recent years.[165]

In April 2009, an accident at the CAC pipeline interrupted Turkmen gas exports to Russia. The failure of Russia and Turkmenistan to agree on the repair of the pipeline led to a fundamental change in Turkmenistan's natural gas strategy: Turkmenistan accelerated the construction of alternative export infrastructures, resulting in pipeline links to Iran and China.[166] Although deliveries to Russia were restored at the end of 2009, they never reached the quantities transported before the accident.[167]

0.4.3.2.1. Gas pipeline from Turkmenistan to Turkey via Iran
Since 1993, the Turkmen authorities have promoted the construction of a gas export pipeline from Turkmenistan via Iran to Turkey. Initially, a pipeline with a capacity of 15bcm per year was planned with a view to supplying the Turkish market. Future upgrades to 28–30bcm annually are envisioned to supply markets farther west. The development of this line is complicated by US sanctions on projects involving Iran. After years of delay, Shell—authorised to form a consortium to build and operate the pipeline—withdrew from the project in April 2003.[168]

0.4.3.2.2. Korpedzhe–Kurt-Kui Gas Pipeline to Iran
In October 1995, certain that the larger pipeline project to Turkey would be slow to develop, the National Iranian Oil Company of Iran decided to begin construction of the Korpedzhe–Kurt-Kui pipeline. In late 1997, Iran completed a 200km, 12bcm[169] per year

---

165 Ibid., pp. 254–255.
166 Horák, Slavomír: Turkmenistan's shifting energy geopolitics in 2009–2011: European perspectives, in: Problems of Post-Communism, 2012 (vol. 59), no. 2, pp 18–30, here p. 22.
167 Ibid., p. 23.
168 International Energy Agency: Caspian oil and gas: the supply potential of Central Asia and Transcaucasus, Paris: IEA, 1998, pp. 40, 256; Olcott, Martha B.: International gas trade in Central Asia: Turkmenistan, Iran, Russia, and Afghanistan, in: Victor, David G./ Jaffe, Amy M./ Hayes, Mark H. (eds): Natural gas and geopolitics: from 1970 to 2040, Cambridge: Cambridge University Press, 2006, pp. 202–233, here pp. 215–216.
169 The information about the pipeline's capacity varies between 8bcm and 13.5bcm per year. Giuli, Marco: Nabucco pipeline and the Turmenistan connumdrum, in: Caucasian Review of International Affairs, 2008 (vol. 2), no. 3, pp. 124–132, here p. 127, http://www.cria-online.

pipeline from Korpedzhe, in southern Turkmenistan, to Kurt-Kui in remote northeastern Iran, where it connects to an existing pipeline to power stations in northwestern Iran and an existing East–West transmission pipeline running from the Khangiran field to Tabriz in northwest Iran. This line may be incorporated into the project to export gas to Turkey. Turkmenistan has seen little revenue from this project for several years, as gas deliveries were used to reimburse Iran for financing 90% of the construction costs of reportedly US$190 million.[170]

The pipeline was opened in December 1997. However the volume of gas transported through the pipeline has fallen short of the Turkmen government's planned goals. Until 2000, Iran had only imported approximately 6bcm through this pipeline;[171] the pipeline remains underutilised.[172]

0.4.3.2.3. Artyk–Lotfabad Gas Pipeline to Iran
In December 2000, another short, low-capacity pipeline connecting Artyk to Lotfabad, in the Iranian province of Khorasan, was put into operation. As in the case of the Korpedzhe–Kurt-Kui pipeline, this line was built through the joint efforts of both countries. According to the Turkmen Minister of Oil and Mines, Khoshgaldi Babaev, the volume of Turkmen natural gas transferred through the pipeline to Iran would reach 15 million cm per year by 2001, which will be increased to 28 million cm in the future.[173]

0.4.3.2.4. Dauletabad–Sarakhs–Khangiran Gas Pipeline to Iran
In early 2010, Turkmenistan launched the third gas link between the two neighbouring countries. The Dauletabad–Sarakhs–Khangiran pipeline to northeastern Iran had

---

org/4_2.html; Pannier, Bruce: Turkmen gas exports to Iran a boon for both countries, in RFE/RL, 6 January 2010, http://www.rferl.org/content/Turkmen_Gas_Exports_To_Iran_A_Boon_For_Both_Countries/1921933.html?page=1#relatedInfoContainer.

170 International Energy Agency: Caspian oil and gas: the supply potential of Central Asia and Transcaucasus, Paris: IEA, 1998, pp. 40, 255; Olcott, Martha B.: International gas trade in Central Asia: Turkmenistan, Iran, Russia, and Afghanistan, in: Victor, David G./ Jaffe, Amy M./ Hayes, Mark H. (eds): Natural gas and geopolitics: from 1970 to 2040, Cambridge: Cambridge University Press, 2006, pp. 202–233, here pp. 213–214; Horák, Slavomír: Turkmenistan's shifting energy geopolitics in 2009–2011: European perspectives, in: Problems of Post-Communism, 2012 (vol. 59), no. 2, pp 18–30, here p. 22.

171 'Opening of Iran–Turkmenistan pipeline welcomed by Turkmen press', IRNA news agency, 7 December 2000, obtained via BBC Monitoring.

172 Olcott, Martha B.: International gas trade in Central Asia: Turkmenistan, Iran, Russia, and Afghanistan, in: Victor, David G./ Jaffe, Amy M./ Hayes, Mark H. (eds): Natural gas and geopolitics: from 1970 to 2040, Cambridge: Cambridge University Press, 2006, pp. 202–233, here p. 214.

173 Ginzburg/ Troschke state a capacity of 6bcm per year. 'Opening of Iran–Turkmenistan pipeline welcomed by Turkmen press', IRNA news agency, 7 December 2000, obtained via BBC Monitoring; Ginzburg, Veniamin/ Troschke, Martina: The exports of Turkmenistan's energy resources: stabilisation without a market economy, in: Central Asia and the Caucasus, 2003, no. 6(24), pp. 108–117, http://www.ca-c.org/journal/2003/journal_eng/cac-06/15.ginen.shtml.

0. Introduction: Export Pipelines in Eurasia 55

an initial capacity of 6bcm of gas per year—subsequently increased to 12bcm[174]—to transport Turkmen gas from the Dauletabad field near the border to the Khangiran gas refinery. Together with the two existing pipelines, the new cross-border link will eventually allow Turkmenistan to supply more than 20bcm of gas to Iran annually. While furthering Ashgabat's efforts to lessen its dependence on Russian-operated export routes, for Iran the new gas supplies will alleviate gas shortages in its northern regions.[175]

0.4.3.2.5. Gas Pipeline to Pakistan and India via Afghanistan (TAPI)
In addition to possible export options via Iran and Turkey, Turkmen authorities also considered tapping the Asian gas market. In 1993, Bridas (Argentina) proposed the construction of a pipeline from Turkmenistan through Afghanistan to Pakistan, with a planned annual capacity of approximately 20bcm and the possibility of extending the line to northern India. Proposed to begin at the Yashlar gas field, the pipeline was to run via western Afghanistan to Chaman on the Afghan–Pakistani border, to the provincial capital of Quetta and finally to Baluchistan in western Pakistan. There, it would connect with Pakistani gas trunk pipelines. The length of the line was to be approximately 1,400km, with construction costs estimated at US$2.5–3 billion.[176]

By late 1995, Bridas lost favour with the Turkmen government. Instead, the Turkmen government signed an agreement with a consortium comprising Unocal (US) and Delta (Saudi Arabia) to build and operate a 1,450km, 20bcm per year pipeline from Turkmenistan's Dauletabad field to Multan in northern Pakistan, where it would be connected to an existing Pakistani pipeline system. The line's route was to be similar to that of the planned Bridas line, with construction costs estimated at US$1.9 billion, with an extension to northern India for an additional US$600 million. The main inno-

---

174 Umbach, Frank: Competing for Caspian energy resources: Russia's and China's energy (foreign) policies and the implications for the EU's energy security, in: Amineh, Mehdi Parvizi/ Guang, Yang (eds): Secure oil and alternative energy: the geopolitics of energy paths of China and the European Union, Leiden: Koninklijke Brill, 2012, pp. 75–114, here p. 90; Mangott, Gerhard: Gestörte Leitungen: Russland und die Gasversorgung der EU, in: Erler, Gernot/ Schulze, Peter W. (eds): Die Europäisierung Russlands: Moskau zwischen Modernisierungspartnerschaft und Großmachtrolle, Frankfurt/ Main: Campus, pp. 189–213, here p. 208.
175 'Another victory for diversification as Turkmenistan opens new gas pipeline to Iran', in: Global Insight, 6 January 2010, http://www.ihs.com/products/global-insight/industry-economic-report.aspx?id=106594710; Pannier, Bruce: Turkmen gas exports to Iran a boon for both countries, in RFE/RL, 6 January 2010, http://www.rferl.org/content/Turkmen_Gas_Exports_To_Iran_A_Boon_For_Both_Countries/1921933.html?page=1#relatedInfoContainer.
176 International Energy Agency: Caspian oil and gas: the supply potential of Central Asia and Transcaucasus, Paris: IEA, 1998, pp. 40, 258; Olcott, Martha B.: International gas trade in Central Asia: Turkmenistan, Iran, Russia, and Afghanistan, in: Victor, David G./ Jaffe, Amy M./ Hayes, Mark H. (eds): Natural gas and geopolitics: from 1970 to 2040, Cambridge: Cambridge University Press, 2006, pp. 202–233, here p. 217.

vation of the Unocal-Delta consortium was its proposal to take delivery of 20bcm per year at the Turkmen–Afghan border and market it at its own risk.[177]

However, the major problem for both pipeline projects is the lack of political stability in Afghanistan, through which both would pass. Questions have also been raised regarding whether future Pakistani demand alone will be sufficient to justify such a pipeline, while any extension to India would entail overcoming political difficulties between the two South Asian neighbours. Eventually, the unresolved nature of the Afghan conflict led to Unocal's decision to withdraw from the TAPI project in 1998.[178]

However, the project has been revitalised since 2009. 'Intensive negotiations took place with representatives of Afghanistan, Pakistan, and India in 2009–2010 on the construction of the TAPI pipeline with an annual capacity of 30bcm', resulting in the signing of some agreements in December 2010.[179] However, Afghanistan still lacks political stability and the security situation in Pakistan has also deteriorated, which puts the viability of the entire project into question.

0.4.3.2.6. Central Asia–China Gas Pipeline

In addition to exports to the south, Turkmenistan has also long explored export options to the east. In 1994, Turkmenistan and China signed a memorandum of understanding to construct a 28bcm gas pipeline from Turkmenistan, via Uzbekistan and Kazakhstan, to China. The main problems were the costs involved in building such a long line, the complexities of dealing with the transit countries, and certain questions regarding Turkmenistan's ability to supply the amount of gas necessary to amortise the project.[180]

Thus, it was not until 2006/2007 that China was able to sign the necessary agreements with Turkmenistan and the transit countries. In the 2006 supply contract, Turkmenistan pledged to supply China with 30bcm of gas annually for a period of 30 years, beginning in 2009. For that purpose, Turkmenistan granted CNPC a license to develop the Bagtyyarlyk deposit in northeastern Lebap province in August 2007, with the aim of contributing 17bcm of that gas. The remaining 13bcm of gas would

---

177 International Energy Agency: Caspian oil and gas: the supply potential of Central Asia and Transcaucasus, Paris: IEA, 1998, p. 258; Olcott, Martha B.: International gas trade in Central Asia: Turkmenistan, Iran, Russia, and Afghanistan, in: Victor, David G./ Jaffe, Amy M./ Hayes, Mark H. (eds): Natural gas and geopolitics: from 1970 to 2040, Cambridge: Cambridge University Press, 2006, pp. 202–233, here pp. 217–218.
178 International Energy Agency: Caspian oil and gas: the supply potential of Central Asia and Transcaucasus, Paris: IEA, 1998, p. 40; Olcott, Martha B.: International gas trade in Central Asia: Turkmenistan, Iran, Russia, and Afghanistan, in: Victor, David G./ Jaffe, Amy M./ Hayes, Mark H. (eds): Natural gas and geopolitics: from 1970 to 2040, Cambridge: Cambridge University Press, 2006, pp. 202–233, here p. 221.
179 Horák, Slavomír: Turkmenistan's shifting energy geopolitics in 2009–2011: European perspectives, in: Problems of Post-Communism, 2012 (vol. 59), no. 2, pp 18–30, here p. 24.
180 International Energy Agency: Caspian oil and gas: the supply potential of Central Asia and Transcaucasus, Paris: IEA, 1998, pp. 40, 259–260.

0. Introduction: Export Pipelines in Eurasia

come from two older gas fields, Samandepe and Altyn Asyr, in the southeastern Mary province of Turkmenistan. To ensure additional gas supplies from Central Asia, Beijing developed energy relationships with Uzbekistan and Kazakhstan.[181] CNPC began construction of the pipeline in August 2007 and provided all of the necessary investment for the pipeline. However, the actual construction work made little progress. It took the supply interruptions with Russia in 2009 for Turkmenistan to eliminate all of the technical and bureaucratic barriers that had hampered the successful completion of the project.[182]

The Central Asia–China gas pipeline runs from Turkmenistan via Uzbekistan and southern Kazakhstan (running parallel to the existing Bukhara–Tashkent–Bishkek–Almaty pipeline), through the Kazakh city of Chimkent on to Xinjiang Province in western China. The 1,830km trunk pipeline consists of two parallel lines (lines A and B) and five compressor stations; its total capacity is 30 bcm per year. The first string of the pipeline (line A) opened in December 2009, and line B became operational in October 2010.[183]

In 2010, the 2006 supply contract was re-negotiated to 40bcm annually through increased pressure in the existing pipeline strings. In September 2011, however, the parties involved agreed to raise annual deliveries to 55bcm by constructing a third line (line C). Construction of the additional string began in December 2011, and completion is expected by 2014. Beginning in Uzbekistan, line C runs parallel to lines A and B and will have a capacity of 25bcm annually, increasing the total transmission capacity of the Central Asia–China pipeline to 55bcm per year by 2015.[184]

---

181 In June 2010, China and Uzbekistan signed an agreement under which Uzbekistan will provide China with 10bcm of gas annually. As part of the agreement, Uzbekistan's gas pipeline grid was connected to the Central Asia–China gas pipeline. 'Central Asia–China gas pipeline, Turkmenistan to China', in: hydrocarbons-technology.com, http://www.hydrocarbons-technology.com/projects/centralasiachinagasp/ (accessed 27 November 2013).

182 Yenikeyeff, Shamil M.: Kazakhstan's gas: export markets and export routes, Oxford: Oxford Institute for Energy Studies, 2008, p. 65, http://www.oxfordenergy.org/wpcms/wp-content/uploads/2010/11/NG25-KazakhstansgasExportMarketsandExportRoutes-ShamilYenikeyeff-2008.pdf; Horák, Slavomír: Turkmenistan's shifting energy geopolitics in 2009–2011: European perspectives, in: Problems of Post-Communism, 2012 (vol. 59), no. 2, pp 18–30, here p. 22.

183 Yenikeyeff, Shamil M.: Kazakhstan's gas: export markets and export routes, Oxford: Oxford Institute for Energy Studies, 2008, p. 64, http://www.oxfordenergy.org/wpcms/wp-content/uploads/2010/11/NG25-KazakhstansgasExportMarketsandExportRoutes-ShamilYenikeyeff-2008.pdf; Horák, Slavomír: Turkmenistan's shifting energy geopolitics in 2009–2011: European perspectives, in: Problems of Post-Communism, 2012 (vol. 59), no. 2, pp 18–30, here p. 22; 'Central Asia–China gas pipeline, Turkmenistan to China', in: hydrocarbons-technology.com, http://www.hydrocarbons-technology.com/projects/centralasiachinagasp/ (accessed 27 November 2013).

184 Kusznir, Julia: TAP, Nabucco West, and South Stream: the pipeline dilemma in the Caspian Sea basin and its consequences for the development of the Southern Gas Corridor, in: Caucasus Analytical Digest, 18 February 2013, no. 47, pp. 2–8, here p. 6, http://www.isn.ethz.ch/Digital-Library/Publications/Detail/?ots591=0c54e3b3-1e9c-be1e-2c24-a6a8c7060233&lng=en&id=160678; 'Central Asia–China gas pipeline, Turkmenistan to China', in: hydrocarbons-technology.com, http://www.hydrocarbons-technology.com/projects/centralasiachinagasp/ (accessed

0.4.3.2.7. Trans-Caspian Gas Pipeline from Turkmenistan

Others have suggested that exports be directed west. Several parties have proposed constructing a gas pipeline beneath the Caspian Sea from Turkmenbashi (formerly Krasnovodsk) in Turkmenistan, near the Kazakh border, to Azerbaijan (it may also include a connection from the Tengiz field in Kazakhstan to Turkmenistan). In Baku, it would be linked to the Baku–Tbilisi–Erzurum gas pipeline to carry natural gas to Turkey before ultimately reaching Europe. If completed, it is expected to be nearly 2,400km long and have a capacity of up to 30bcm per year. The project was designed to accommodate 16bcm of gas for the Turkish market and 14bcm for European consumers. However, the Turkmen government has not completely endorsed this scheme, citing higher costs compared to a route via Iran, a conflict between Turkmenistan and Azerbaijan over their gas shares in the proposed pipeline, the division of Caspian hydrocarbon fields and problems related to the unsettled legal status of the Caspian Sea and thus the opposition of Russia and Iran to the construction of a submarine pipeline in the Caspian Sea, citing environmental risks. Subsequently, the project became stalled and was ultimately abandoned.[185]

However, the US actively lobbied for the project in the 1990s. Recently, renewed European and US interest in the Trans-Caspian pipeline has been associated with the Nabucco project.[186]

0.4.3.2.8. Pre-Caspian Gas pipeline from Turkmenistan to Russia via Kazakhstan

In addition to diversifying its gas export options, Turkmenistan has co-operated with Russia in the construction of new export pipelines. In December 2007, Russia, Turkmenistan and Kazakhstan signed an agreement on the construction of the Pre-Caspian gas pipeline, i.e., on the modernisation of the existing gas transportation net-

---

27 November 2013); 'Construction on third line begins for Central Asia–China gas pipeline', in: Pipelines International, March 2012, http://pipelinesinternational.com/news/construction_on_third_line_begins_for_central_asia-china_gas_pipeline/066998/ (accessed 27 November 2013).

185 International Energy Agency: Caspian oil and gas: the supply potential of Central Asia and Transcaucasus, Paris: IEA, 1998, pp. 40, 257; Yenikeyeff, Shamil M.: Kazakhstan's gas: export markets and export routes, Oxford: Oxford Institute for Energy Studies, 2008, p. 68, http://www.oxfordenergy.org/wpcms/wp-content/uploads/2010/11/NG25-KazakhstansgasExportMarketsandExportRoutes-ShamilYenikeyeff-2008.pdf; Krauer-Pacheco, Ksenia: Turkey as a transit country and energy hub: the link to its foreign policy aims, Bremen: Forschungsstelle Osteuropa, 2011, p. 33, http://www.forschungsstelle.uni-bremen.de/UserFiles/file/fsoAP118.pdf.

186 Yenikeyeff, Shamil M.: Kazakhstan's gas: export markets and export routes, Oxford: Oxford Institute for Energy Studies, 2008, p. 69, http://www.oxfordenergy.org/wpcms/wp-content/uploads/2010/11/NG25-KazakhstansgasExportMarketsandExportRoutes-ShamilYenikeyeff-2008.pdf; Krauer-Pacheco, Ksenia: Turkey as a transit country and energy hub: the link to its foreign policy aims, Bremen: Forschungsstelle Osteuropa, 2011, p. 33, http://www.forschungsstelle.uni-bremen.de/UserFiles/file/fsoAP118.pdf.

0. Introduction: Export Pipelines in Eurasia

work and the construction of new export capacity for Central Asian gas.[187] They sought to construct a new 1,700km Caspian littoral gas pipeline from the Belek compressor station outside Turkmenbashi in western Turkmenistan to Alexandrov Gai in Russia's Saratov region. The Pre-Caspian gas pipeline is planned to run over 500km across Turkmenistan and some 1,200km across Kazakhstan, parallel to the CAC pipeline, which is also scheduled to receive upgrades. The construction for the 10bcm pipeline was scheduled to begin in early 2009, with completion projected in 2010. However, due to the 2009 gas conflict between Russia and Turkmenistan, the Pre-Caspian pipeline has been shelved.[188]

Table 0-8: Turkmenistan's Gas Export Pipelines

Pipeline	Route	Year commissioned	Designed capacity	Operator
Artyk–Lotfabad	Artyk–Lotfabad	2000	28 million cm	Turkmengaz, National Iranian Gas Company
Central Asia–China	Saman-Depe-Olet–Chimkent-Khorgos	2010	30bcm	Turkmengaz, Asia TransGas, Intergaz, CNPC
Dauletabad–Sarakhs–Khangiran	Dauletabad–Sarakhs–Khangiran	2010	12bcm	Turkmengaz, National Iranian Gas Company
Korpedzhe–Kurt-Kui	Korpedzhe–Ekarem–Kurt-Kui	1997	12bcm	Turkmengaz, National Iranian Gas Company
Pre-Caspian	Turkmenbashi–Bekdash–Uzen–Beineu–Aleksandrov Gai	Under consideration	10bcm	Turkmengaz, Intergaz, Gazprom
TAPI	Turkmenistan–Afghanistan, Pakistan–India	Under consideration	20bcm	N/A
Trans-Caspian gas pipeline	Turkmenistan–Azerbaijan	Under consideration	30bcm	N/A
Turkmenistan–Iran–Turkey	Turkmenistan–Iran–Turkey	Under consideration	30bcm	N/A

Sources: see text.

---

187 Cf. Gazprom, http://www.gazprom.com/about/production/projects/pipelines/; http://www.gazpromquestions.ru/?id=2#c536 (accessed 23 October 2013).
188 Yenikeyeff, Shamil M.: Kazakhstan's gas: export markets and export routes, Oxford: Oxford Institute for Energy Studies, 2008, pp. 60, 62, http://www.oxfordenergy.org/wpcms/wp-content/uploads/2010/11/NG25-KazakhstansgasExportMarketsandExportRoutes-ShamilYenikeyeff-2008.pdf: Horák, Slavomír: Turkmenistan's shifting energy geopolitics in 2009–2011: European perspectives, in: Problems of Post-Communism, 2012 (vol. 59), no. 2, pp 18–30, here p. 25.

0.4.4. Azerbaijan

Azerbaijan has been an important oil-producing region since the late 19th century; however, its importance has declined rapidly since the middle of the 20th century. As the centre of Soviet oil production had moved to Siberia, production and transport facilities in Azerbaijan had been neglected and, as a result, were deteriorating. The oil (and few gas) deposits were increasingly depleted, and although Soviet authorities were aware of potential reserves of oil and gas, no new deposits had been brought online for many years. Thus, the country suffered, as did Kazakhstan and Turkmenistan, from a dire need to modernise its energy sector and attract foreign investments to finance these improvements.

0.4.4.1. Oil

Azerbaijan had traditionally refined most of its own crude oil production and some crude imported from its neighbours while primarily exporting oil products. Being confident in its reserve base, Azerbaijan planned to export a much higher percentage of future output; however, it has faced a number of problems, including a lack of export pipelines.[189]

In 1994, the Azerbaijani state oil company SOCAR signed a US$7.4 billion, 30-year production contract for the Azeri, Chirag and deepwater Guneshli oil field complex with a consortium of major international oil companies called the Azerbaijan International Operating Company (AIOC).[190] As part of its production contract, AIOC is responsible for helping the Azeri government to develop export routes. The construction of pipelines has been divided into an 'early oil' (pre-peak production) phase, for production up to approximately 15mt per year, and a 'main oil' (peak production) phase, to handle a throughput of 35–40mt per year.[191]

Azerbaijan could use Soviet-era pipelines traversing Russia. From a commercial perspective, relying on the existing pipelines for the new projects would have been more feasible, as small modifications or the construction of new parts are less expensive than is constructing a completely new pipeline. However, the AIOC consortium was reluctant to opt for the exclusive use of existing pipelines. Instead, it pursued a 'mul-

---

189 International Energy Agency: Caspian oil and gas: the supply potential of Central Asia and Transcaucasus, Paris: IEA, 1998, p. 162.
190 AIOC was founded in September 1994 under the leadership of BP and originally had twelve shareholders: BP (17.13%), Amoco (17.01%), Unocal (10.05%), LUKOIL and SOCAR (10% each), Statoil (8.56%), Exxon (8%), TPAO (6.75%), Pennzoil (4.82%), Itochu (3.92%), Ramco (2.08%), and Delta Nimir (1.68%). The composition of the consortium has changed over time; it now includes BP (34.14%), Chevron (10.28%), SOCAR and Inpex (10% each), Statoil (8.56%), ExxonMobil (8%), TPAO (6.75%), Devon Energy (5.63%), Itochu (3.92%), and Hess (2.72%).
191 International Energy Agency: Caspian oil and gas: the supply potential of Central Asia and Transcaucasus, Paris: IEA, 1998, pp. 162–163.

0. Introduction: Export Pipelines in Eurasia   61

tiple pipelines' approach, seeking to reduce Russia's position as a transit country for Caspian energy supplies and diversify Azerbaijani energy export options.[192]

In October 1995, AIOC decided to export its 'early oil' via two export routes: a northern route from Baku to the Russian Black Sea port of Novorossiisk and a western route from Baku to the Georgian Black Sea port of Supsa. It was essentially the reconstruction and refurbishment of two existing pipelines.[193] Because the main oil pipeline would only be needed in 2003, AIOC announced that it would make its decision by October 1998.[194]

0.4.4.1.1. Baku–Novorossiisk Pipeline

In 1996, Azerbaijan signed an intergovernmental agreement with Russia to construct the Baku–Novorossiisk pipeline. The 1,347km long pipeline began operations in December 1997 with a total capacity of 7mt annually, of which 5mt are reserved for Azeri oil (the remaining capacity is used by oil from Kazakhstan and Turkmenistan delivered via tanker to Makhachkala in Dagestan). The northern route primarily consists of upgraded (and in some cases rebuilt) sections of pipeline from Baku to Novorossiisk via Grozny (in Chechnya) and Tikhoretsk. The section within Azerbaijan was originally employed to import Russian oil for processing in Azeri refineries and had to be reversed. The upgrading and replacement of the Russian section was the responsibility of Transneft, which also financed the work.[195]

Some 153km of the northern route pass through the secessionist region of Chechnya in the Russian Caucasus. Continued unrest in this region posed considerable concerns, as much of this section was damaged during the first Chechen war of 1994–1996. Thus, the Russian government built a 283km bypass between Khasavyurt (in Dagestan) and Tikhoretsk that would be used principally as a back-up in the event of problems on the route via Chechnya. The bypass appears to be influenced by Russia's determination to make the northern route an attractive option for the main oil pipe-

---

192   Badalyan, Lusine: Interlinked energy supply and security challenges in the South Caucasus, in: Caucasus Analytical Digest, 12 December 2011, no. 33, pp. 2–5, here p. 3, http://www.isn.ethz.ch/Digital-Library/Publications/Detail/?ots591=0c54e3b3-1e9c-be1e-2c24-a6a8c7060233&lng=en&id=135318.
193   International Energy Agency: Caspian oil and gas: the supply potential of Central Asia and Transcaucasus, Paris: IEA, 1998, pp. 38, 163; United Nations Development Programme/ World Bank: Cross-border oil and gas pipelines: problems and prospects, Washington DC: Energy Sector Management Assistance Programme, 2003, p. 79, http://siteresources.worldbank.org/INTOGMC/Resources/crossborderoilandgaspipelines.pdf.
194   International Energy Agency: Caspian oil and gas: the supply potential of Central Asia and Transcaucasus, Paris: IEA, 1998, pp. 163–164.
195   Ibid., p. 163; Energy Charter Secretariat: Bringing oil to the market: transport tariffs and underlying methodologies for cross-border crude oil and products pipelines, Brussels: Energy Charter Secretariat, 2012, p. 71.

line. However, upgrading the current northern route to serve as the main oil route would cost an estimated US$2–2.4 billion.[196]

Notwithstanding the pipeline's commercial viability, the northern route was not a reliable option, as military and terrorist activities have caused supply interruptions, not only in Chechnya but also throughout the North Caucasus.[197] When the current delivery contract expires in February 2014, it will not be renewed and the Azeri section of the pipeline will be closed; however, the section from Makhachkala to Novorossiisk will remain available for Kazakh and Turkmen oil.[198]

0.4.4.1.2. Baku–Supsa Pipeline

The 920km western route from Baku to the Georgian Black Sea port of Supsa with an annual capacity of 6mt was inaugurated in April 1999. Similar to the northern route, it combines existing stretches of pipeline with new sections. AIOC, which is financing the western route, allocated US$315 million for the project, of which US$60 million were used to build a terminal and storage facilities for 240,000 tonnes of oil in Supsa. The Supsa terminal has an annual capacity of 10mt, with the potential to eventually be upgraded to 50–70mt.[199]

Some sections of the Baku–Supsa line were newly constructed. This includes the 150km section from Akstafa on the Georgian–Azeri border to the Georgian village of Samgori (near Tbilisi). After initial upgrading work on other sections, it became evident in early 1998 that much of the existing pipeline in Azerbaijan would likely have to be replaced rather than refurbished.[200] Thus, the initial costs of the Baku–Supsa pipe-

---

196 International Energy Agency: Caspian oil and gas: the supply potential of Central Asia and Transcaucasus, Paris: IEA, 1998, p. 163.
197 With the opening of the Baku–Tbilisi–Ceyhan pipeline (see below), the volumes of Azeri oil transported to Novorossiisk remained low; in 2010, only 2.3mt were delivered annually. United Nations Development Programme/ World Bank: Cross-border oil and gas pipelines: problems and prospects, Washington DC: Energy Sector Management Assistance Programme, 2003, p. 79, http://siteresources.worldbank.org/INTOGMC/Resources/crossborderoilandgaspipelines.pdf; Badalyan, Lusine: Interlinked energy supply and security challenges in the South Caucasus, in: Caucasus Analytical Digest, 12 December 2011, no. 33, pp. 2–5, here p. 3, http://www.isn.ethz.ch/Digital-Library/Publications/Detail/?ots591=0c54e3b3-1e9c-be1e-2c24-a6a8c7060233&lng=en&id=135318; Energy Charter Secretariat: Bringing oil to the market: transport tariffs and underlying methodologies for cross-border crude oil and products pipelines, Brussels: Energy Charter Secretariat, 2012, p. 71.
198 'Russia to close 250-km section of Baku–Novorossiisk pipeline', in: NewsBase, FSU Oil & Gas Monitor, no. 49, 11 December 2013, p. 7.
199 International Energy Agency: Caspian oil and gas: the supply potential of Central Asia and Transcaucasus, Paris: IEA, 1998, p. 164; United Nations Development Programme/ World Bank: Cross-border oil and gas pipelines: problems and prospects, Washington DC: Energy Sector Management Assistance Programme, 2003, p. 79, http://siteresources.worldbank.org/INTOGMC/Resources/crossborderoilandgaspipelines.pdf.
200 International Energy Agency: Caspian oil and gas: the supply potential of Central Asia and Transcaucasus, Paris: IEA, 1998, p. 164; United Nations Development Programme/ World Bank: Cross-border oil and gas pipelines: problems and prospects, Washington DC: Energy Sector

0. Introduction: Export Pipelines in Eurasia   63

line nearly doubled: 'Most of the overrun is attributable to AIOC's decision to replace large sections of the pipeline rather than refurbish it'.[201]

0.4.4.1.3. Baku–Tblisi–Ceyhan Pipeline (BTC)

As it was clear from the onset that these two pipelines for the 'early oil' would not be sufficient to transport the entirety of peak oil production, a main oil export pipeline had to be designated. Regional and international powers—such as Russia, Iran, Turkey and the United States (US)—attempted to influence AIOC's decision regarding its main export pipeline, leading to a politicised debate. The regional powers of Russia, Iran, and Turkey were all interested in becoming transit countries for Azeri oil and, thus, lobbied for the main export pipeline to be built through their territory.[202]

Since the dissolution of the Soviet Union, the US as an external power had increasingly attempted to influence the planning and construction of export pipelines in the Caspian Sea area, a region that was considered a viable supply alternative to reduce US dependence on oil from the Middle East. One of the US policy goals regarding the construction of export pipelines was to ensure that neither Russia nor Iran would develop a monopoly over pipelines from the Caspian region, which could be employed as political and economic leverage against the countries of the region.[203] As a consequence, the pipelines that had become the centre of diplomatic efforts by the US in the region were guided more by political than economic considerations.[204]

Management Assistance Programme, 2003, p. 80, http://siteresources.worldbank.org/INTOGMC/Resources/crossborderoilandgaspipelines.pdf.
201 United Nations Development Programme/ World Bank: Cross-border oil and gas pipelines: problems and prospects, Washington DC: Energy Sector Management Assistance Programme, 2003, p. 84, http://siteresources.worldbank.org/INTOGMC/Resources/crossborderoilandgaspipelines.pdf.
202 Cf. Miles, Carolyn: The Caspian pipeline debate continues: why not Iran?, in: Journal of International Affairs, 1999 (vol. 53), no. 1, pp. 325–346; Tayfur, M. Fatih/ Göyman, Korel: Decision making in Turkish foreign policy: the Caspian oil pipeline issue, in: Middle Eastern Studies, 2002 (vol. 38), no. 2, pp. 101–122; Bolukbasi, Suha: Jockeying for power in the Caspian region: Turkey versus Iran and Russia, in: Akiner, Shirin / Aldis, Anne (eds): The Caspian: politics, energy and security, London: Routledge, 2004, pp. 198–207; Makhmudov, Rustam: Russia and Iran: attempts to implement new strategic steps in the Caspian-Central Asian oil and gas industry, in: Central Asia and the Caucasus, 2008 (vol. 53), no.5, http://www.ca-c.org/journal/2008-05-eng/08.shtml.
203 Cf. Jofi, Joseph: Pipeline diplomacy: the Clinton administration's fight for Baku Ceyhan, Princeton, NJ: Woodrow Wilson School, 1999 (Case Study 1/99), http://www.dtic.mil/dtic/tr/fulltext/u2/a360382.pdf; Morningstar, Richard: From pipe dream to pipeline: the realization of the Baku–Tbilisi–Ceyhan pipeline, Harvard University, Belfer Center, Event Report, 8 May 2003, http://belfercenter.ksg.harvard.edu/publication/12795/from_pipe_dream_to_pipeline.html; Starr, S. Frederick/ Cornell, Svante E. (eds): The Baku–Tbilisi–Ceyhan pipeline: oil window to the West, Stockholm/ Washington, DC: Central Asia – Caucasus Institute & Silk Road Studies Program, 2005.
204 Stauffer, Thomas R.: Caspian fantasy: the economics of political pipelines, in: Caspian Studies, 2000 (vol. 7), no. 263, http://www.caspianstudies.com/article/caspian_fantasy.htm.

Experts considered the southern export option through Iran to be the most economically effective. Iran offered two inexpensive and quick routes to transport oil from the Caspian to world markets through swap deals. In both cases, Caspian oil would have been imported into northern Iran and processed in Iranian refineries to meet domestic demand, whereupon comparable volumes of Iranian oil would have been delivered on behalf of the Caspian producers to buyers via the Persian Gulf.[205]

While the economics have favoured an Iranian export route, US sanctions against Iran have been the major obstacles to this option. To deter foreign investment in the Iranian energy sector, the US implemented a law to 'impose sanctions on persons making certain investments directly and significantly contributing to the enhancement of the ability of Iran or Libya to develop its petroleum resources, [...]', the so-called Iran and Libya Sanctions Act of 1996.[206] The law's definition of 'investment' does not specifically include oil or gas purchases from Iran. However, the position of every US administration has been that the construction of energy pipelines to or through Iran would constitute sanctionable activity because such infrastructures would develop Iran's petroleum resources.[207]

With the Iranian option rendered impossible,[208] in July 1997, AIOC ultimately announced that it had narrowed the possible routes for the so-called 'main oil' pipeline to three: expanded versions of the two routes used for 'early oil' or a third route to the Turkish Mediterranean port of Ceyhan.[209]

In addition to geopolitical considerations in the case of Russia, both the US and Turkey advanced another argument against an expansion of both 'early oil' pipelines, which was proposed by AIOC and supported by Russia's Transneft: the US and Turkey were concerned by the environmental impact that these pipelines would have due to the increase in tanker traffic from the Russian port of Novorossiisk and Georgia's port

---

205   Ibid. The first route involves an existing but unused gas pipeline running from Iranian Azerbaijan to Baku. This line could be upgraded and reversed and the oil then pumped to refineries in Tabriz and Teheran. The second route entails the proposed expansion of another existing pipeline. Oil can be sent via tanker from Kazakhstan, Turkmenistan, or Azerbaijan to the Iranian port of Neka, which is connected by a small pipeline with the Teheran refinery.
206   Cf. http://www.gpo.gov/fdsys/pkg/BILLS-104hr3107enr/pdf/BILLS-104hr3107enr.pdf (accessed 28 November 2013).
207   Katzman, Kenneth: The Iran sanctions act (ISA), Washington, DC: Congressional Research Service, 2007, pp. 1, 4 (Order Code RS20871), http://www.fas.org/sgp/crs/row/RS20871.pdf.
208   In what seems to be a face-saving argument, BP officials have stated that the combined cost of the pipeline and transit fees would come close to the cost of other export options. As the Iranian pipelines were already in place, this argument sounds unconvincing. Cornell, Svante E./ Tsereteli, Mamuka/ Socor, Vladimir: Geostrategic implications of the Baku–Tbilisi–Ceyhan pipeline, in: Starr, S. Frederick/ Cornell, Svante E. (eds): The Baku–Tbilisi–Ceyhan pipeline: oil window to the West, Stockholm/ Washington, DC: Central Asia – Caucasus Institute & Silk Road Studies Program, 2005, pp. 17–38, here p. 19.
209   International Energy Agency: Caspian oil and gas: the supply potential of Central Asia and Transcaucasus, Paris: IEA, 1998, pp. 163–164.

0. Introduction: Export Pipelines in Eurasia 65

of Supsa through the Turkish Straits (Bosporus and Dardanelles), an already heavily congested waterway.[210] Ankara stated the possible environmental and human risks associated with the transportation of oil through the straits because of their location near the densely populated city of Istanbul. They argued instead for a pipeline that would bypass the environmentally sensitive Bosporus Strait and run to Ceyhan.[211]

However, the Turkish option was considered economically infeasible and engendered substantial opposition in both business and governmental circles in the US and the Caspian region:[212] There were concerns that the Baku–Ceyhan project does not connect to the large, future volumes of oil from the Caspian, it runs through politically unstable regions of Georgia and Turkey, as well as environmental concerns because the route would direct the pipeline through a seismically active region of Turkey with the potential risks of earthquakes. Additionally, it would require billions of dollars of investment that could not be attracted without massive subsidies from some extra-regional power.[213]

Nevertheless, even without a strong business model, the Turkish option has received strong endorsement from the governments of Turkey, Azerbaijan and the US, as it would be the first major oil pipeline in this region that circumvents Russia. To make investments in the project more feasible, the US government made financing available from governmental agencies, such as the US Export-Import Bank. The Turkish government also guaranteed that the construction cost for the Turkish section would not exceed the original estimate. Once these and other financing options were in place, AIOC broke ground on the Baku–Tbilisi–Ceyhan (BTC) pipeline in September 2002.[214]

---

210 Morningstar, Richard: From pipe dream to pipeline: the realization of the Baku–Tbilisi–Ceyhan pipeline, Harvard University, Belfer Center, Event Report, 8 May 2003, http://belfercenter.ksg.harvard.edu/publication/12795/from_pipe_dream_to_pipeline.html.
211 International Energy Agency: Caspian oil and gas: the supply potential of Central Asia and Transcaucasus, Paris: IEA, 1998, p. 164; Krauer-Pacheco, Ksenia: Turkey as a transit country and energy hub: the link to its foreign policy aims, Bremen: Forschungsstelle Osteuropa, 2011, p. 32, http://www.forschungsstelle.uni-bremen.de/UserFiles/file/fsoAP118.pdf.
212 AIOC originally preferred an extended version of the Baku–Sapsa pipeline as its main export pipeline, which would have been less expensive and faster to build. Cf. 'Decision on Azeri oil pipeline will not take "a long time"—president', in: Azerbaijani TV, Channel One, Baku, 31 October 1998, obtained via BBC Monitoring.
213 Stauffer, Thomas R.: Caspian fantasy: the economics of political pipelines, in: Caspian Studies, 2000 (vol. 7), no. 263, http://www.caspianstudies.com/article/caspian_fantasy.htm; Morningstar, Richard: From pipe dream to pipeline: the realization of the Baku–Tbilisi–Ceyhan pipeline, Harvard University, Belfer Center, Event Report, 8 May 2003, http://belfercenter.ksg.harvard.edu/publication/12795/from_pipe_dream_to_pipeline.html.
214 International Energy Agency: Caspian oil and gas: the supply potential of Central Asia and Transcaucasus, Paris: IEA, 1998, pp. 164–165; Peimani, Hooman: The Caspian pipeline dilemma: political games and economic losses, Westport, CT: Praeger Publishers, 2001, p. 2; Morningstar, Richard: From pipe dream to pipeline: the realization of the Baku–Tbilisi–Ceyhan pipeline, Harvard University, Belfer Center, Event Report, 8 May 2003, http://belfercenter.ksg.harvard.edu/publication/12795/from_pipe_dream_to_pipeline.html; Cornell, Svante E./ Tsereteli, Mamuka/ Socor, Vladimir: Geostrategic implications of the Baku–Tbilisi–Ceyhan pipeline, in: Starr, S. Frederick/

The 1,768km-long BTC begins at the Azeri–Chirag–Guneshli oil fields and connects to the Turkish Mediterranean port of Ceyhan via Tbilisi; it began operations in July 2006, with an initial capacity of 50mt per year.[215] In 2009, its capacity was successfully augmented, reaching 59.8mt.[216] It 'was designed initially to transport Azeri oil, but has enough capacity to also ship crude from other Caspian producers, especially Kazakhstan'.[217] In 2006, Kazakhstan made a pledge to provide the BTC an additional 7.2mt of oil annually from the Kashagan oil field carried by tanker across the Caspian Sea. However, the exploitation of the Kashagan field has encountered problems, and the start of production had to be postponed until September 2013.[218]

Table 0-9: Azerbaijan's Oil Export Pipelines

Pipeline	Route	Year commissioned	Designed capacity	Operator
Baku–Novorossiisk	Baku–Khasavyurt–(Grozny)–Tikhoretsk–Novorossiisk	1997	7mt	Transneft
Baku–Supsa	Baku–Akstafa–Tbilisi–Supsa	1999	6mt	AIOC
Baku–Tbilisi–Ceyhan	Baku–Tbilisi–Horasan–Erzurum–Erzincan–Ceyhan	2006	50mt	AIOC

Sources: see text.

Cornell, Svante E. (eds): The Baku–Tbilisi–Ceyhan pipeline: oil window to the West, Stockholm/Washington, DC: Central Asia – Caucasus Institute & Silk Road Studies Program, 2005, pp. 17–38, here p. 24.

215 It was commissioned by the Baku–Tbilisi–Ceyhan Pipeline Company (BTC Co.), which is a joint venture company of eleven shareholders and managed by BP. The eleven BTC Co. shareholders are BP (30.1%), SOCAR (25%), Chevron (8.9%), Statoil (8.71%), TPAO (6.53%), Eni and Total (5% each), Itochu (3.4%), Inpex and ConocoPhillips (2.5% each), and Hess Corporation (2.36%) (cf. BP, Baku–Tbilisi–Ceyhan Pipeline, http://www.bp.com/sectionge nericarticle.do?categoryId=9 006669&contentId=7015093, accessed 28 April 2011).

216 Krauer-Pacheco, Ksenia: Turkey as a transit country and energy hub: the link to its foreign policy aims, Bremen: Forschungsstelle Osteuropa, 2011, p. 31, http://www.forschungsstelle.uni-bremen.de/UserFiles/file/fsoAP118.pdf.

217 Perovic, Jeronim/ Orttung, Robert W./ Perovic, Jeronim (eds): Energy and the transformation of international relations: toward a new producer–consumer framework, Oxford: Oxford University Press, 2009, pp. 117–157, here p. 119.

218 So far, only small volumes of oil from Kazakhstan and Turkmenistan are delivered to Baku via tankers. Badalyan, Lusine: Interlinked energy supply and security challenges in the South Caucasus, in: Caucasus Analytical Digest, 12 December 2011, no. 33, pp. 2–5, here p. 4, http://www.isn.ethz.ch/Digital-Library/Publications/Detail/?ots591=0c54e3b3-1e9c-be1e-2c24-a6a8c7060233&lng=en&id=135318; Krauer-Pacheco, Ksenia: Turkey as a transit country and energy hub: the link to its foreign policy aims, Bremen: Forschungsstelle Osteuropa, 2011, pp. 31–32, http://www.forschungsstelle.uni-bremen.de/UserFiles/file/fsoAP118.pdf; Energy Charter Secretariat: Bringing oil to the market: transport tariffs and underlying methodologies for cross-border crude oil and products pipelines, Brussels: Energy Charter Secretariat, 2012, p. 67.

0. Introduction: Export Pipelines in Eurasia   67

0.4.4.2. Gas

From 1992–1995, Azerbaijan was a net gas importer (chiefly from Turkmenistan). In March 1996, the Azeri government announced that it would discontinue all imports of gas and concentrate on developing its own resources in the Caspian Sea. The exploitation of new fields and an overhaul of the gas supply system, including the upgrading of compressor stations, were intended to make the country self-sufficient in gas.[219] Principally, this means exploiting the Shah Deniz field, the largest discovery of a natural gas since 1978, which is located offshore in the Caspian Sea near Baku. It is estimated to contain 35 trillion cm of natural gas. The first phase of its exploitation was completed in 2006 and completion of the second phase is expected in 2016.

0.4.4.2.1. Mozdok–Makhachkala–Kazi Magomed Gas Pipeline
In 2007, Azerbaijan was finally able to halt gas imports that had been transported via the Mozdok–Makhachkala–Kazi Magomed gas pipeline. The pipeline from Mozdok in North Ossetia through Chechnya and Dagestan to Azerbaijan was built in 1983. The Azerbaijani section is known as Kazi Magomed or Baku–Novo Filya, and it connects Baku with the Novo Filya gas metering station in Russia. After 2007, the pipeline became largely inactive. In 2010, however, the flow direction of the Mozdok–Makhachkala–Kazi Magomed pipeline was reversed and, since 2011, has been used to transport Azerbaijani gas to Russia. With a designed capacity of 13bcm, the pipeline currently has an annual throughput capacity of 5bcm. In 2012, Azeri gas exports amounted to 1.5bcm.[220]

0.4.4.2.2. Baku–Tbilisi–Erzurum Pipeline (BTE)
However, the main export pipeline for the peak production of the Shah Deniz gas field is the Baku–Tbilisi–Erzurum (BTE) gas pipeline (also known as the South Caucasus Pipeline). It carries natural gas from the Shah Deniz field through Tbilisi in Georgia and links to the Turkish national gas pipeline network in Erzurum. The BTE has a total annual capacity of 6bcm; it became operational in December 2006, delivering the first gas to Georgia and in July 2007, to Turkey. The 692km-long BTE runs parallel to the BTC oil pipeline.[221]

---

219 International Energy Agency: Caspian oil and gas: the supply potential of Central Asia and Transcaucasus, Paris: IEA, 1998, p. 173.
220 Wilson, David/ Drayton, Geoffrey: Soviet oil and gas to 1990, Cambridge, MA: Abt Books, 1982, p. 45; 'Azeri gas safe from Dagestan unrest', in: United Press International, 21 May 2010, http://www.upi.com/Business_News/Energy-Resources/2010/05/21/Azeri-gas-safe-from-Dagestan-unrest/UPI-96861274453399/; Gazprom, http://www.gazprom.com/about/production/projects/pipelines/; http://www.gazpromquestions.ru/?id=2#c536 (accessed 30 October 2013).
221 Krauer-Pacheco, Ksenia: Turkey as a transit country and energy hub: the link to its foreign policy aims, Bremen: Forschungsstelle Osteuropa, 2011, p. 32, http://www.forschungsstelle.uni-bremen.de/UserFiles/file/fsoAP118.pdf.

This line was constructed by the South Caucasus Pipeline Company,[222] which is a joint venture of seven international shareholders and is operated by BP and Statoil. It is planned to triple its capacity to over 20bcm per year. This would be in line with the second exploitation phase of the Shah Deniz deposit and involve the laying of a new pipeline string across Azerbaijan and the construction of two new compressor stations in Georgia.[223]

0.4.4.2.3. Nabucco: Its Competitors and Its Successor

The Nabucco pipeline project, a 3,900km pipeline from Turkey to Austria via Bulgaria and Hungary, as well as its competitor projects such as the Interconnection Turkey–Greece–Italy (ITGI) pipeline, the Trans-Adriatic Pipeline (TAP) and the Trans-Anatolia Pipeline (TANAP), and its successor project, Nabucco West, are not included in this overview because these pipeline projects are not intended to begin in one of the Eurasian producer countries under study but rather in Turkey. Detailed information on these projects can be found in the chapters by Rustamov and Kustova.

Table 0-10: Azerbaijan's Gas Export Pipelines

Pipeline	Route	Year commissioned	Designed capacity	Operator
Mozdok–Makhachkala–Kazi Magomed	Baku–Novo Filya–Makhachkala	1983	13bcm	SOCAR, Gazprom
Baku–Tbilisi–Erzurum	Baku–Tbilisi–Horasan–Erzurum	2006	6bcm	South Caucasus Pipeline Company

Sources: see text.

0.5. Summary

Despite many newly built export pipelines, the Soviet pipeline system remains in use and constitutes the backbone of the Russian export infrastructure; it also remains important for other FSU countries. Most Russian gas destined for Western Europe still transits through a narrow Ukrainian–Slovak–Czech corridor that was originally created in 1967 for the Bratstvo pipeline: '[...] all subsequent export pipelines—except the lines to Finland, Greece, and Turkey—had been built in parallel with this line'.[224] Although

---

222 The shareholders are BP (technical operator, 25.5%), Statoil (commercial operator, 25.5%), SOCAR, LUKOIL, NICO, Total (10% each), and TPAO (9%) (cf. http://www.bp.com/genericarticle.do?categoryId=9006615&contentId=7018471, accessed 30 October 2013).
223 Krauer-Pacheco, Ksenia: Turkey as a transit country and energy hub: the link to its foreign policy aims, Bremen: Forschungsstelle Osteuropa, 2011, p. 32, http://www.forschungsstelle.uni-bremen.de/UserFiles/file/fsoAP118.pdf.
224 Högselius, Per: Red gas: Russia and the origins of European energy dependence, New York: Palgrave Macmillan, 2013, p. 212.

the construction of the Yamal–Europe, Blue Stream, Nord Stream, and South Stream pipelines will lessen the importance of the Bratsvo system to some extent, it will remain an essential component of the Russian gas export infrastructure. The Druzhba oil pipeline system also remains in operation; however, its role has been reduced by the expansion of exports by tanker from Russian Baltic Sea ports.

Kazakhstan and Turkmenistan were both able to build export outlets to enter new consumer markets. However, both countries continue to extensively employ the Soviet-era pipeline grid and/or Russia as a transit country for their exports. Nevertheless, they have been able to reduce their dependence on Russia to some extent.

Only Azerbaijan has been able to significantly reduce its dependence on Russian infrastructure; the country was able to build two pipelines to world markets independent of Russia (one for oil and one for gas).

0.6. Structure of the Book

This book presents the papers discussed at the 8th Changing Europe Summer School on 'Export pipelines from the CIS region' organised by the Central Asian Foundation for the Development of Democracy, Almaty, and the Research Centre for East European Studies at the University of Bremen, Germany and with financial support from the Volkswagen Foundation.

The book is divided into four, geographically focussed, parts: the first analyses Russia's foreign energy policy.

In his chapter, 'Russian foreign energy policy under Vladimir Putin: Norms, ideas, and determining factors', András Molnár applies neoclassical realism to analyse the political, economic, and ideological motives and objectives behind the creation and structural development of Russian energy foreign policy since the collapse of the Soviet Union. The Russian state attempts to expand its increasing influence over energy resources as a potential tool to achieve its strategic aims. Molnár considers states to be rational actors. However, in the case of energy resources, this does not always coincide with economic rationality. Instead, Russia subordinates economic considerations to developing state power and influence abroad. Thus, contemporary Russian energy policy represents a diffusion of several (geo-)political-economic considerations that have led to the emergence of a multidimensional policy.

In the following chapter, Inna Chuvychkina applies actor-centred institutionalism to Russian pipeline policies and analyses the influence of institutions on the preferences and perceptions of actors and their forms of interaction. Once institutions are created, institutional reform is costly and requires substantial political will. Under Vladimir Putin, informal institutions have emerged that foster the power-seeking and profit-making interests of his closest circle of friends and associates, such as informal practices, the management of resource revenue, and corruption schemes. These informal institutions also affect the economic outcomes of Russian state-controlled oil and gas companies and contribute to high construction costs for pipelines and kickbacks from pipeline construction. Pipeline projects are primarily driven by political considerations, the near-sighted, egoistic use of rents and the short-termism of Russian policy, which does not consider the economic justification of their construction. Chuvychkina concludes that these informal institutions lead to a loss of market shares and profits for Russia as an energy exporter.

This is followed by Niels Smeets' chapter, 'Opening up the black box: Russia's energy security concept'. Using the concept of energy security and distinguishing between the security of supply and the security of demand, Smeets demonstrates that Russia, although an energy producer, faces considerable challenges concerning the security of supply. Similar to consumer countries, the availability, acceptability, accessibility, and affordability of energy all play an important role for Russia in securing stable domestic supplies to its end-consumer. Applying securitisation theory, he conducts a comparative discourse analysis of the 2003 and 2009 versions of the Russian energy strategy. Evidence of both aspects of energy security was abundantly observed, but twice as much attention is devoted to the security of supply. This finding modifies the statement that producer countries will primarily be concerned with security of demand issues. Smeets concludes that security of supply remains the most important concern of the Russian elite.

Energy policy in Central Asia is scrutinised in the second part of the book.

In 'Challenges along the way towards a maximally secure Central Asian gas system', Farkhod Aminjonov analyses the volatile regional energy relations in the Central Asian energy system (CAES), a complex framework within which various energy actors interact and affect one another's security. A secure the CAES means that all Central Asian states enjoy sufficient and sustainable energy supplies for the needs of both the population and the economy. Due to asymmetric power relations within the CAES, larger powers attempt to maintain their influence over energy policy choices, and weaker powers are constantly searching for ways to increase their leverage. The dependence of Central Asian exporters' on the Russian pipeline network to export their natural gas places them in a vulnerable position. Russia effectively exploits this dependence to promote its economic and political interests. However, this Russian monopoly has been challenged by Iranian and Chinese energy pipeline infrastructures that have transformed the CAES. This indicates partially successful efforts to diversify away from Russia and possibly obtain a higher price for the exported gas. However, attempts to fulfil obligations to supply the agreed amount of natural gas to external customers may result in an unstable regional energy market in which less gas is available for Central Asian consumer countries.

In the following chapter, 'The geo-economics of Eurasian gas: the evolution of Russian–Turkmen relations in natural gas (1992–2010)', Boris Barkanov studies this dyad

0. Introduction: Export Pipelines in Eurasia 71

to understand the future prospects for Russia's dominant position on the Eurasian market, the tools at Russia's disposal to control the flow of Turkmen gas to Europe, and the trade-offs and potential limitations Russia faces in their successful deployment. He employs the concept of 'conditions of trade', which include the gas price, form of payment, contract duration, and key features related to bargaining dynamics to measure their bilateral relations and examine how and why Russian relations with Turkmenistan have changed. Barkanov's principal findings are the following: (1) the supply-demand dynamics for Russian gas were not sufficient to change the relationship; (2) the Central Asia–China gas pipeline was sufficient to produce the largest price increases since 2006, while the Trans-Caspian gas pipeline was necessary but not sufficient for earlier changes; and (3) the political factors associated with state-firm relations in Russia were necessary for price changes beginning in 1999 and changes in the form of payment, the time-frame of the relationship and bargaining dynamics.

In his chapter, 'Pipelines and hegemonies in the Caspian: A Gramscian appraisal', Paolo Sorbello analyses the dynamic relationship between energy and foreign policy. The chapter specifically focuses on a case study of Kazakhstan and the circumstances that led to the construction of two oil pipelines, the Caspian Pipeline Consortium and the Kazakhstan–China pipeline. Using a Gramscian approach and his concepts of historical bloc and hegemony, Sorbello assesses whether a change in the material conditions of the Kazakh energy sector produced a shift in foreign policy or, more specifically, the new configuration in the oil industry triggered the realisation of the multi-vector foreign policy designed in the 1990s. He concludes that a shift in the material conditions led to new relations in terms of upstream operations and the success of intergovernmental negotiations over the construction of the Kazakhstan–China pipeline. Sorbello argues that the transnational corporations and Russia have lost their material ability to trigger Kazakh consent. The rise of Kazakhstan as an independent actor that is capable of shaping the social configurations (from contracts to technical expertise) that concern its energy policy had a critical impact on foreign policy.

Part three of the book focuses on the Caucasus and the Southern Energy Corridor.

In the chapter, 'The Trans-Adriatic Pipeline and Nabucco West pipeline projects: advantages and disadvantages for Azerbaijan', Rufat Rustamov identifies future possibilities for Azerbaijan in the event that any of these pipelines are constructed and what types of challenges and difficulties the country could face. He concludes that international gas pipeline projects have broadened Azerbaijan's room for manoeuvre in foreign policy and simultaneously decreased its dependence on Russia. As a result, Azerbaijan gained economic, political and, partly, geopolitical benefits. Energy pipelines have strengthened the economic security of both Azerbaijan and Georgia, helping the two countries move away from the Russian sphere of influence and more firmly orient themselves towards the West, as well as attracting the attention of international institutions to the region.

While the geopolitical perspective is important in accounting for the configuration of gas export routes in the region, Irina Kustova assesses developments internal to the EU in her chapter: 'EU energy policy towards the Caspian region: Assessing the Southern Gas Corridor'. She seeks to shed light on additional aspects of the EU's limited success that are not captured by the geopolitical approach. Instead of analysing the stakes of the parties involved in geopolitical games, she focuses on the internal inconsistencies of EU policies, both in terms of integration processes in the energy domain and the commercial interests of energy companies. Using a combination of theoretical approaches (historical institutionalism, securitisation theory, and the concept of supranational spill-over), Kustova argues that the EU's internal inconsistencies in decision-making and its disregard for the commercial aspects of pipeline projects have seriously impeded the successful implementation of the Southern Gas Corridor. It additionally undermines the EU's credibility as an important regional actor. Regarding the prospects for the Southern Gas Corridor, Kustova concludes that absent a common EU position, the EU is unlikely to be able to pursue any effective policies in the Caspian region.

In her chapter, 'The EU's democracy promotion in its Eastern Neighbourhood: Consistency in the shadow of energy trade and the EU's ability', Lusine Badalyan addresses the EU's democracy-promotion policies in its Eastern Partnership and seeks to explore the conditions under which the EU's efforts to promote democracy are effective and to what extent. Badalyan, using a critical approach to the Normative Power Europe concept, argues that the EU may deviate from a consistent pattern in cases where its strategic objectives, such as security or economic interests, are prioritised over the transfer of norms of democratic rule. In its Eastern Neighbourhood, the EU's energy supply security interests are considered one of the key concerns that can mitigate its attempts to promote democratic rule and good governance. A clear example of how the EU's security considerations can override its support for democracy and human rights reforms can be found in the case of Azerbaijan. Here, despite continuing violations of human rights and democratic principles, the EU has voiced little criticism, which is primarily linked to strategic considerations. To empirically assess these arguments, an Ordinary Least Squares regression analysis is conducted. The results suggest that the EU's consistent assessments of the target countries are strongly associated with more diversified trade structures and in cases that do not involve energy interests.

The final part of this book analyses Ukraine's energy policy with a focus on Russian–Ukrainian gas relations.

In his chapter, 'Gas disputes between Russia and Ukraine: Patterns of escalation and Russian stakes (2006–2009)', Eric Pardo Sauvageot focuses on three disputes in January 2006, March 2008, and January 2009. Ukraine is both a consumer of Russian

0. Introduction: Export Pipelines in Eurasia

gas and a transit country for the latter's gas exports to Central and Western Europe. For Russia, this complex triangular relationship implies that any dispute with Ukraine could eventually disrupt its export market, posing a serious dilemma for Russia: it has to decide whether damage to its image as a reliable supplier is acceptable for the sake of prevailing in disputes with transit countries. Comparing the escalation patterns, Pardo Sauvageot considers the explanatory power of a possible changing valuation of the status quo by Russia to be an explanation for the different patterns of escalation. Considering insights from cognitive psychology provided by prospect theory, the analysis focuses on how actors rationally (even if the selection of a point of reference may not be rational) behave in response to risks and how their responses may differ depending on whether they value the status quo positively or negatively. Having analysed these three disputes, it seems obvious to assume that the main risk factor for Gazprom, and thus for Russia, was the effect of gas cut-offs to Ukraine on its reputation as a reliable supplier. The empirical analysis revealed that the changes and derived divergences in the patterns of escalation were primarily due to decisions made by the Russian side for which, as predicted by prospect theory, the status quo clearly worsened during the January 2009 dispute.

Since the dissolution of the Soviet Union, Russia has attempted to gain control of Ukraine's gas transit system (GTS), but to no avail. In her chapter, 'The struggle over Ukraine's gas transit pipeline network through the lenses of securitisation theory', Katerina Malygina answers the question of how Ukraine managed to sustain the pressure throughout this period. By reconstructing the main events surrounding the GTS privatisation debate in 1991–2009, she reveals the subtle power game between the Ukrainian government, which was attempting to increase its scope of activity, and the opposition, which was attempting to discredit the government and improve its own image. While the former employed a 'multipolarity' strategy, the latter used securitisation. The case of gas transit in Ukraine reveals the limits of Russia's energy leverage and rent seeking in Ukraine. For more than 20 years, Ukraine's GTS has remained under state control, despite numerous attempts to change its status. Undoubtedly, the role of the parliament in blocking GTS privatisation was and remains large. In addition to legislative means, the Ukrainian opposition heavily relied on public discourse in its power struggles with the government. The securitisation of GTS as an indispensable part of Ukraine's independence hindered its ultimate privatisation and gave Ukraine the strength to successfully resist Russia's pressure, leading, in some cases, to irrational and ineffective decisions with detrimental consequences for Ukraine's economy.

This collective volume concludes an annex that includes figures on the reserve bases and oil and gas production levels of the four main Eurasian producer countries under study and a bibliography with relevant literature on Eurasian export pipelines.

Part I. Russia's Foreign Energy Policy

András Molnár

1. Russian Foreign Energy Policy Under Vladimir Putin: Norms, Ideas, and Determining Factors

1.1. Introduction

Every country attempts to make use of its political relations, its economic and military potential, and every possible resource to gain prestige and strengthen its influence in the international system. What makes Russia special in this area is its loss of status as a superpower after the collapse of the Soviet Union in 1991, which led to the disintegration of the country and to a deep political, ideological, and economic crisis. As a result of these transformational difficulties, the gross domestic product (GDP) of the country dropped by almost half, and the decline of the Russian economy lasted until 1998.

The credit for the country's recovery in the 2000s was not due to its industrial achievement or to foreign investments but rather to the fast-paced increase in the world market prices of fossil energy sources. The increasing revenues from energy exports to the Russian state strengthened its national identity, which, in turn, encouraged the political leadership to make an effort to regain Russia's lost superpower status. However, relying solely on the country's military power did not seem possible. The technical obsolesce of Russia's huge weapon arsenal from the Soviet times made it inadequate as a 'hard power' tool to efficiently increase Moscow's influence in world politics. Therefore, Russia needed a better tool in the 21st century's increasingly multipolar system to regain its powerful position in the international arena.

Given the country's endowments, only the energy sector seemed to fit this requirement. However, because of ideological considerations and the lack of control over the energy sector during President Boris Yeltsin's term, Moscow put even less emphasis on energy policy in foreign policy contexts. This situation changed radically in 2000, when Vladimir Putin came to power.

Under President Putin, with a gradual shift from 'big pipeline', 'big business' became the determining element of the Russia's energy policy. The strategy of export maximising gave way to Moscow's aim of gradually strengthening its bargaining position in international political and economic relations and support for Russian foreign policy goals with the help of energy policy. With this move, Russia brought its 'soft power' to the foreground rather than its military and hard power politics. This was an understandable and tactical step because the 'energy weapon' as a foreign policy tool can be used in peacetime, as opposed to military force.

The primary aim of this study is to present and analyse the political, economic, and ideological motives and objectives behind the creation and structural development of

Russian energy foreign policy from the collapse of the Soviet Union until the present day.

1.2. Theoretical Background: Neoclassical Realism

Foreign policy analysis, in terms of generating a scientific background for research on energy politics, can be divided into two groups. One group of paradigms reflects geopolitical concepts of political power. The other analytical method's perspective is influenced by economy policy. Whereas the former considers energy a strategic and security factor,[1] the latter sees energy sources as internationally traded goods with which a country can gain political and economic benefits in the international system.[2] However, these two concepts—the 'securitisation' and 'commodification' of energy resources—cannot always explain the behaviour of Russia's external energy priorities and objectives. Therefore, Moscow's foreign energy policy cannot be explained through (geo)political or (geo)economic concepts exclusively.

The main reason for this is that Russia plays several simultaneous roles in international energy policy (producer, consumer, importer, exporter, and transit country). Consequently, taking only (geo)political or (geo)economic factors into consideration leads to misinterpretation and cannot provide a satisfactory explanation of Russia's complex and manifold system of foreign energy policy. Thus, to analyse the foreign energy policy of Russia, I will use the neoclassical realist theoretical background of international politics, given that this approach makes it possible to consider not only the international distribution of power (independent variable) and foreign policy decisions of states (dependent variable) but also the relevant variables of internal politics. As a result, this theoretical approach is the most appropriate tool to describe and analyse the complex structural, political, economic, and technological elements of Russian foreign energy policy.

Neoclassical realism is a diffuse approach based on classical realism and neorealism. It takes into account domestic variables and analyses their significance based on the assumption that these influence the political mechanisms of states.[3] Neoclassical realism incorporates foreign policy as a dependent variable to obtain an image of the relative power capabilities of a given state and how leaders perceive it. Accordingly, neoclassical realism sees state leaders rather than states as the most important actors

---

1   Blank, Stephen: Russia's new gas deal with China: background and implications, in: Northeast Asia Energy Focus 2009, (vol. 6), no. 4. pp. 16–30; Mankoff, Jeffrey: Russian foreign policy: the return of great power politics, Lanham, MD: Rowman & Littlefield, 2009.
2   Yergin, Daniel: The fundamentals of energy security, in: Committee on Foreign Affairs, US House of Representatives, 22 March 2007, http://democrats.foreignaffairs.house.gov/110/yer032207.htm; Finon, Dominique/ Locatelli, Catherine: Russian and European gas interdependence: could contractual trade channel geopolitics, in: Energy Policy, 2007 (vol. 36), no. 1, pp. 423–442.
3   Elman, Colin: Realism, in: Griffiths, Martin (ed.): International Relations theory for the twenty-first century: an introduction, New York: Routledge, 2007, pp. 11–20.

1. Russian Foreign Energy Policy Under Vladimir Putin

in the international system. As Zakaria emphasises, 'statesmen, not nations confront the international system and construct the architecture of the international arena'.[4] Nevertheless, statesmen's access to state power is limited; they can only use a given aspect (e.g., military, economic, or energy resources) extracted by a state apparatus to realise foreign and internal policy aims. As Zakaria describes the hypothesis of neoclassical realism, 'nations try to expand their national interests abroad when central decision-makers perceive a relative increase in state power'.[5]

Neoclassical realism distinguishes two types of power: national power and state power. 'The measure of national power is sometimes confined to military strength, but often [...] it is gauged using aggregated material indicators such as GDP, percentage of world trade, and population'.[6] The same type of power, in Waltz's interpretation, is highly convertible; economic power, for instance, can be transformed into military power, and vice versa.[7] Nevertheless, there are certain limitations. It is not true for all types of power that one can be converted into the other.

Zakaria defines state power as the 'function of national power and state strength and the ability of the state apparatus to extract national power for its purposes'.[8] This definition suggests that different states have different capabilities in international relations and that they are mainly determined by the broadness of the state, meaning that the state has sufficient power to achieve its aims. If a state has ambitions abroad, it is vital to have national power that certain state institutions can access and use for a given purpose. Energy resources are not part of foreign policy until it is possible to extract them for state purposes. The fact that governments and not nations influence foreign policy and have the potential to choose the appropriate instruments to implement politics supports the idea that neoclassical realist theory sees state power rationally and objectively. Given that the focal point of examination in neoclassical realism is material power and, accordingly, the state domestic structure and how statesmen perceive the international system, it provides an excellent theoretical background for the study of energy resources in foreign policy. These aspects provide an opportunity to put the role of energy resources in a given state's foreign policy into perspective. Moreover, they make it possible to examine differences in how energy resources are perceived in foreign policy even over periods when no change can be observed in national energy power. The example of Russia fits perfectly because energy resources became more important in foreign policy after changes in the presidency from Yelsin to Putin.

---

4   Zakaria, Fareed: From wealth to power: the unusual origins of America's world role, Princeton, NJ: Princeton University Press, 1998, p. 35.
5   Ibid., p. 42.
6   Ibid., p. 19.
7   Waltz, Kenneth N.: Realism and international politics, London: Routledge, 2008.
8   Zakaria, Fareed: From wealth to power: the unusual origins of America's world role, Princeton, NJ: Princeton University Press, 1998, p. 38.

Neoclassical realism also offers a simple means of analysis without the need to analyse domestic politics in detail. When changes in decision-making groups or changes in perception are found, they can be used to explain changes in state interests in the international system and changes in foreign policy. This is of vital importance when analysing energy resources because it may serve as an appropriate explanation for significant differences between states belonging to the same group, such as exporters, consumers, and transit states.

Taking advantage of energy resources to achieve specific goals abroad is characteristic of states where there is a closer relation between the energy sector and the government. Accordingly, if the energy sector is independent of the government, the transfer of energy resources from national power to state power is more problematic. This situation explains why states that are more distant from democracy use energy resources in foreign policy more commonly and more efficiently when expanding the state's interests abroad. Consequently, these states have a more advantageous position concerning the inclusion of energy resources in foreign policy and energy security questions, especially in the case of consumer or transit states.

Democracy in the governance structure manifests in decentralisation in decision-making processes and the involvement of competitiveness, which results in an increase in the number of procedures. Non-democratic states, however, have the advantage of transforming energy power to state power more efficiently by applying fewer procedures and non-democratic processes while expanding their interests abroad. The gained state power, then, can be used for foreign policy purposes, either as a 'petro-carrot' or as a 'petro-stick'. If we take Russia as an example, various cases could be enumerated for both uses of power.[9] Moreover, this way of wielding power has been gaining importance with the emergence of 'petro states' all over the world.

1.3. The Roots of Russia's Energy Influence

The roots of today's Russian oil and gas influence date to the country's Soviet past. From the early 1950s, it was a deliberate endeavour of Moscow to make the energy resources of the USSR crucial for both the East and the West of Europe. During the late 1950s and 1960s, the Soviet Union expanded its efforts to link itself economically to its Warsaw Pact allies. The measures of the Council for Mutual Economic Assistance (CMEA) concerning economic cooperation aimed to strengthen the military and ideological ties within the Soviet Bloc. The USSR's leading role in raw materials and energy supply to other countries of the bloc formed a principal part of the evolving idea of the

---

9   Goldman, Marshall I.: Petrostate: Putin, power, and the new Russia, Oxford: Oxford University Press, 2008.

'socialist division of labor'.[10] Subsidised fuel became indispensable for the emerging big energy consumer industries, such as petrochemical and fertiliser production or iron smelters.[11] The pattern of using heavily subsidised oil and prices as a reward for allies in return for their political compliance is evident. Critics of the Kremlin, such as Romania, have needed to rely on the much pricier world market for fuel.[12] This pattern has recently been applied by Russia as well.

As a result, a new aspect was added to the Soviet energy influence, namely the idea that further expansion of East European pipeline system towards the West made it possible to export large amounts of oil and gas to the capitalist Western countries.[13] The USSR expected that this would provide it with valuable hard currency and a unique opportunity in politics to guide Western Europe away from the influence of the US.

Throughout the 1960s, it was a serious concern of US leaders that the Soviets would gain economic influence over Western Europe through the new oil pipeline Druzhba (Friendship), which was under construction at the time running from Siberia to East Germany. The stage was set for the USSR to benefit from its energy resources in two ways: on the one hand, by offering subsidised prices to its allies; on the other hand, by achieving high profits by trading with enemies at the world market price. Either way, Moscow gained power. However, the world market price for oil collapsed in 1986. Numerous experts, including the former Russian Prime Minister Yegor Gaidar, agree that this situation contributed significantly to the collapse of the USSR.[14]

The years under President Yeltsin did not favour the regaining of the old Soviet 'petro power' by an independent Russia. The tendency towards decrease dragged on. Several reasons can be cited for this situation. First, the price of oil remained very low, usually ranging between US$15 and US$25 in the period between 1991 and 1999.[15] Second, as a consequence, the entire Russian economy took a nosedive, aggravating

---

10 Kramer, John: Soviet-CEMA energy ties, in: Problems of Communism, 1985 (vol. 34), no. 4, pp. 32–47; Newnham, Randall: Soviet oil and gas trade with Eastern Europe: developments in the 1980s, Santa Monica, CA: California Seminar on International Security and Foreign Policy, 1990 (Discussion Papers, no. 112).
11 Shadrina, Elena: Russia's foreign energy policy: norms, ideas and driving dynamics, Turku: Pan-European Institute, 2010, p. 110, http://www.utu.fi/fi/yksikot/tse/yksikot/PEI/raportit-ja-tietopa ketit/Documents/Shadrina_final_netti.pdf.
12 In 1979, Hungary received 97% of its energy imports from the USSR, Bulgaria 92%, and East Germany 90%. Romania, in contrast, received only 16% of its imports from the Moscow, having to buy the rest on the world market. Kramer, John: Soviet-CEMA energy ties, in: Problems of Communism, 1985 (vol. 34), no. 4, pp. 32–47.
13 Jentleson, Bruce: Pipeline politics: the complex political economy of East–West energy trade, Ithaca, NY: Cornell University Press, 1986.
14 Gaidar, Yegor: Collapse of an empire: lessons for modern Russia, Washington, DC: Brookings Institution Press, 2007.
15 British Petroleum: BP Statistical Review of World Energy, June 2012, http://www.bp.com/liveas sets/bp_internet/globalbp/globalbp_uk_english/reports_and_publications/statistical_energy_ review_2011/STAGING/local_assets/pdf/statistical_review_of_world_energy_full_report_2012. pdf.

the government's urgent need for hard currency exports. Third, the sharp decline in the economy unavoidably entailed a fallback in oil and gas production, which again hindered the recovery of the Russian economy. Finally, the privatisation of energy resources in Russia placed a significant part of the remaining 'petro-power' in the hands of businessmen instead of politicians.[16]

The disadvantages of the serious financial fallback that hit the Russian 'petro-power' were twofold. On the one hand, the urgent need for hard currency resulted in Russia's moving towards a policy of 'export at all costs'. On the other hand, to fill deep budget gaps, the Kremlin privatised state assets, including oil and gas fields, often at a fire sale price. Under the Yeltsin presidency, privatisation gradually removed state influence from the oil and gas sectors. By the end of the Yeltsin era, the state controlled barely 10% of Russia's oil production. The situation was somewhat better in the gas sector, but the power of the Kremlin weakened there as well. Most natural gas production remained under Gazprom, the state-owned energy giant created just before the fall of the USSR. Nevertheless, the state's stake decreased to approximately 38%; the rest was in the hands of domestic and foreign buyers. Parallel to the decline in the state's influence, Gazprom's actions were increasingly based on economic and not political calculations.[17] Its usefulness as a tool of state leverage was declining. Towards the end of the Yeltsin era, there were bold speculations about the full privatisation of Gazprom, which would have excluded the Kremlin from the energy sector completely. In the final years of the Yeltsin presidency, control over the oil and gas industry slipped out of the hands of the Kremlin to the extent that, in many aspects, the industry controlled it. This situation is not comparable to the forceful and self-conscious foreign energy policy characteristic of today's Russia.

1.4. The Putin Era

With the beginning of Vladimir Putin's presidency in 2000, Russian foreign policy received a new perspective, the hallmark of which was Putin's statement that the collapse of the Soviet Union was the 'greatest geopolitical catastrophe of the twentieth century'.[18] In particular, Putin argued that the liberalisation of the energy sector in the 1990s was not a well-based decision, and the most appropriate tool in the hands of the Russian government for the restoration of the state's economic and societal stability was mineral resources. Accordingly, the principal objective of Russia's foreign policy in the Putin years has been the restitution of Russia as a global player in world

---

16   Newnham, Randall: Oil, carrots, and sticks: Russia's energy resources as a foreign policy tool, in: Journal of Eurasian Studies, 2011 (vol. 2), no. 2, pp. 134–143.
17   Ibid., p. 137.
18   'Putin deplores collapse of USSR', in: BBC News, 25 April 2005, http://news.bbc.co.uk/2/hi/4480745.stm.

energy policy.[19] To achieve this, Putin recentralised political power and renationalised the crucially important energy sector.

Thus, the level of governmental control over the energy sector and its foreign policy significance increased significantly. According to Hanson, Putin's policy change was characterised by

> the move for control of parts of the economy—both by direct state ownership and by ensuring that politically compliant businessmen are running things—would on this view be a move to ensure that no significant base of independent social and political power exists.[20]

Based on Putin's dissertation, completed in 1997, and his article published in the Journal of Sankt Petersburg Mining Institute,[21] the following main goals in foreign energy policy can be identified.

The central idea is that the government should play a key role in major decisions concerning energy and natural endowments. There is no need for total control; rather, there is a need for a 'managed' market with various possible forms of ownership. Although the significance of market forces and private property is recognised, it is obvious that the decisive role of the state in Russia's energy sector is non-negotiable.[22] That is, Putin believes that even though private property would still exist, the property owner's rights would not be absolute, and the state's interests could overwrite of those of private firms and market regulations.[23] As Putin sees it, governmental control over mineral resources would contribute significantly to economic security and thus serve the interests of society.[24] In the same article, Putin claims that to catch up with the leading economies of the world, an annual economic growth rate between 4% and 6% is necessary.[25] In Balzer's opinion, this means that mineral resources, if used effectively, can ensure Russia's position in the world economy.[26] That is, the raw materials sector is vital for key aspects of the state, such as supporting industry or providing

---

19 Lennon, Alexander T./ Kozlowski, Amanda: Global powers in the 21st century, Cambridge, MA: MIT Press, 2008.
20 Hanson, Philipp: Managing the economy, in: White, Stephen/ Sakwa, Richard/ Hale, Henry (eds): Developments in Russian politics, London: Palgrave Macmillan, 2010, pp. 188–206, here p. 197.
21 Balzer, Harley: The Putin thesis and Russian energy policy, in: Post-Soviet Affairs, 2005 (vol. 21), no. 3, pp. 210–225.
22 Hober, Kaj: Law and policy in the Russian oil and gas sector, in: Journal of Energy and Natural Resources Law, 2009 (vol. 27), no. 3, pp. 420–432.
23 Olcott, Martha B.: Vladimir Putin and the geopolitics of oil, Houston, TX: Rice University, Institute for Public Policy, 2004, http://bakerinstitute.org/files/2778/.
24 Balzer, Harley: The Putin thesis and Russian energy policy, in: Post-Soviet Affairs, 2005 (vol. 21), no. 3, pp. 210–225.
25 Ibid.
26 Ibid.

the basis for the modernisation of Russia's military-industrial complex. Furthermore, it enhances social stability and elevates the quality of life of the population.[27]

Considering natural resources, the most effective tool for the reconstruction of the Russian economy and regional influence, Putin identified the three most important steps in his 1999 'Russia at the turn of the Millennium' speech[28]:
- The economic modernisation of Russia;
- The pursuit of stability in domestic politics;
- The enhancement of Russian national security.

The 'oil boom' at the end of the 1990s provided Russia with an opportunity to benefit greatly from its natural resource endowments. According to Goldman, the recently acquired wealth in connection with energy prices 'provide[d] Putin [...] with foreign policy leverage that is unprecedented in Russian, even Soviet, history'.[29] Putin recognised the key importance of Russia's abundance in natural resources in both the modernisation of economy and in ensuring the country's international position.[30] Additionally, the Putin years brought a shift in foreign policy from traditional 'hard' power vision to 'soft' power tactics.[31] Trenin's words indicate the value and the measure of this policy; according to him, 'fluctuating energy prices, not nuclear warheads, really matter to Moscow'.[32]

Therefore, it is not surprising that the main goal for Putin in his first presidential term was to restore central authority and thus to strengthen the Russian state. Bugajski claims that as a result, a system characterised by 'managed democracy' came into being, with a top-down controlled government.[33] The energy sector was no exception to this. Putin recognised the strategic value inherent in the natural gas sector and acted swiftly to bring the corporation into the hands of the government.

Putin's first measures aimed to restructure the energy sector according to the aforementioned ideas. By the end of Yeltsin's presidency, the Russian energy sector

---

27 Pleines, Heiko: Developing Russia's oil and gas industry, in: Perovic, Jeronim/ Orttung, Robert W./ Wenger, Andreas (eds): Russian energy power and foreign relations, London: Routledge, 2009, pp. 71–87.
28 Putin, Vladimir: Russia at the turn of the millennium, 30 December1999, http://pages.uoregon.edu/kimball/Putin.htm.
29 Goldman, Marshall I.: Moscow's new economic imperialism, in: Current History, 2008 (vol. 107), no. 711, pp. 322–329, here p. 326.
30 Olcott, Martha B.: Vladimir Putin and the geopolitics of oil, Houston, TX: Rice University, Institute for Public Policy, 2004, http://bakerinstitute.org/files/2778/.
31 Tsygankov, Andrey P.: Vladimir Putin's vision for Russia as a normal great power, in: Post Soviet Affairs, 2008 (vol. 21), no. 2, pp. 132–158; Hill, Fiona: Energy empire: oil, gas and Russia's revival, London: The Foreign Policy Centre, 2004; Stent, Angela E.: Restoration and revolution in Putin's foreign policy, in: Europe-Asia Studies, 2008 (vol. 60), no. 9, pp. 1089–1106.
32 Trenin, Dmitri: Russia redefines itself and its relations with the West, in: The Washington Quarterly, 2007 (vol. 30), no. 2, pp. 95–105, p. 95.
33 Bugajski, Janusz: Expanding Eurasia: Russia's European ambitions, Washington, DC.: Centre for Strategic and International Studies, 2008.

1. Russian Foreign Energy Policy Under Vladimir Putin

was dominated by thirteen major vertically integrated oil companies, eight of which were in private hands and three that were under governmental influence. At the end of Putin's second term, however, the number of oil companies was reduced to five. Pleines distinguishes two phases of the process, which are parallel to Putin's two presidential terms:[34]

- 1999–2004: In this period, the number of major oil companies was reduced from thirteen to eight. The only oil company that remained under governmental control was Rosneft, by reducing governmental ownership of the oil sector below 15%.
- 2004–2008: Five major companies dominated the energy industry of Russia, namely Gazprom, Rosneft, LUKOIL, Surgutneftegaz, and TNK-BP. The first two were under governmental control, and the rest were in private hands.

For example, Putin replaced the chairman of Gazprom, Rem Vyachirev, with one of his long-time associates, Alexei Miller, in 2001. In 2004, as part of the renationalisation of the company, the government regained the majority of the shares.[35] The final objective of renationalising Gazprom was the diffusion of state (or, rather, presidential) interests and the capabilities of the private sector.[36] According to Heinrich, the motive behind regaining state authority over Gazprom was the idea that governmental control over strategic sectors made it possible to protect Russian interests and increase the state's influence abroad.[37]

Another measure taken in the first term of Putin's presidency was to return Russian gas fields developed by foreign companies to the hands of the government. Putin managed to reverse concessions made to several foreign energy companies, achieving majority stakes in foreign development programmes.[38]

Indeed, the significance of the energy sector as a foreign policy tool soon became apparent, as the re-emerging Russian political and economic interests were posed to challenge by the 'colour revolutions' in Georgia in 2003 and in Ukraine in 2004.

Ukraine, like most countries of the former Soviet Union (FSU), benefited from heavily subsidised gas prices.[39] Nevertheless, a type of interdependence characterises the

---
34 Pleines, Heiko: Developing Russia's oil and gas industry, in: Perovic, Jeronim/ Orttung, Robert W./ Wenger Andreas (eds): Russian energy power and foreign relations, London: Routledge, 2009, pp. 71–87.
35 Olcott, Martha B.: Vladimir Putin and the geopolitics of oil, Houston, TX: Rice University, Institute for Public Policy, 2004, p. 3, http://bakerinstitute.org/files/2778/.
36 Goldman, Marshall I.: Moscow's new economic imperialism, in: Current History, 2008 (vol. 107), no. 711, pp. 322–329.
37 Heinrich, Andreas: Under the Kremlin's thumb: does increased state control in the Russian gas sector endanger European energy security?, in: Europe-Asia Studies, 2008, (vol. 60), no. 9, pp. 1539–1574, here p. 1563–1565.
38 Orttung, Robert: Energy and state-society relation: socio-political aspects of Russia's energy wealth, in: Perovic, Jeronim/ Orttung, Robert W./ Wenger Andreas (eds): Russian energy power and foreign relations, London: Routledge, 2009, pp. 51–70.
39 Viter, Olena/ Pavlenko, Rostyslav/ Honchar, Mykhaylo: Ukraine: post-revolution energy policy and relations with Russia, London: GMB Publishing, 2006, pp. 19–20.

bilateral relationship between the two countries. Ukraine depends on Russia in terms of gas imports for domestic consumption, and Russia, in turn, depends on Ukraine due to the country's role as a transit state for deliveries to the European Union (EU). After the collapse of the Soviet Union in 1991, Ukraine's leaders slowly began to build stronger ties with Western institutions such as the EU and NATO (North Atlantic Treaty Organization). Parallel to Ukraine's drifting away from Russian influence, Moscow began to foster ideas of introducing 'market prices' in natural gas export dealings with Kiev. This involved a price hike, often a threefold increase over a one-year period. At the beginning of 2008, it became apparent that a gas conflict was developing between Ukraine and Russia. Upon Ukraine's unwillingness to pay the elevated price for its gas imports, Russia's answer was first to limit deliveries to Ukraine by the amount the country consumed. Then, when it turned out that Ukraine had been using the gas destined for the European costumers of Russia, Moscow decided to cut all deliveries to Ukraine.[40]

Of course, the root cause of the conflict was seen differently by the two sides. The Ukrainians saw the gas cut as revenge for the 'Orange Revolution' and as the manifestation of Russia's intention to obstruct Ukraine's drifting towards the West, namely its NATO membership bid. From Russia's perspective, Ukraine was guilty of not paying its debts for gas consumption and using gas deliveries destined for the EU.

Russia's foreign policy trends that began in the Putin years continued during Dmitrii Medvedev's presidency (2008–2012). Russia's assertive use of natural gas for foreign policy goals in the 2009 and 2010 gas 'wars' with the former Soviet republics of Belarus and Ukraine supports this view. In June 2010, Gazprom cut off the flow of gas to Belarus. According to Sergeyev, the source of the conflict was Belarus' inability to keep up with gas payments, accumulating to a debt of US$192 million.[41] This situation reflected the souring of the once close relations between Belarus and Russia. The ties between the two states were also becoming increasingly strained because of Belarus' initiatives to build closer relationships with the EU and its refusal to join the Russian-backed customs union with Kazakhstan.

Concerning the experiences and the developments of Russian foreign policy in the last decade, the question arises of what conclusions can be drawn concerning its future trends and goals by means of neoclassical realism.

---

40 Westphal, Kirsten: Russian gas, Ukrainian pipelines, and European supply security: lessons of the 2009 controversies, Berlin: SWP, 2009, p. 9, http://www.swp-berlin.org/fileadmin/contents/products/research_papers/2009_RP11_wep_ks.pdf.
41 Sergeyev, Dimitry: Russia, Belarus, gas talks fail cut off looms, in: Reuters, 19 June 2010, http://www.reuters.com/article/idUSLDE65I0C720100619.

1.5. Strategic Goals of Russia's Foreign Energy Policy: the Nexus Between Energy Policy and National Priorities

Western countries see Russia's energy-related decisions that do not match their interests as Russia's use of the 'energy weapon' or 'flexing energy muscles' to realise foreign policy goals.[42] However, the role of energy policy has often been one of a multiplier rather than a cause. Most commonly, the analysis of contemporary Russian foreign policy is based on the 'West vs East' paradigm. The traditional distinction between the East and the West dates back to the Slavophiles' initiatives to define Russian national identity and the country's place in the world. The most frequently debated question between the two sides was whether Russia should choose integration or isolation—in other words, whether it should remain inward-oriented or assimilate and adapt to the West and thus become an integral part of it.[43]

Concerning Russia's foreign energy policy, a general phenomenon that can be observed is characterised by multipolarity in the shifting balance between Europeanism[44] and Eurasianism.[45] As a result of the strong commercial relations with Europe—a perfect example of which is energy trade—Europeanism has been the prevailing view in Russian energy policy. According to European-oriented researchers and politicians, Russia's lagging behind Europe in terms of economy and technology not only indicates a different path of development but also hinders the state from making full use of its position as a bridge between Europe and Asia.[46]

The significant upsurge of the external use of oil, gas, and other raw materials from the 2000s on facilitated the diffusion of a pragmatic way of thinking in today's Russian foreign energy policy-making. Although energy is frequently seen as an ace in the hole for Russia's foreign policy goals and as a tool that unavoidably entails coercion and blackmail,[47] it seems to be a highly simplistic approach.

---

42 Smith Stegen, Karen: Deconstructing the 'energy weapon': Russia's threat to Europe as case study, in: Energy Policy, 2011 (vol. 39), no. 10, pp. 6505–6513.
43 Buzan, Barry/ Wæver, Ole: Regions and powers: the structure of international security, Cambridge: Cambridge University Press, 2003, pp. 398–405.
44 According to Shadrina, 'Europeanism holds that to reach a level of economic development somewhat comparable to the West's, Russia needs to discontinue its traditional Great Power politics and concentrate on solving domestic economic problems' (Shadrina, Elena: Russia's foreign energy policy: norms, ideas and driving dynamics, Turku: Pan-European Institute, 2010, p. 36, http://www.utu.fi/fi/yksikot/tse/yksikot/PEI/raportit-ja-tietopaketit/Documents/Shadrina_final_netti.pdf).
45 According to Shadrina, '[…] since 1991 the geopolitical situation changed drastically literally challenging Russia's security at all fronts. Thus, Eurasianism embracing both the West (Euro-Atlantic) and the East (Asia-Pacific) stands to be the only policy mode to adequately tackle a vast array of lingering uncertainties and potential threats' (ibid.).
46 Ibid., p. 26.
47 Liuhto, Kari: Energy in Russia's foreign policy, Turku: Pan-European Institute, 2010, http://www.utu.fi/fi/yksikot/tse/yksikot/PEI/raportit-ja-tietopaketit/Documents/Liuhto_final_netti.pdf.

The strategic goals in Russian foreign energy policy, described in the following, are based on two suppositions:
1. Russia's inherited desire to become a great power by gaining leverage in the post-Soviet space.
2. Russia's energy policy is likely to shift in the direction of 'state energy capitalism'.

Of course, the strategic goals of Russian foreign policy are subordinated to more general Russian foreign policy objectives. Bugajski enumerates six broader objectives in Russian foreign policy:[48]

- Strengthening foreign policy influences to re-establish at least part of the one-time power of the USSR.
- Putting Russia's energy resources and capital investments under as much governmental control as possible.
- Restricting Western influence in Russia's zone of interest.
- Regaining global influence by building upon Eastern Europe and the post-Soviet space to expand Russia's field of influence and great power.
- Expanding the monopolisation of the economy through pre-determined foreign investments and strategic infrastructure buyouts.
- Eliminating US predominance by de-emphasising links between Europe and the US.

Based on the previous two suppositions and six core objectives of Russia's foreign policy, the following five conclusions can be drawn concerning Russia's strategic foreign energy policy aims:

1. Russia uses the energy dependencies of FSU countries to gain greater influence over them and to develop Russia into a post-Soviet regional power

Belarus's dependence on Russia is already at a critical phase, and, assuming that the current trend continues, Ukraine is likely to follow soon. Kazakhstan may remain under Russia's influence because of the Customs Union, but Central Asia seems to fall within the reach of China,[49] forcing Russia to implement a CoCoCo strategy in Central Asia: to co-operate, co-exist, and compete with China.

2. Russia intends to push up gas prices

It is in Russia's interest to obstruct the spread of unconventional gas (shale gas) production and consumption in the EU so that gas imports and prices remain high. Although Russia remains outside of OPEC (Organization of Petroleum Exporting Countries), it

---

48 Bugajski, Janusz: Cold peace: Russia's new imperialism, London: Prager, 2004.
49 Russia must pay attention to the increasing influence of China and its resurgent interest in the possible energy sources in Central Asia. The Easter Siberian and Central Asian Russian population is shrinking, and the migratory influx in the border areas continues to grow.

supports the Gas Exporting Countries Forum[50] to increase gas prices. Furthermore, Russia aims to break the connection between oil and gas pricing. Political pricing is likely to persist in strategically important countries, such as Belarus.

3. *Strategic foreign energy policy objectives of Russia continue to be influenced by individual interests*

Russia's energy policy is still strongly shaped by influential political actors and business clans despite the exclusion of the oligarchs from politics in the 2000s and stronger governmental control over the oil and gas industry. Russia's actual foreign energy policy is influenced by many factors, such as the presidential administration, the government, and the state-run energy corporations. Occasionally, this may lead to unnecessary uncertainties and short-term actions, whereas in other cases, the personal interests of the Russian politico-economic elite may better match those of their Western counterparts than the national interests of the Russian government.

4. *In the absence of sustainable rules in East–West energy co-operation, the probability of Russia's foreign energy policies shifting towards a more conservative direction is very high*

If conservative trends increase in Russian energy policy, its foreign influences may be manifold. First, foreign energy companies in Russia will be placed in a more disadvantageous situation; as a result, Western firms will be hesitant to join Russian energy savings programmes or unconventional gas production. Second, Russia will implement its foreign energy policy arsenal more often, especially in the FSU, with increasingly aggressive measures. Third, the decrease in the production of hydrocarbons will entail changes in the mentality of energy collaboration as well; energy will become a subject of bargaining rather than an integrative bridge.

1.6. Conclusion

Changing circumstances in the international system put increasing emphasis on non-military aspects of power. Because there is no acting global supranational organisation to handle energy resource-related issues in foreign policy, states are the main actors for energy resources in foreign policy. Given that energy resources are material objects, a materialistic ontology is needed to conduct well-based research on these resources. Thus, a realistic paradigm is best suited for this analysis. Energy resources in foreign policy should be examined in a realistic paradigm because its focal point is material power or hard power.

---

50 The member states are Algeria, Bolivia, Egypt, Equatorial Guinea, Iran, Libya, Nigeria, Norway, Oman, Qatar, Russia, Trinidad and Tobago, UEA, and Venezuela. Iraq, Kazakhstan, and the Netherlands have observer status.

In addition to maintaining its influence over energy resources, the state also makes efforts to expand this influence as a potential tool for strategic aims. Moreover, no universal energy market exists; thus, it can be concluded that the liberal paradigm should be waived.

States are rational actors. However, in the case of energy resources, this does not always coincide with economic rationality. States may subordinate economic considerations to building state power and influence abroad. Dating back to the intellectual debate concerning Russia's role in the world and ideal path of development, the 'East or West?' question has affected Russian foreign energy policy in various ways. However, the current energy policy of Russia represents neither side; rather, a diffusion of several (geo-)political-economic considerations has led to the emergence of a multidimensional policy.

As mentioned, Russian foreign energy policy is shaped by a wide variety of external and internal factors. Olcott provides a straightforward description of the crucial point as the follows:

> The primacy of the Russian state in the country's energy sector is non-negotiable. While Vladimir Putin recognizes the importance of market forces and the need to protect private property, he believes that both must be managed to ensure that neither takes precedence over the interests of the state, which exercises its control in the name of the Russian people.[51]

Russia's long-standing export dependence is the basis of its major energy security concerns. An extended export disruption or a significant fallback in its volumes would endanger the capability of the Russian government and industry to maintain economic development, putting national security at risk in the long run.

On these grounds, it can be concluded that three major mechanisms of geographic, economic, and political realms will provide the basis for policy transformations. The idea of Eurasianism is likely to contribute to the shaping of Russia's relations with both the East and the West. Pragmatist considerations, which were the first to influence Russia's foreign policy shift, will persist as background motives in every respective geographical dimension. Lastly, the idea of multipolarity, which focuses on guaranteeing a secure environment for domestic sustainable development by maintaining regional stability and enhancing international cooperation, also continues as an important influence on Russia's foreign energy policy.

---

51   Olcott, Martha B.: Vladimir Putin and the geopolitics of oil, Houston, TX: Rice University, Institute for Public Policy, 2004, p. 3, http://bakerinstitute.org/files/2778/.

Inna Chuvychkina

2. An Actor-Centred Institutionalist Approach to Russia's Pipeline Policies

2.1. Introduction

Oil and gas pipeline construction is one of the central pillars of Russian energy policy as well as an important dimension of the specific economic and political system designed by Russian President Vladimir Putin. The development of oil and gas export pipeline projects was caused not only by a shortage of transport capacity and the aim of overcoming Russia's dependence on transit countries but also by the desire to diversify Russia's export direction (i.e., promoting the eastern vector of energy export). On the one hand, the development of a pipeline infrastructure is impartially required to supply energy resources from the new oil and gas fields to a consumer market. On the other hand, pipeline policy serves the different interests of the political elite. Therefore, investing in some pipeline projects is related to the pursuit of (geo)political goals and power-seeking interests, whereas economic considerations of the pipeline project are ignored by state actors and economic elites. Pipeline construction is also influenced by the material context and profit-seeking strategies of the interest groups. For instance, given that every pipeline project has unique technical characteristics, capital expenditures for pipeline construction in Russia are many times higher than in other developed countries. Oil and gas companies, pipeline construction companies, manufacturers of steel pipes, and pipe trading companies lobby for energy infrastructure projects. They are interested in the rapid development of the construction and infrastructure sector because it can generate high levels of kickbacks from the building of pipelines.

This chapter analyses how institutional arrangements affect the formulation of Russia's pipeline policy as well as its political and economic outcomes. It focuses on the analysis of pipeline politics after 2000, when a new framework for institutional interactions in the relationship between business and state was established. This shift refers to the weakening of Russian business tycoons, the redistribution of power resources, and the centralisation of state control over the energy sector. This chapter considers only oil and gas pipelines that are already constructed or under construction. Planned pipelines are not included.

The chapter proceeds as follows. In the first section, the institutional framework is introduced, in which different state and non-state actors are represented. Following this, the constellation of actors and their interests are discussed. The political and economic outcomes are specified by examining pipeline policy making. The conclusion discusses the influence and implications of institutional arrangements on the preferences and interaction of actors.

2.2. Theoretical Considerations

Actor-centred institutionalism describes the influence of institutions on the preferences and perceptions of actors and their forms of interaction. The central analytical categories are institutions, actors, and their constellations as well as the forms and orientations of interaction. The approach begins from the assumption that 'the analysis of structures without reference to actors remains as deficient as the analysis of player behavior without reference to structures'.[1] Actor-centred institutionalism is thus based on the assumption that policy decisions are made under institutional framework conditions. The preferences, resources, and action strategies of actors shape their constellations, which can be either cooperative or conflicting.

To evaluate the results of these interactions, the approach makes use of game theory, which analyses interactions under conditions of mutual dependence. Constructing the game matrices may help to explain the outcomes. For instance, the theory of non-cooperative games can be a useful tool to explain the unilateral actions of actors. In this case, the payoff matrix can identify the profit maximisation policy of one actor and the vulnerability of outcome due to the response actions of another actor.

The clash of actors in solving problems or achieving goals occurs in an environment that is characterised by a combination of cooperation and the promotion of divergent factors. The institutional context influences both the preference order of the players and the forms of interaction.

Institutions are defined as 'systems of rules that structure the courses of actions that a set of actors may choose'.[2] Institutions are considered both dependent and independent variables that facilitate and restrict the actions of actors. At the same time, they have no determinative effect. Thus, institutions form a context for action and influence behaviour. They constitute the actors and their constellations through resources and competences. However, they may be altered by the actions of actors. North emphasises that 'institutions define and limit the choice sets of actors and thus serve as the framework for interaction'.[3]

In contrast to institutions, actors are capable of acting and of producing goals, standards, and actions. Actor-centred institutionalism is based on the assumptions of

---

1   Mayntz, Renate/ Scharpf, Fritz W.: Der Ansatz des akteurszentrierten Institutionalismus, in: Mayntz, Renate/ Scharpf, Fritz W. (eds): Gesellschaftliche Selbstregelung und politische Steuerung, Frankfurt am Main: Campus, 1995, pp. 40–72, here p. 46.
2   Scharpf, Fritz W.: Games real actors play: actor-centered institutionalism in policy research, Bolder, CO: Westview Press, 2007, p. 38.
3   North, Douglass C.: Institutions, institutional change, and economic performance, Cambridge: Cambridge University Press, 1990, p. 4.

2. An Actor-Centred Institutionalist Approach to Russia's Pipeline Policies

the rational choice approach and ascribes rational behaviour to actors that aims to maximise profits. The actors' skills, perceptions, and preferences predetermine their actions. The concept of actor constellations includes strategies, perceptions, and evaluations of achievable results and refers to resources for action and the orientations of the actors. The interaction modes can be identified as follows: individualistic, competitive, solidarity, altruism, and hostile.

The modes of interaction are influenced by institutional rules and the institutional context. Four orientations are particularly relevant to explain the policy decisions of actors: unilateral action, negotiated agreement, majority vote, and hierarchical direction.[4]

'Unilateral action' involves a high degree of cooperation and coordination losses and occurs in the case of minimal institutions and the minimal presence (or even the absence) of rules and obligations. 'Negotiated agreement', as a second form of interaction, refers to arrangements between actors that make factual but temporary decisions. This form of interaction is associated with high transaction costs because it is necessary to overcome the negotiating dilemma. The actors are dependent on finding optimal solutions and resolving conflicts, which requires cooperative interaction orientation, effective communication, and mutual trust. 'Majority vote' involves lower transaction costs because the decisions are made by majority decisions or are caused by hierarchical control. In the fourth form of interaction, decisions are imposed hierarchically. 'Hierarchical direction' is a way to eliminate conflicting interests and to change the payoff matrices in the prisoner's dilemma. It overcomes the problem of collective action when

> the members of a large group rationally seek to maximize their personal welfare, they will not act to advance their common or group objectives unless there is coercion to force them to do so, or unless some separate incentive, distinct from the achievement of the common or group interests, is offered to the members of the group individually on the conditions that they help bear the costs or burdens involved in the achievement of the group objectives.[5]

Through the hierarchy, selective incentives are allocated and/or monitoring is performed. This process allows the stabilisation of collective action. Under the condition that potential sanctioning can be used, welfare-maximising decisions are thus dependent on success.

---

4   Scharpf, Fritz W.: Games real actors play: actor-centered institutionalism in policy research, Bolder, CO: Westview Press, 2007, p. 46.
5   Olson, Mancur: The logic of collective action: public goods and the theory of groups, Cambridge, MA: Harvard University Press, 1971, p. 2.

2.3. Influence of Institutional Design on the Policy-Making Process

The current political system in Russia is characterised by the presence of neopatrimonial rules and patron-client relations as well as high levels of corruption. According to the typology of Acemoglu and Robinson, Russia is an extractive type of state. The main features are the monopolisation and centralisation of power, the concentration of economic resources in the hands of the ruling elite, the suppression of competition, and the preservation of existing modes of production as well as a lack of mechanisms for changing elites. In contrast to 'extractive institutions', inclusive political institutions tend to protect individual rights, the rule of law, and an equal distribution of income.[6]

Resource nationalism is a key factor in the development of the Russian energy complex. Resource-rich countries seek to protect their economic benefits and rents from the exploration of natural resources and to achieve political gains by supply restrictions. This situation may not only affect the operations and financial sources of energy companies working in these countries but may also destabilise the global energy markets. The Resource Nationalism Index covers 197 countries and rates the stability, transparency, and robustness of political and legal institutions, respect for property rights, and economic factors, such as indebtedness and dependence on natural resources. Russia ranked 15th in the Index in 2012 and is considered a country with a high resource nationalism risk.[7] In this context, some studies show that

> rising oil prices have triggered a new wave of nationalization in the oil and gas sector, particularly in 'petro-authoritarian' countries with weak political institutions. The more valuable oil and gas assets, the stronger the incentives seem to be to expropriate them. Nationalization and cartelization trends have in turn contributed to higher oil prices, since the state control of production almost inevitably leads to inefficiency and corruption which raise export prices. Instead of investing in the exploration, production and refining sectors, the revenues from state-owned companies are more often than not diverted from social welfare programmes and government operating budgets to arms purchases and foreign aid.[8]

Huge profits from the export of energy resources are obstacles to the modernisation of the energy sector and the economy as a whole. Collier and Godernis found a correlation between energy prices and economic growth. They emphasise that high energy prices affect economic growth positively in the short term but can retard it in the long

---

6 Acemoglu, Daron/ Robinson James A.: Why nations fail: the origins of power, prosperity and poverty, London: Profile Books, 2012.
7 '44% of global oil production taking place in countries with a "high risk" of resource nationalism—new report', in: www.maplecroft.com, 1 March 2012, http://maplecroft.com/about/news/resource_nationalism_index_2012.html.
8 Umbach, Frank: Energy security in Eurasia: clashing interests, in: Dellecker, Adrian/ Gomart, Thomas (eds): Russian energy security and foreign policy, London: Routledge, 2011, pp. 23–38, here p. 24.

2. An Actor-Centred Institutionalist Approach to Russia's Pipeline Policies

run. Considering the question of why income from the export of energy resources in different countries affects producers differently, the authors note the relationships between the resource curse and political processes and institutions. The influx of windfall profits allows authoritarian regimes to focus on the distribution of resource rents and to pursue populist and inefficient policies rather than investing in the development of the economy. However, this is not the case in countries with strong political institutions.[9]

A study by Van der Ploeg confirms these assumptions and shows how natural resources affect politics, economic and institutional development, rent-seeking, and the occurrence of civil conflicts. Natural resource abundance per se is not a curse and can have a positive effect on economic growth if states have stable political and legal institutions. However, some resource-rich countries are not able to implement their resource rents for prosperity. These negative effects of resource wealth are reflected in unstable countries with poor institutions and a lack of the rule of law, high levels of corruption, and underdeveloped financial systems. The wealth of resources also increases 'rent grabbing and civil conflict especially if institutions are bad, induces corruption especially in non-democratic countries, and keeps in place bad policies'.[10]

The impact of natural resource abundance on political regimes is examined in a study by Caselli and Tesei. In democratic states, increasing resource rents do not affect the level of democracy. In authoritarian states, however, an increase in resource rents has significant political consequences and causes countries to be increasingly autocratic. These authors note that in 'autocracies, spending on self-preservation is an increasing and concave function of the resource rents'.[11] The distribution of rents among some political groups allows them to remain in power. This situation slows the long-term development of the economy.

In summary, a better institutional environment enhances economic growth, whereas corruption has a negative impact on effectiveness, competitiveness, and economic development. The study by Akindinova et al. notes that 'improving [the] quality of the main institutes, characterizing protection of personal security and own-

---

9 Collier, Paul/ Goderis, Benedikt: Commodity prices, growth, and the natural resource curse: reconciling a conundrum, Oxford: University of Oxford, Department of Economics, mimeo, 2008, (MPRA Paper no. 17315, http://mpra.ub.uni-muenchen.de/17315/).
10 Van der Ploeg, Frederick: Natural resources: curse or blessing?, in: Journal of Economic Literature, 2011 (vol. 49), no. 2, pp. 366–420.
11 Caselli, Francesco/ Tesei, Andrea: Resource windfalls, political regimes, and political stability, London: London School of Economics and Political Science, Centre for Economic Performance, mimeo, 2011 (CEP Discussion Paper No 1091, http://cep.lse.ac.uk/pubs/download/dp1091.pdf), here p. 9.

ership rights, obeyance of laws and legal framework, effectiveness of bureaucracy are positively connected with the rates of economic growth".[12]

2.4. Actors and Actor Constellations

Tkachenko distinguishes between a number of actors within Russian energy policy, including the president and the presidential administrations, ministries of the Russian federal government, regional administrations in the Northwestern Federal District of Russia, legislative assemblies of Russian regions, leading energy companies, and other business and financial actors. He also notes that the Russian oil and gas industry is in private hands, but its connection with the government remains strong. This situation leads to competing interests in the energy sector: public and private, national and foreign, conservative and liberal, federal and regional. Tkachenko emphasises that the president is the real centre of power, and it is in his interest to use the energy resources as an instrument of state power.[13]

In addition to the political-administrative elite, business elites have the potential to influence and play a role in political decision-making.[14] The involvement of the business elite and contacts with the state actors take usually place when there is a potential for tangible benefits and greater profits. The business elite's means of political influence is lobbying.[15]

Lobbying is a widespread practice of business communication with government representatives. To influence government policy, business actors employ both direct and indirect methods. The main lobbying techniques involve direct influence through personal contacts and consultations to achieve goals. Indirect lobbying takes the form of bribery or the financing of electoral activity and is hidden from the public eye.

Given the state of affairs in the Russian lobbying system, it should be noted that this system represents an interpenetration of business elite lobbying and politics conducted predominantly through informal relationships. The lobbying practices are based on personal ties to government deputies and informal cooperation between friends, family enterprises, partners, and close associates. Informal networks are becoming more important; they are 'always personal and link up individuals or groups that share

---

12 Akindinova, N.V./ Aleksashenko, S.V./ Yasin, E.G.: Scenarios and challenges of macroeconomic policy: report at the XII international academic conference on economic and social development, Moscow, 5–7 April 2011, Moscow: HSE Publishing House, 2011.
13 Tkachenko, Stanislav L.: Actors in Russia's energy policy towards the EU, in: Aalto, Pami (ed.): The EU-Russian energy dialogue: Europe's future energy security, Aldershot: Ashgate, 2008, pp. 163–192.
14 Tschirikowa, Alla: Regionale Eliten und regionale Machtstruktur, in: Goszka, Gabriele/ Schulze, Peter (Hg.): Russlands Weg zur Zivilgesellschaft, Bremen: Edition Temmen, 2000, p. 127-139.
15 Kusznir, Julia: Der politische Einfluss von Wirtschaftseliten in russischen Regionen, Stuttgart: ibidem Verlag, 2008, p. 55.

2. An Actor-Centred Institutionalist Approach to Russia's Pipeline Policies

similar interests, allegiances and identification. In the Russian case, there is a clear tendency for the elite networks to identify themselves with the state, which makes it very difficult to distinguish between the two'.[16]

Among the lobby patterns that distinguish the main aims and favours of oil and gas companies are (1) obtaining governmental budgets to promote their own corporate activities (government contracts, subsidies), (2) obtaining state privileges (tax breaks, a monopoly or favourable export quotas), and (3) eliminating government regulations that hinder their own corporate activities (foreign trade restrictions, safety or environmental regulations, license requirements).[17]

The system of interest intermediation is organised by various lobbying associations. The Russian Union of Industrialist and Enterprises (RSPP) represents the interests of business circles from the energy sector as well as from other economic sectors. The Russian Gas Society (RGO) is a non-commercial partnership of companies in the oil and gas industry; its members are Russian gas producers, gas transport companies, regional gas distribution, sales organisations, and gas research institutes. Other channels of interest representation are the Intergovernmental Commission on Fuel and Energy Complex and the Council for Competitiveness and Entrepreneurship under the Government of the Russian Federation. These channels are mostly coordinative and consultative institutions. Their structures not only serve to coordinate between representatives of the legislature but also contribute to the exchange of views with representatives of the business community.

Although there are a number of industry lobbying associations, the system of communication between government agencies and non-state actors remains at a low level. Until 2004, a number of leading tycoons participated in the activities of the RSPP, and the coordinated actions were mutually beneficial. The business elite have had access to policy makers, and their interests have been primarily represented. The state benefited from the information on 'where measures need to be adopted and to determine whether planned policies have sufficient support from those directly involved to ensure that they will be implemented'.[18]

Currently RSPP 'continue[s] to be active in formulating policy advice at the more technical end of the scale, and in managing some inter-firm disputes. [...] What the RSPP no longer does is defend the shared interests of big business against the state, even the common interest in making perfectly legal use of tax loopholes'.[19] However,

---

16 Kononenko, Vadim/ Moshes, Arkady: Russia as a network state: what works in Russia when state institutions do not? Basingstoke: Palgrave Macmillan, 2011, p. 6.
17 Pleines, Heiko: Wirtschaftseliten und Politik im Russland der Jelzin-Ära (1994–99), Münster: Lit, 2003, p. 64.
18 Hanson, Philip/ Teague, Elizabeth: Big business and the state in Russia, in: Europe-Asia Studies, 2005 (vol. 57), no. 5, pp. 657–680, here p. 658.
19 Ibid, here p. 664.

both the industry's lobbying associations and some government agencies are merely a façade. For example, the Ministry of Energy of the Russian Federation does not play a role in decision-making; its activities are mainly reduced to the preparation of strategic documents (for instance, elaborating the Energy Strategy of Russia until 2030) and the compilation of statistical data.

In addition, there is intra-elite struggle for control over decision-making in the energy sector. During Putin's term as prime minister (2008–2012), the centre of the decision-making process on the fuel and energy complex was the Russian government (headed during this period by Igor Sechin and after 2012 by Arkadii Dvorkovich). However, in 2012, the President's Commission for the Strategic Development of the Fuel and Energy Complex and Environmental Safety was established with Putin as chairman and Sechin as responsible secretary. The main functions of the Commission are price controls on the internal market, fiscal policy, and decisions on strategies for developing the sector as a whole.[20] After the establishment of the Commission, the decision-making centre of the energy sector shifted again to the presidential administration, and the influence of the prime minister on energy policy has weakened.

As noted above, the management and control of the energy sector are within the purview of the president, who is also the arbiter in case of intra-elite conflicts. Thus, there is a one-man political dominance and no alternative political pole in Russia.

A study by Dyatlikovich and Chapkovsky aimed to identify the personal ties of the ruling elite, the upper layer of the executive authorities of Russia (i.e., members of the government, the leadership of the presidential administration, intelligence and security service, and heads of state-owned companies). The authors traced the personal connections between the members of government in 2000 and 2011. They concluded that there was a consolidation and strengthening of small groups of nomenclature. Most of the staff recruiting and even the political processes of the past twenty years are easily explained by personal relationships without the involvement of other factors. The entire political system is built on personal loyalty. In addition, the system reproduces itself; the key positions are replaced by familiar and loyal people. Putin plays a mediating role among various interest groups.[21]

Rutland observes that 'lobbying mainly takes the form of individual approaches to presidential and governmental leaders who can issue decrees or instructions granting the required exemption. This lobbying flows through networks of personal contacts'.[22]

---

20   Paszyc, Ewa: The 'energy tandem': Putin and Sechin control the Russian energy sector, in: OSW Eastweek, 20 June 2012, http://www.osw.waw.pl/en/publikacje/eastweek/2012-06-20/energy-tandem-putin-and-sechin-control-russian-energy-sector.
21   Dyatlikovich, Viktor/ Chapkovskii, Filipp: Kto est' kto i pochemu v rossiiskoi elite. "Sotsial'naya set'" federal'nykh chinovnikov, in: www.rusrep.ru, 7 September 2011, http://rusrep.ru/article/2011/09/07/who_is_who.
22   Rutland, Peter: Business lobbies in contemporary Russia, in: The International Spectator, 2007 (vol. 32), no. 1, pp. 23–37, here p. 32.

2. An Actor-Centred Institutionalist Approach to Russia's Pipeline Policies

Given the role of the president as the mediator between competing elite groups and the economic elite, the decision-making process is associated with high transaction costs that require the identification of an optimal solution and the resolution of conflicting interests. The interaction modus increasingly takes the form of negotiated agreement in terms of actor-centred institutionalism. Acting as the centre of real power, the president can solve the problem of collective action through hierarchical direction and control.

2.5. Economic Outcomes of the Pipeline Policy

The interplay between the actors in the energy sector is characterised by high interaction density. The oil and gas industry accounts for the largest share of government revenue, is the monetary potential of the country, and determines the development of other sectors of the economy through investment programmes. This situation leads to continual consultations, strategic planning, and discussion of options for action.

The state has an immediate impact on the oil and gas industry. On one hand, it affects energy exports through the tax and customs policy. On the other hand, it controls the most important actors in the energy sector (Gazprom, Rosneft). The oil pipelines are managed by the state company Transneft, whereas the gas supply pipelines are controlled by Gazprom. Government influence on the energy sector is thus exercised less through ownership than through control of transport routes and export flows.[23]

In general, the degree of government control and the politicisation of the state-controlled companies differs significantly from country to country. Some companies are considered efficient and competitive firms that follow an entrepreneurial logic. In other countries, ideological orientations and political targets determine the business decisions. The type of institutional environment has a greater effect on the efficiency of company operations than the actual form of ownership. State-owned companies in the oil and gas sector can be 'both efficient and incorruptible' and can perform impressively for a long time, most notably Norway's state oil company Statoil.[24] However, these companies require competent management that is shielded from political meddling, and they are embedded in strong institutional settings with secure property rights, clear rules for the market players, and functioning regulatory agencies. In addition, they are exposed to market pressures and competition. Thus, the institutional setting and the political attitude of the government in charge are important cri-

---

23 Götz Roland: Russlands Öl und Europa, Bonn: Friedrich-Ebert-Stiftung, 2006 (FES-Analyse: Russland).
24 Stiglitz, Joseph: What is the role of the state?, in: Humphreys, Macartan/ Sachs, Jeffrey/ Stiglitz, Joseph (eds): Escaping the resource curse, New York: Columbia University Press, 2007, pp. 23–52, here p. 30.

teria for the efficiency of the oil and gas sector. In a weak institutional environment, companies tend to be less efficient.[25]

The main tasks of state-owned energy companies in an extractive state such as Russia are 'to supply the rentier state with funds for its budget, which includes in particular the alimentation of corrupt clientele systems'.[26]

The state's presence in the fuel and energy complex has increased steadily since 2003.[27] For example, the share of state-controlled companies in Russian oil production in that year was 13%, and it grew to 24% in 2005 and 40% in 2011.[28] The state controls the oil and gas companies through the involvement of politicians in the management of companies. During his presidency, Dmitry Medvedev pursued a policy of reducing the number of public officials on the boards of directors. However, soon after Vladimir Putin began his third term as the Russian president, government officials still held appointments or were re-elected on the boards of directors. The involvement of public officials creates a lack of transparency and poor corporate governance in Russia, and it allows these officials to receive benefits and personal gains from their jobs. According to the Peterson Institute for International Economics, 'Gazprom posted nominal profits of $46 billion in 2011, [but] it lost $40 billion to corruption and inefficiency'.[29]

Gazprom is 'not a coherent entity but a conglomerate of interest'.[30] It is understood that

> since 2001, Gazprom's management has been dominated by three groups: CEO Alexei Miller's young St. Petersburg economists, a group of St. Petersburg KGB officers—both closely linked to Putin—and a third group of old Gazprom officials. Putin himself has arbitrated between these three factions, preventing any one of them from gaining the upper hand.[31]

---

25   Yakovlev, Andrei: Spros na pravo v sfere korporativnogo upravleniya, Moscow: GU VShE, 2004, pp. 148–155; Stiglitz, Joseph: What is the role of the state?, in: Humphreys, Macartan/ Sachs, Jeffrey/ Stiglitz, Joseph (eds): Escaping the resource curse, New York: Columbia University Press, 2007, pp. 23–52, here p. 34–38.
26   Dirmoser, Dietmar: Energy security: new shortages, the revival of resource nationalism and the outlook for multilateral approaches, Bonn: Friedrich-Ebert-Stiftung, 2007, p. 14.
27   For the gas sector, see Heinrich, Andreas: Under the Kremlin's thumb: does increased state control in the Russian gas sector endanger European energy security?, in: Europe-Asia Studies, 2008 (vol. 60), no. 9, 1539–1574.
28   Compiled by the Research Centre for East European Studies at the University of Bremen with data from companies, the Russian Federal Service for Statistics and NewsBase (www.newsbase.com).
29   'Gazprom: Russia's wounded giant', in: The Economist, 23 March 2013, http://www.economist.com/news/business/21573975-worlds-biggest-gas-producer-ailing-it-should-be-broken-up-russias-wounded-giant.
30   Kivinen, Markku: Public and business actors in Russia's energy policy, in: Aalto, Pami (ed.): Russia's energy policies: national, interregional and global levels, Cheltenham: Edward Elgar, 2012, pp. 45–62, here p. 49.
31   Åslund, Anders: Why Gazprom resembles a crime syndicate, in: The Moscow Times, 28 February 2012, http://www.themoscowtimes.com/opinion/article/why-gazprom-resembles-a-crime-syndicate/453762.html.

2. An Actor-Centred Institutionalist Approach to Russia's Pipeline Policies 101

Moreover, Gazprom works with many affiliated companies. The prospects of the largest pipe suppliers for Gazprom and contractors in the pipeline construction depend on the development of pipeline politics. In this case, trading companies play a significant role.

Gazprom buys pipes from intermediaries. Most of them are supplied by the 'North European Pipe Project', owned by Boris and Arkadii Rotenberg.[32] According to an investigation of the Federal Antimonopoly Service in 2013, the pipe trader was placed in an exclusive position in the market, leading to a restriction of competition. Gazprom was suspected of limiting access to the market for the pipe manufacturers and imposing unfavourable terms. Tenders were formed so that the pipe manufacturers could not participate in them on their own.[33] This situation indicates a privileged position for the 'North European Pipe Project'.

Furthermore, most of the contracts for the construction of gas pipelines were won or received on a non-competitive basis by Rotenberg's company Stroigazmontazh. Established in 2007, contracts with Gazprom made the company the industry's second-largest player by revenue in 2009. Its largest client is Gazprom, representing approximately 65% of its turnover.[34] Stroigazmontazh is involved in the construction of the Dzhubga–Lazarevskoe–Sochi pipeline, the Russian section of the Nord Stream pipeline, and the Sakhalin–Khabarovsk–Vladivostok pipeline.

It should be noted that the main contractor until 2006 was Stroitransgas. The shareholders of this company were relatives of the former Prime Minister Viktor Chernomyrdin and the former head of Gazprom, Rem Vyakhirev. The principal owner of Stroitransgaz is now the oil trader Gennady Timchenko, and the major customers of the company are Rosneft and Transneft.[35]

The tenders for the construction of both the Sakhalin–Khabarovsk pipeline and the Dzhubga–Lazarevskoe–Sochi pipeline were received by Stroigazmontazh on a non-competitive basis.

The Sakhalin–Khabarovsk–Vladivostok gas pipeline is owned and operated by Gazprom. The transported gas is intended for domestic consumption in Russia and, potentially, for export to the Asia-Pacific region. The main resource base for the

---

32 Petlevoi, Vitalii: FAS rekomenduet "Gazpromu" pokupat' truby napryamuyu u proizvoditelei, in: Vedomosti, 18 March 2013, http://www.vedomosti.ru/companies/news/10132391/fas_rasputala_delo_o_trubah.
33 Terent'eva, Aleksandra: FAS vozbudila delo protiv "Gazproma" i kompanii Rotenbergov, in: Vedomosti, 3 April 2013, http://www.vedomosti.ru/companies/news/10729561/fas_vozbudila_delo_protiv_gazproma_i_kompanii_rotenbergov.
34 Mazneva, Yelena: Gazprom keeps projects in Rotenberg's pipeline, in: The Moscow Times, 23 July 2010, http://www.themoscowtimes.com/business/article/gazprom-keeps-projects-in-rotenbergs-pipeline/410878.html.
35 Mazneva, Elena: "Gazprom" svoich ne brosaet, in: Vedomosti, 10 September 2009, http://www.vedomosti.ru/newspaper/article/2009/09/10/213606.

development of the gas transportation system is the offshore fields of the Sakhalin I and Sakhalin II. From 2014, this pipeline will carry gas from the Sakhalin III project. The costs of the project are estimated to be US$15.6 billion. This pipeline is a prestige project because the construction of the first stage was dedicated to the APEC 2012 Summit.

Another prestige project is the Dzhubga–Lazarevskoe–Sochi pipeline. This pipeline is designed to provide a reliable gas supply for Sochi and the Olympic constructing facilities. The construction costs amount to approximately US$1.28 billion. The construction of prestige projects eliminates questions about the feasibility and reasonableness of costs.

The Nord Stream project is a gas pipeline beneath the Baltic Sea that directly connects Russia and Germany and that was implemented by the Nord Stream AG.[36] There is an onshore section of the pipeline in Russia and Western Europe that includes two transmission pipelines in Germany, OPAL (Ostsee-Pipeline-Anbindungs-Leitung) and NEL (Nordeuropäische Erdgasleitung). Shareholders provide 30% of the funding in proportion to the project. The remaining 70% was received on the capital markets. External funding to control costs as well as the creation of a joint-stock company with foreign partners reduces the possibility of kickbacks in the pipeline construction.

However, the building cost of one kilometre of the pipeline on the Russian side was approximately three times higher than the cost of the Western European part of the pipeline. Despite the fact that each pipeline project (or parts of it) has unique characteristics and depends on climate, geography, and landscape features, the unit cost of construction of one kilometre of the Russian onshore section totalled €5.8 million. One kilometre of the land section of the OPAL pipeline is €2.1 million, and the NEL pipeline is €2.3 million.[37]

The Dzhubga–Lazarevskoe–Sochi pipeline was built without foreign loans at a cost per kilometre of US$7.25 million. The unit cost of the construction of one kilometre of the Sakhalin–Khabarovsk–Vladivostok pipeline amounted to US$8.5 million.[38] A comparison of the costs of gas pipeline construction in Russia raises many questions with regard to economic justification, the competitiveness of routes, and inefficient spending.

The Eastern Siberia–Pacific Ocean (ESPO I and ESPO II) oil pipeline supplies Siberian oil to Asian markets. The construction and operating company is Transneft. The construction of the ESPO pipeline cost a total of US$23 billion. The unit costs of building one kilometre of the first section (from Taishet to Skovorodino) were approximately US$4 million,[39] and the unit costs of the second phase (from Skovorodino to Kozmino)

---

36 The shareholding structure is as follows: Gazprom (51 per cent), Wintershall Holding (BASF subsidiary) and E. ON Ruhrgas (15.5 per cent each), Gasunie and GDF Suez (9 per cent each).
37 East European Gas Analysis, http://www.eegas.com.
38 'Zolotye truboprovody', in: Neftegazovaja vertikal', no. 12, 2011, pp. 36–40, here p. 39, http://www.ngv.ru/upload/iblock/355/355502b9ea944e2813c17b6c72d66af7.pdf.
39 Skornyakova Anna: Zolotye nefteprovody, in: Profil', 5 March 2007, http://www.profile.ru/items_22163.

2. An Actor-Centred Institutionalist Approach to Russia's Pipeline Policies

amounted to US$5.8 million.[40] Another relevant oil pipeline built by Transneft is the Baltic Pipeline System-II. The construction was implemented in cooperation with Stroitransgaz. The unit costs of building one kilometre of the Baltic Pipeline System-II amounted to US$3.6 million.[41]

In 2010, it was reported that the damage caused by the construction of the ESPO pipeline totalled at least US$4 billion. Blogger and anti-corruption activist Alexei Navalny published documents containing information about enormous embezzlement during the construction of the ESPO project. Contractors were involved without tenders; there were unreasonably high estimated costs of project work, falsification of documents, and the removal of funds through front companies in offshore accounts. The disclosure of corrupt practices did not result in any significant effects for any Transneft managers.

Money invested in the construction of the pipeline does not provide additional income to Gazprom's shareholders. On the contrary, profits decline, and the company receives a subsidy to compensate for losses from the operation of the new pipeline. The federal budget of the Russian Federation for 2011–2013 provides Gazprom with a subsidy to cover the difference between the purchase prices of the operator of the Sakhalin II project and the set feed-in price to the gas transportation system to supply the power stations of the Far Eastern region. The subsidy amounts were 1.9 million Roubles in 2011, 11.2 million Roubles in 2012, and 11.5 million Roubles in 2013.[42]

Within the framework of 'extractive institutions', the state-owned companies are too self-confident in their monopoly positions. The monopoly on pipeline transportation and the export of energy resources undermines free competitions and eliminates private investments. Corruption in state-controlled Russian companies leads to higher costs and prices of products. The political elite is allowed to benefit from corruption in exchange for its loyalty; the spoils of this corruption are primarily distributed among Vladimir Putin's closest circle, including friends, collaborators, and business partners. This situation explains the selective use of the rule of law and the impunity for the large scale of embezzlement.

2.6. Institutional Influence on Political Outcomes

The export of Russian energy resources is oriented primarily towards the European Union (EU), but Russia has recently begun to develop the Asian direction of its energy

---
40 'Zolotye truboprovody', in: Neftegazovaja vertikal', no. 12, 2011, pp. 36–40, here p. 40, http://www.ngv.ru/upload/iblock/355/355502b9ea944e2813c17b6c72d66af7.pdf.
41 Ibid.
42 Federal'nyj zakon ot 13 dekabrya 2010 goda. N 357-FZ "O federal'nom byudzhete na 2011 god i na planovyj period 2012 i 2013 godov", pp. 444, 1116, http://www1.minfin.ru/ru/budget/federal_budget/index.php?pg4=3.

policy. One of the key elements of the Russian export pipeline policy is harmonised cooperation between the state and the pipeline lobby, in which state and private interests coincide. Pipelines contribute to the establishment of long-term relationships between producers and consumers, which provide oil and gas companies with profits in the long run. Construction pipeline companies, steel pipe suppliers, and pipe manufacturers win tenders in the pipeline projects, which contribute to the profit flow.

Under the influence of 'extractive institutions', the main concern of the oil and gas companies is 'to ensure and control external security of demand as much as possible in order to sustain a rent-based system in Russia'.[43] Many state-controlled companies also exercise specific tasks related to the geopolitical goals of their governments. Thus, the Russian gas and oil giants Gazprom and Rosneft are regarded as 'willing henchmen' of the Kremlin in expanding the Russian influence in Europe and Asia and in the positioning of Russia as a great power.[44] Infrastructure projects in the European direction are seen as strengthening the presence of the Russian companies on the lucrative European energy market and are intended to increase their market share. The construction of pipelines in the Asian direction aims to diversify export options by strengthening the eastern vector of energy exports and reducing dependence on a single consumer market.

The established informal institutions in Russia are responsible for the rational-egoistic and power-seeking preferences and motivations of actors aimed at maximising short-term benefits. Following game theory, the pursuit of short-term relative gains becomes obvious in the example of the Russia-Ukraine gas disputes. To promote the Nord Stream pipeline, between 2006 and 2009, gas supplies were repeatedly interrupted to compromise the supply route through Ukraine.[45] If Russia chooses the non-cooperative strategy while the EU is oriented towards sustainability and cooperation, the worst case situation develops for the EU: interruption in the supply or delivery cuts. However, Russia has all of the advantages on its side in the short term (i.e., short-term political gains, a gas contract with Ukraine that is beneficial for Russia). In the long run, however, Russia must anticipate reputational damage, financial loss, and the loss of market share.

On the one hand, the gas supply interruptions led to a rethinking of the European energy policy towards Russia and the search for new energy supply routes. The EU developed the Southern Gas Corridor programme, which included several pipeline projects and identified a number of partner countries. This initiative should be con-

---

43 Faber van der Meulen, Evert: Gas supply and EU-Russia relations, in: Europe-Asia Studies, 2009 (vol. 61), no. 5, pp. 833–856, here p. 848.
44 Dirmoser, Dietmar: Energy security: new shortages, the revival of resource nationalism and the outlook for multilateral approaches, Bonn: Friedrich-Ebert-Stiftung, 2007, p. 14.
45 Kaczmarski, Marcin/ Konończuk, Wojciech/ Paszyc, Ewa: The Russia-EU summit took place on 21–22 May in Khabarov, in: OSW Eastweek, 26 May 2009, http://www.osw.waw.pl/en/publikacje/eastweek/2009-05-27/ukraine-background-russia-eu-summit.

2. An Actor-Centred Institutionalist Approach to Russia's Pipeline Policies

ductive to enhancing the security of the energy supply. On the other hand, the gas disputes caused a zero-sum game in the pipeline policies between Russia and the EU. The profits of a single interested party mean losses for the other party in a zero-sum game. Russian losses in the case of the construction of one Southern Corridor project may mean a declining share in the European market. EU losses mean greater dependence on Russia and undiversified gas supply sources. As a result, the competition for energy resources and transit routes can intensify the crisis potential of regional constellations.

Thus, the decision to build pipelines is affected by political considerations and is influenced by power-seeking interests. Power politics thinking discounts the future and prioritises short-term relative gains over absolute gains. That is, the actors favour short-terms profits over longer-term priorities. Pursuing absolute gains may be conductive to increasing benefits for all Russian and European actors through the establishment of legally binding norms (institutions).

It should also be noted that the gas disputes were one of the critical junctures in European de-carbonisation policy, which, in conjunction with high oil and gas prices, aimed to replace the fossil energy sources with renewable energy and to create a competitive energy market.

The inclusive political institutions are aware of the role of different innovation mechanisms. Sustainable growth and development can be achieved through policy-induced innovations. The 'extractive institutions', in contrast, have little interest in technological change and the wealth of the society. They avoid creative destruction and prevent changes to the current energy infrastructure.

The Russian ruling elite benefits from the existing informal and weak institutions and is not interested in revising the achieved state of equilibrium. Due to short-term policy, the elite resists any change. These factors create lock-in situations that are difficult to change. Enormous profits from the export of energy resources also affect investments in the promotion and modernisation of the energy infrastructure. A close link between business and the state negatively affects the investment climate, and both domestic and foreign investors have a low level of confidence in investing in Russian projects.

The institutional lock-in relates to the 'dominance of [an] inefficient policy instrument, in the presence of superior institutional arrangements'.[46] Non-democratic regimes are often characterised by loyalty versus competence trade-offs in their principal-agent relationships that can lead to negative effects by blocking change and/or hampering adaption to change. Leaders of non-democratic regimes often face the dilemma of which loyalty prevails at the expense of competence: the loyal but less competent are actors in key positions.[47]

---

46   Woerdman, Edwin: The institutional economics of market-based climate policy, Amsterdam: Elsevier, 2004, p. 57.
47   Egorov, Georgy/ Sonin, Konstantin: Dictators and their viziers: endogenizing the loyalty-competence trade-off, in: Journal of European Economic Association, 2011 (vol. 9), no. 5, pp. 903–930.

For instance, this situation is reflected in the fact that actors, trapped in the established informal institutions, are unable to develop new methods of energy transportation. This is why Gazprom initially did not actively seek to acquire LNG (liquefied natural gas) technology and a share in the global LNG market; the company was hesitant about changes and trends on the gas markets. The possibility of capturing a key position and market share at the early stage was missed, so Russia is not well represented in the LNG segment today.[48]

2.7. Conclusion

Once institutions are created and actors have come to rely on these institutions' influence on their preferences and constellations, institutional reform is costly and requires substantial political will. Power-seeking interests and profit-making business for his closest circle of friends and associates were the main reasons for the emergence of informal institutions in Vladimir Putin's Russia. The informal rules of the game structure actors' interactions and orientations. Actors' incentives and expectations are shaped by informal practices, the management of resource revenue, and corruption schemes. Actors' orientations are determined by the near-sighted, egoistic use of rents and the short-termism of their policy.

Under the influence of exclusive political institutions, the state-controlled oil and gas companies operate at an inefficient level. The political outcome of exclusive institutions in the Russian case relates to short-term relative gains, a zero-sum game, lock-in situations, and the avoidance of creative destruction. The weak institutional arrangement also affects economic outcomes and contributes to high building costs for pipelines and kickbacks from the building of pipelines. The pipelines are driven primarily by political reasons that do not consider the economic justification of their construction. However, power politics thinking is not acceptable in the case of pipeline construction and maintenance. Pipelines connect suppliers and consumers in a long-term relationship and are affected by political interrelations. Pipelines are also highly vulnerable to security instability. Thus, they require long-term contractual obligations and reciprocal coordination.

The environment for natural gas and crude oil suppliers is changing due to technological innovations. The extractive political institutions are not capable of creatively destroying prior economic and political order. In contrast to inclusive institutions, they cannot succeed in the long run. This situation leads to losses in market share and a loss of profits for Russia as an energy exporter.

---

48   Kardaś, Szymon: Russia activates the LNG sector, in: OSW EastWeek, 16 January 2013, http://www.osw.waw.pl/en/publikacje/eastweek/2013-01-16/russia-activates-lng-sector.

Niels Smeets

3. Opening Up the Black Box: Russia's Energy Security Concept

> Let's draw up a new energy charter or a new version of the Energy Charter. But what should it be like? It should not benefit just the consumers. Yes, a consumer is a vulnerable party. But sometimes we need to think about the producers as well (Russian President Dmitri Medvedev, March 2009).[1]

3.1. Russia's Energy Security Concept: A Black Box?

Although the International Energy Agency (IEA) claims to report on the *world's* energy security, which by definition includes both consumer and producer countries, the security interests of producer countries are mostly neglected. In its 2007 World Energy Outlook, the IEA defines energy security as adequate, affordable and reliable *supplies* of energy.[2] The IEA's history, in which the organisation saw the light in reaction to the 1970s oil shocks, and its membership of major industrialised consumer states explains its focus on consumer countries' interests. The most important challenge of energy security is safeguarding an uninterrupted flow of energy supplies at affordable prices.

On the European Union (EU) level, it is unsurprising that energy security and security of supply are considered synonymous. Except for Denmark, the EU-28 countries are net-importers of oil and gas. Moreover, the EU's waves of eastward enlargement in 2004, 2007 and 2013 increased the number of EU countries that are dependent on a single gas supplier. Bulgaria, the Czech Republic, Slovakia, and the Baltic states are entirely dependent on Russian gas. Before the enlargements, only Finland, which joined the EU in 1995, was fully dependent on Russian gas imports.[3]

Because of its dependency on external suppliers, the EU became highly susceptible to supply shocks. The 2006 and 2009 energy crises between Russia and Ukraine emphasised the importance of supply security and put it on top of the EU's agenda. Some countries in Southeastern Europe, for which gas is the main source for heat generation, suffered gas supply interruption for 20 days in the dead of winter 2009. These characteristics of the EU as consumer countries explain why both concepts of energy security and security of supply (SOS) are used interchangeably.

---

1 'Europe needs new Energy Charter—Medvedev', in: Russia Today, 1 March 2009, http://rt.com/politics/official-word/europe-needs-new-energy-charter-medvedev/.
2 IEA: World Energy Outlook, Paris: International Energy Agency, 2007, here pp. 160–161.
3 Kaczmarski, Marcin: Bezpieczeństwo energetyczne Unii Europejskiej, Warsaw: Wydawnictwa akademickie i profesjonalne, 2010.

In general, security of supply of net energy importers is overrepresented in both political and scholarly debate. In the academic literature on energy security, it has become common practice to treat energy security and security of supply as synonymous.[4] Energy security is operationalised as an uninterrupted supply of energy at affordable prices. However, from a theoretical point of view, this does not imply that energy security and security of supply can be used interchangeably. This chapter addresses two shortcomings that result from this bias. First, producer countries are disregarded as having no supply security issues due to the abundant availability of energy reserves. The focus on one aspect of supply security draws attention away from producer countries' concerns about supply security. Second, the demand side of the energy equation has been largely neglected as a result of overemphasising importers' interests that equate energy security with security of supply. As a result, both security of supply and demand of exporting countries are rarely considered.

The producer perspective adds an alternative, country-specific interpretation of energy security.[5] Whereas importing countries are mainly concerned about stable energy supplies at affordable prices, producer states seek to secure stable export volumes at high prices to reliable consumers, generating a stable inflow of energy revenues. Energy exporters are equally concerned about spreading risks by penetrating alternative markets and protecting their energy sector from foreign influences. At the same time, producer countries must ensure a stable supply for domestic end-consumers. To address this dual bias, the Russian Federation has been chosen as a case study. As a major energy producer that must meet both domestic and international demand, Russia provides insights into the complexity of the concept of energy security. This chapter first unveils Russia's domestic SOS concerns that are comparable to consuming countries struggles with the four A's of energy (i.e., availability, accessibility, acceptability and affordability).[6] The main difference in the Russian case lies within the virtual absence of a geopolitical element that adheres to international supplies.

Second, the chapter turns its attention to an alternative way of considering energy security. The analysis indicates that the global financial crisis prompted a sense of urgency, increasing Russia's attention to security of demand (SOD).

---

4   Cf. e.g., Chester, Lynne: Conceptualising energy security and making explicit its polysemic nature, in: Energy Policy, 2010 (vol. 38), no. 2, pp. 887–895; Vivoda, Vlado: Diversification of oil import sources and energy security: a key strategy or an elusive objective?, in: Energy Policy, 2009 (vol. 37), no. 11, pp. 4615–4623; Kruyt, Bert/ van Vuuren, D.P./ de Vries, H.J.M./ Groenenberg, H.: Indicators for energy security, in: Energy Policy, 2009 (vol. 37), no. 6, pp. 2166–2181; Cohen, Gail/ Joutz, Frederick/ Loungani, Prakash: Measuring energy security: trends in the diversification of oil and natural gas supplies, in: Energy Policy, 2011 (vol. 39), no. 9, pp. 4860–4869.
5   Chester, Lynne: Conceptualising energy security and making explicit its polysemic nature, in: Energy Policy, 2010 (vol. 38), no. 2, pp. 887–895.
6   First introduced by: Asia Pacific Energy Research Centre: A quest for energy security in the 21st century: resources and constraints, Tokya: APERC, 2007.

3. Opening Up the Black Box: Russia's Energy Security Concept

The EU's SOS is not independent from Russia's SOD decisions. The shift in Russia's energy export policy from the EU to the Chinese consumer market is the most revealing example.[7] This policy change has already resulted in the construction of the Eastern Siberia–Pacific Ocean oil pipeline, while the Altai–China gas pipeline, which would connect Russia directly to mainland China, has been proposed. Given the declining production of West Siberian oil and gas fields that serve the EU market, the redirection of new fields in Eastern Russia towards China could decrease the availability of energy supplies for Europe within decades. To open up this black box, this chapter deconstructs Russia's SOD concept. Four constituent parts were separated analytically: rents, recovery, resource nationalism and reliability.

This chapter consists of three sections. The first section presents a literature review highlighting the biased conceptualisation of energy security towards consuming countries' security of supply. The second section focuses on the theoretical and methodological framework for viewing the energy security concept as comprised of SOS and SOD. Securitisation theory is used to explain the shift towards a geopolitical approach to energy security.[8] The subsequent empirical section analyses the different aspects of Russia's energy security concept (SOS and SOD) by applying a computer-based discourse analysis to the Russian energy strategy documents (i.e., Energy Strategy until 2020; Energy Strategy until 2030). Both SOS and SOD are deconstructed to produce an alternative conception of energy security from a producer country's point of view.

3.2. The Dual Bias Towards Consuming Countries' Security of Supply

A dual bias towards security of *supply* of *consuming* countries can be observed in the academic literature on energy security.[9] The consumer-centric approach to energy security has two shortcomings.

---

7   Ministerstvo Energetiki Rossiiskoi Federatsii: Energeticheskaya strategiya Rossii na period do 2030 goda. Utverzhdena rasporyazheniem Pravitel'stva Rossiiskoi Federatsii ot 13 noyabrya 2009g. no. 1715-r, 13 November 2009, http://minenergo.gov.ru/aboutminen/energostrategy/.
8   Buzan, Barry: People, states and fear: an agenda for international security studies in the post-cold war era, 2nd ed., New York: Harvester Wheatsheaf, 1991; Buzan, Barry/ Wæver, Ole: Regions and powers: the structure of international security, Cambridge: Cambridge University Press, 2003; Buzan, Barry/ Wæver, Ole/ de Wilde, Jaap: Security: a new framework for analysis, Boulder, CO: Lynne Rienner, 1998.
9   Cf. e.g., Kruyt, Bert/ van Vuuren, D.P./ de Vries, H.J.M./ Groenenberg, H.: Indicators for energy security, in: Energy Policy, 2009 (vol. 37), no. 6, pp. 2166–2181; Luciani, Giacomo: Security of supply for natural gas markets: what is it and what is it not?, Milan: Fondazione Eni Enrico Mattei, 2004; Tonjes, Christoph/ de Jong, Jacques J.: Perspectives on security of supply in European natural gas markets, The Hague: Clingendael International Energy Programme, 2007; Weisser, Hellmut: The security of gas supply: a critical issue for Europe?, in: Energy Policy, 2007 (vol. 35), no. 1, pp. 1–5; Vivoda, Vlado: Diversification of oil import sources and energy security: a key strategy or an elusive objective?, in: Energy Policy, 2009 (vol. 37), no. 11, pp. 4615–4623; Sovacool,

First, security of supply of energy consumers and energy security are often used as synonyms.[10] The focus on the interests of consumer countries overlooks the alternative perspective of producing countries, which also securitise demand issues. Even authors who plead for a more holistic approach to energy security nevertheless equate the concept with security of supply. For example, Sovacool and Mukherjee seek key phrases such as 'energy security' and 'security of supply,' leaving out the security of demand perspective.[11] Bradshaw adds a geographical dimension to the concept and considers emerging economies.[12] The forecasted rise in energy demand and the geographical shift towards Asian importing countries are interpreted as problems. However, from a producer perspective, both trends could imply new opportunities to improve SOD.

Some recent studies diagnose the complex nature of energy security in relation to producer and consumer interpretations.[13] However, a clear definition of the SOD concept and its theoretical or methodological relationship to SOS has not been consistently developed.

Mansson et al. mention the interrelationship of supply and demand security but confine the issue to mutual interdependence.[14] Their interpretation is based on the reasoning that both supplier and consumer states are interested in a stable energy flow since the consumer needs the energy, whereas the producer state is equally dependent on the energy revenues.[15] The interdependence argument is further developed by reciprocal trade relations: 'the EU is Russia's main export market while Russia is the EU's main external supply source'.[16]

However, this approach overestimates the commonalities of both parties' interests and downplays the contrasts between SOD and SOS strategies. For instance, Mansson

---

Benjamin K./ Mukherjee, Ishani: Conceptualizing and measuring energy security: a synthesized approach, in: Energy, 2011 (vol. 36), no. 8, pp. 5343–5355; Chester, Lynne: Conceptualising energy security and making explicit its polysemic nature, in: Energy Policy, 2010 (vol. 38), no. 2, pp. 887–895.

10 Kruyt, Bert/ van Vuuren, D.P./ de Vries, H.J.M./ Groenenberg, H.: Indicators for energy security, in: Energy Policy, 2009 (vol. 37), no. 6, pp. 2166–2181.

11 Sovacool, Benjamin K./ Mukherjee, Ishani: Conceptualizing and measuring energy security: a synthesized approach, in: Energy, 2011 (vol. 36), no. 8, pp. 5343–5355, here p. 5344.

12 Bradshaw, Michael J.: Global energy dilemmas: a geographical perspective, in: Geographical Journal, 2010 (vol. 176), no. 4, pp. 275–290.

13 Dellecker, Adrian/ Gomart, Thomas: Russian energy security and foreign policy, London: Routledge, 2011; Boussena, Sadek/ Locatelli, Chatherine: Energy institutional and organisational changes in EU and Russia: revisiting gas relations, in: Energy Policy, 2013 (vol. 55), pp. 180–189.

14 Mansson, A./ Johansson, B. / Nilsson, L. J.: Methodologies for characterising and valuing energy security: a short critical review, paper prepared for the 9th International Conference on the European Energy Market, Florence, Italy, 9–12 May 2012.

15 Umbach, Frank: Energy security in Eurasia: clashing interests, in: Dellecker, Adrian/ Gomart, Thomas (eds): Russian energy security and foreign policy, London: Routledge, 2011, pp. 23–38.

16 Boussena, Sadek/ Locatelli, Chatherine: Energy institutional and organisational changes in EU and Russia: revisiting gas relations, in: Energy Policy, 2013 (vol. 55), pp. 180–189, here p. 180.

3. Opening Up the Black Box: Russia's Energy Security Concept 111

et al. do not analyse the crucial difference between producer and consumer countries with regard to price shocks.[17] They do not discuss the possibility that producer countries might improve their demand security through higher export revenues as a result of higher energy prices.

The second shortcoming of previous literature concerns its failure to address the SOS of producing countries. For instance, Sovacool and Mukherjee highlight the scarce attention that energy security concerns of non-industrialised countries receive.[18] Nevertheless, the research interviews were conducted exclusively in consumer countries such as China, India and Japan.

Although producer states have ample resources, they may still face supply issues. Although Russia possesses substantial energy resources, the supply chain between extraction and delivery to the end-consumer is long and winding. Internal obstacles such as underinvestment in domestic energy infrastructure, deficiencies in the integration of regional pipeline systems and uncertain liberalisation initiatives that affect access to the pipeline network complicate delivering a stable stream of supplies to consumers.[19]

Moreover, the academic literature that considers producer's concerns tends to overemphasise security of demand. For example, Umbach states: 'Whereas consumer nations are primarily interested in security of supply, producer countries are more focused on security of demand from foreign markets,'[20] implying that energy producers are not mainly interested in security of supply issues. However, as the analysis below will demonstrate, SOS issues remain the main energy security issue even for energy-rich countries such as Russia.

In sum, producer countries' SOD and SOS largely remain unnoticed in the literature, and if they are mentioned, they remain a black box. This chapter is an attempt to comprehensively deconstruct both SOD and SOS using Russia as a case study.

---

17 Mansson, A./ Johansson, B. / Nilsson, L. J.: Methodologies for characterising and valuing energy security: a short critical review, paper prepared for the 9th International Conference on the European Energy Market, Florence, Italy, 9–12 May 2012.
18 Sovacool, Benjamin K./ Mukherjee, Ishani: Conceptualizing and measuring energy security: a synthesized approach, in: Energy, 2011 (vol. 36), no. 8, pp. 5343–5355, here p. 5353–5354.
19 Predsedatel' Pravitel'stva Rossiiskoi Federatsii: Energeticheskaya strategiya Rossii na period do 2020 goda. Utverzhdena rasporyazheniem Pravitel'stva Rossiiskoi Federatsii ot 28 avgusta 2003g. no. 1234-r, 28 August 2003, http://www.rg.ru/2003/09/30/energeticheskajastrategija. html; Ministerstvo Energetiki Rossiiskoi Federatsii: Energeticheskaya strategiya Rossii na period do 2030 goda. Utverzhdena rasporyazheniem Pravitel'stva Rossiiskoi Federatsii ot 13 noyabrya 2009g. no. 1715-r, 13 November 2009, http://minenergo.gov.ru/aboutminen/energostrategy/.
20 Umbach, Frank: Energy security in Eurasia: clashing interests, in: Dellecker, Adrian/ Gomart, Thomas (eds): Russian energy security and foreign policy, London: Routledge, 2011, pp. 23–38.

3.3. Energy Trade Versus Energy Security

Two main perspectives can be distinguished in previous research on energy security: a market-based and a geopolitical approach.[21] Whereas the former emphasises the tendency towards an internationalisation of energy markets,[22] the latter focuses on the increasing role of the state enabling the use of energy as a political tool.[23] The market-based approach is a consequence

> of the liberalisation of energy markets, energy security is a market outcome, determined by the operation of the market and thus can only be defined in market terms—particularly supply (physical availability) and price.[24]

The economic interpretation of SOS is actually a logistical problem. Given the limited reserves, transport infrastructure and production rates, how can the market demand be met?

After the interruptions of the Russian gas supply to Europe in 2006 and 2009, the geopolitical approach gained new momentum.[25] Similar to the political reaction to the 1973 oil shock,[26] the energy dispute between Russia and Ukraine in January 2006 fostered the perception of a permanently threatened energy supply: Russia could cut off supplies at any time. Based on the Copenhagen School definition of securitisation as the process of presenting an issue in security terms or, in other words, as an existential threat, this change can be interpreted as a renewed securitisation of energy trade relations.[27] The supply risk becomes an issue that could threaten the fundamental operations of a country's economy and, in extreme cases, the existence of the state.

However, the global financial crisis fostered the perception of a threatened energy demand in Russia (see section 3.4.2). SOD is equally fundamental to ensuring regime survival for energy producing countries such as Russia. As a means of maintaining power, Russian elites might want to ensure stable energy revenues that can be used

---

21   Youngs, Richard: Energy security: Europe's new foreign policy challenge, London: Routledge, 2009; Chester, Lynne: Conceptualising energy security and making explicit its polysemic nature, in: Energy Policy, 2010 (vol. 38), no. 2, pp. 887–895.
22   Mitchell, John V./ Morita, Koji/ Selley, Norman/ Stern, Jonathan: The new economy of oil: impacts on business, geopolitics and society, London: Royal Institute of International Affairs/ Earthscan, 2001.
23   Baev, Pavel K.: Russian energy policy and military power: Putin's quest for greatness, London: Routledge, 2009.
24   Chester, Lynne: Conceptualising energy security and making explicit its polysemic nature, in: Energy Policy, 2010 (vol. 38), no. 2, pp. 887–895, here p. 889.
25   Youngs, Richard: Energy security: Europe's new foreign policy challenge, London: Routledge, 2009, here p. 9.
26   Cohen, Gail/ Joutz, Frederick/ Loungani, Prakash: Measuring energy security: trends in the diversification of oil and natural gas supplies, in: Energy Policy, 2011 (vol. 39), no. 9, pp. 4860–4869.
27   Buzan, Barry/ Hansen, Lene: The evolution of international security studies, Cambridge: Cambridge University Press, 2009, here p. 214; Smeets, Niels: Eurasian energy security and sustainable development: the bias towards security of supply, in: Digest of world politics of the XXI century, 2013 (vol. 6), pp. 197–211, here pp. 198–199.

3. Opening Up the Black Box: Russia's Energy Security Concept

to promote economic growth. Energy revenues are significant for the state budget and to finance modernisation programmes.[28] The former can be used to build a strong security service and to buy political support within the elite, whereas the latter can be used to boost living standards, generating popular support.[29]

The securitisation process thus adds a geopolitical layer to the economic trade relationship. Energy security is about political interpretations of economic variables. For instance, the EU tends to securitise energy, which is mainly supplied by non-democratic states.

The geopolitical approach has a dual impact on conceptual measurement. First, economic variables are not the most effective means of measuring security. They are the inputs that are (re)interpreted, adapted and used within a political discourse to defend certain geopolitical decisions. A securitised context always involves subjectivity.[30] Political perceptions not only influence the perceived level of energy security but also have real consequences due to the elite's political power in policy and law making.

Second, by focusing on political discourse, the measuring instruments—political perceptions—are uniform. In contrast, economic operationalisations that do attempt to include geopolitical variables struggle with the relative weights attached to economic and geopolitical indicators. As Kruyt et al. explain, 'indices that cover multiple dimensions of energy security are inherently subjective, as there is no fundamental basis to assign weights on'.[31] This problem is avoided by assuming that energy security is a geopolitical construct. Therefore this chapter maps Russia's elite discourse to capture a producer state's conceptualisation of energy security rather than to focus on economic data.

3.4. Russian Interpretations of Energy Security

This section aims to specify the underlying aspects of Russia's complex conception of energy security. The energy security concept is analysed at the strategic, geopolitical level rather than an as economic operationalisation. As Chester explains: 'The salient point is that energy security is a concept with strategic intent. Energy security is not a policy. Specific policy measures are implemented by governments to achieve

---

28 Denchev, K.: Mirovaya energeticheskaya bezopasnost': istoriya i perspektivy, in: Novaya i Noveyshaya Istoriya, 2010 (vol. 2), pp. 1–43, here p. 23.
29 Øverland, Indra/ Kjærnet, Heidi/ Kendall-Taylor, Andrea: The resource curse and authoritarianism in the Caspian petro-states, in: Øverland, Indra/ Kjærnet, Heidi/ Kendall-Taylor, Andrea (eds): Caspian energy politics: Azerbaijan, Kazakhstan and Turkmenistan, London: Routledge, 2010, pp. 1–12, here p. 4.
30 Aalto, Pami: The EU-Russian energy dialogue and the future of European integration: from economic to politico-normative narratives, in: Aalto, Pami (ed.): The EU-Russian energy dialogue: Europe's future energy security, Aldershot: Ashgate, 2008, pp. 23–41, here p. 30.
31 Kruyt, Bert/ van Vuuren, D.P./ de Vries, H.J.M./ Groenenberg, H.: Indicators for energy security, in: Energy Policy, 2009 (vol. 37), no. 6, pp. 2166–2181, here p. 2177.

the objective(s) of energy security'.[32] Political perceptions of energy security are the object of analysis.

Russia is selected as a case study since it is a major energy producer facing broad internal and external demands for energy. Consequently, Russia faces both SOS and SOD issues. Furthermore, Russia's semi-authoritarian regime prioritises elite survival, which is highly dependent on energy revenues to fuel patronage networks; Russia's regime allows the elite to securitise issues and to take extraordinary measures to ensure energy security (e.g., the *de facto* nationalisation of the Russian oil company, Yukos, in 2003). Moreover, Russia is a major oil and gas exporter to the EU, increasing the impact of Russia's demand decisions on the EU's security of supply.

The methodology consists of a computer-based[33] Critical Discourse Analysis[34] of Russia's energy strategies in 2003 and 2009 (i.e., Energy Strategy until 2020; Energy Strategy until 2030).[35] These documents are selected because of their comprehensiveness in both thematic and institutional scope, which allows us to consider Russia's energy security concept from different perspectives. The energy strategy covers diverse issues, such as quantitative demand, price and production forecasts, qualitative assessments of ecologic and social issues and a discussion on regional challenges. This broad thematic approach is a reflection of the policy making process: the document, which is distributed to all relevant ministries, is the result of a joint governmental project. The authors of both texts are considered to be the Russian elite consistent with Kryshtanovskaya's and White's definition of elite as 'the ruling group in a society, consisting of the people who take decisions of national significance'.[36]

The taxonomy of Russia's energy security concept has been developed both deductively and inductively. The theoretical discussion above provides us with the general deconstruction of the concept into SOS and SOD aspects. Statements were deductively coded into the predefined sub-categories of SOS: availability, acceptability, accessibility and affordability.[37]

---

32 Chester, Lynne: Conceptualising energy security and making explicit its polysemic nature, in: Energy Policy, 2010 (vol. 38), no. 2, pp. 887–895.
33 The software used is NVivo 10, which supports qualitative and mixed-methods research (for more information see http://www.qsrinternational.com/).
34 Fairclough, Norman: Discourse and social change, Cambridge: Polity Press, 2000.
35 Predsedatel' Pravitel'stva Rossiiskoi Federatsii: Energeticheskaya strategiya Rossii na period do 2020 goda. Utverzhdena rasporyazheniem Pravitel'stva Rossiiskoi Federatsii ot 28 avgusta 2003g. no. 1234-r, 28 August 2003, http://www.rg.ru/2003/09/30/energeticheskajastrategija. html; Ministerstvo Energetiki Rossiiskoi Federatsii: Energeticheskaya strategiya Rossii na period do 2030 goda. Utverzhdena rasporyazheniem Pravitel'stva Rossiiskoi Federatsii ot 13 noyabrya 2009g. no. 1715-r, 13 November 2009, http://minenergo.gov.ru/aboutminen/energostrategy/.
36 Kryshtanovskaya, Olga/ White, Stephen: From Soviet nomenklatura to Russian elite, in: Europe-Asia Studies, 1996 (vol. 48), no. 5, pp. 711–733, here p. 712.
37 Kruyt, Bert/ van Vuuren, D.P./ de Vries, H.J.M./ Groenenberg, H.: Indicators for energy security, in: Energy Policy, 2009 (vol. 37), no. 6, pp. 2166–2181.

3. Opening Up the Black Box: Russia's Energy Security Concept   115

A comparable taxonomy of SOD is currently not available in the literature. Therefore, an exploratory discourse analysis of Russia's interpretation of SOD is applied to both energy strategies to develop this alternative concept. Both documents have been coded inductively within NVivo. Based on this discourse analysis, comparable statements were thematically grouped into four categories: recovery, rents, reliability, and resource nationalism (see Figure 3-2). In the remainder of this chapter, Russia's SOS and SOD concepts are deconstructed correspondingly.

3.4.1. Russia's Security of Supply

The statement that producer countries are mainly concerned about SOD is misleading since it underemphasises the importance of the distribution of domestic energy flows to reach the end-consumer in a timely manner, at affordable prices, under ecologically acceptable conditions, and without disruptions. Although Russia has vast resources at its disposal, it still faces enormous challenges to transport energy over its spacious territory. Major energy fields are located at the periphery in northwest Siberia (Yamal) and the Far East (Sakhalin), whereas well over two-thirds of the population lives in the European part of Russia. According to the projections of Russia's latest Energy Strategy, the Central and Volga Federal Districts will continue to face serious energy deficits, even at the end of the Strategy's 2030 horizon.[38] Moreover, underinvestment in the energy infrastructure has caused supply disruptions of electricity, gas and heating. Energy Minister Aleksandr Novak identified 5,300 disruptions during winter 2012–2013.[39] Every industrialised country essentially struggles with this same logistic problem. However, the EU's SOS is more complicated since the EU depends mostly on non-democratic suppliers adding geopolitical considerations. Russia's SOS concerns are concentrated on the domestic sphere; geopolitical considerations apply to a much lesser extent since Russia imports only small amounts of gas and oil from Turkmenistan, Kazakhstan, and Uzbekistan. However, due to the growing awareness of their independence, these countries might pose similar geopolitical challenges in the future. Because the amounts of imported gas and oil represent a small share of Russia's energy demand, the Russian economy as a whole remains self-sufficient and SOS issues remain primarily domestic.

The 2009 Energy Strategy describes energy security as consisting of three elements: sufficient resources, economic accessibility and ecological and technologi-

---

38 Ministerstvo Energetiki Rossiiskoi Federatsii: Energeticheskaya strategiya Rossii na period do 2030 goda. Utverzhdena rasporyazheniem Pravitel'stva Rossiiskoi Federatsii ot 13 noyabrya 2009g. no. 1715-r, 13 November 2009, http://minenergo.gov.ru/aboutminen/energostrategy/.
39 'Rabochaya vstrecha s Ministrom energetiki Aleksandrom Novakom', 20 May 2013, http://www.kremlin.ru/news/18143.

cal acceptability.[40] This definition corresponds almost perfectly to the four suggested elements of energy security: availability, acceptability, affordability and accessibility.[41] In Russia's definition, the latter two categories have been combined (to 'economic accessibility').

Both Russian energy strategies have been deductively coded on all four elements of this classification, reflecting the vital importance of SOS to exporting states. Comparable to consuming countries, Russia's primary concern with regard to energy security rests on SOS issues, coded 1,359 times against 651 instances of SOD (see Figures 3-1 and 3-2). This finding challenges the stark opposition between consuming and producing countries with regard to the importance of SOS. The deductively predefined categories relating to SOS are defined as follows:

- Availability: the physical existence of finite resources, production levels and demand
- Accessibility: access of consumers and producers to pipeline infrastructure
- Affordability: costs of energy production and regulated retail price
- Acceptability: social and environmental constraints

The subsequent qualitative analysis discloses the SOS challenges behind the numbers. The **availability** dimension covers the balancing act between available resources, production capacity and investments, contrasted with ever increasing domestic and foreign demand for energy. Two diversification strategies that address this relationship relate to the broadening of the available energy mix and the exploration of new energy fields. A shift from gas and oil to renewable energies and coal would broaden the scope of alternative energy resources, whereas the exploration of new energy fields, particularly in the Arctic, would greatly enhance the available energy resources.

Accessibility is closely related to the process of developing new energy fields in the periphery, which widens the spatial gap between production sites and the end-consumer. One priority of Russia's energy strategy is the development of an integrated energy infrastructure, including electricity grids, railroads to transport coal and oil pipelines, to connect remote areas with consumers in western Russia.[42] Other aspects of the strategy concern the accessibility of independent suppliers to Gazprom's pipeline network and the liberalisation of Russia's monopolistic energy sector.

---

40   Ministerstvo Energetiki Rossiiskoi Federatsii: Energeticheskaya strategiya Rossii na period do 2030 goda. Utverzhdena rasporyazheniem Pravitel'stva Rossiiskoi Federatsii ot 13 noyabrya 2009g. no. 1715-r, 13 November 2009, http://minenergo.gov.ru/aboutminen/energostrategy/.
41   Cf. e.g., Kruyt, Bert/ van Vuuren, D.P./ de Vries, H.J.M./ Groenenberg, H.: Indicators for energy security, in: Energy Policy, 2009 (vol. 37), no. 6, pp. 2166–2181.
42   Predsedatel' Pravitel'stva Rossiiskoi Federatsii: Energeticheskaya strategiya Rossii na period do 2020 goda. Utverzhdena rasporyazheniem Pravitel'stva Rossiiskoi Federatsii ot 28 avgusta 2003g. no. 1234-r, 28 August 2003, http://www.rg.ru/2003/09/30/energeticheskajastrategija. html; Ministerstvo Energetiki Rossiiskoi Federatsii: Energeticheskaya strategiya Rossii na period do 2030 goda. Utverzhdena rasporyazheniem Pravitel'stva Rossiiskoi Federatsii ot 13 noyabrya 2009g. no. 1715-r, 13 November 2009, http://minenergo.gov.ru/aboutminen/energostrategy/.

3. Opening Up the Black Box: Russia's Energy Security Concept 117

The **affordability** dimension of Russia's SOS relates to the price of energy products for households and businesses. Because energy deposits become more difficult to exploit in ever remote areas, production costs are set to rise. Moreover, the Energy Strategy of 2003 articulated a policy to gradually increase the regulated energy prices. Russia's promise to further increase domestic gas and oil prices was formalised as an accession condition during Russia's negotiations with the World Trade Organization (WTO). Additionally, the practice of cross subsidising should end, implying that private consumers' energy bills will increase in favour of industrial energy consumers, which currently overpay for energy.

Price increases and redistributions for energy, however, create challenges to its social **acceptability**. Because prices for energy are a large part of the household budget, particularly during Siberian winters, socially weak households should be protected. Regarding ecological acceptability, both energy strategies are imbued with energy efficiency and conservation measures to reduce the negative influence of fossil fuels on the environment.

The analysis indicates that the presence of massive gas and oil reserves does not guarantee Russia's exemption from supply issues. However, in contrast to the EU, Russia is far less dependent on foreign suppliers, which makes it a less securitised aspect of energy security in the international arena.

3.4.2. The Global Financial Crisis: Increasing Urgency in Russia's Security of Demand

Although the majority of the literature focuses on SOS, energy security can also be approached from a demand-side perspective. The Russian SOD strategy aims to overcome threats, such as declining demand for energy products, and counteract dependence on the EU as Russia's main export market.

The economic crisis increased the sense of urgency.[43] The plunge in energy prices from US$96.71 per barrel in September 2008 to US$30.7 per barrel on Christmas Day 2008, affected the Russian economy significantly since most economic growth and state revenues originates from the energy sector. A sudden feeling of dependency on world oil prices shocked the country. The large EU export market became a liability rather than an asset. More than 80% of Russia's energy exports are EU-oriented. However, the global financial crisis and subsequent Euro crisis resulted in a production slowdown that led to faltering demand for Russian energy resources. Russia found itself unable to switch between markets, particularly in the case of gas, which is dependent on a rigid pipeline infrastructure. The energy-hungry Asian economies could not

---

43 Buzan, Barry/ Wæver, Ole/ de Wilde, Jaap: Security: a new framework for analysis, Boulder, CO: Lynne Rienner, 1998, here p. 29.

compensate for the decline in demand in Europe. The model of economic growth on the basis of energy revenues was under existential threat, at least in 2009.

The effects of the economic crisis are reflected in the energy strategy. In the 2009 version, the impact of the economic crisis is mentioned 16 times. When qualitatively analysing these statements relating to the economic crisis, it becomes clear that they are discursively connected with SOD. Most attention focuses on fears of economic downfall as a result of contracting markets and the decrease in energy prices and different diversification strategies that were developed to counter the consequences of the crisis.[44]

Economic collapse as a result of declining demand and prices for Russian energy resources would challenge the Putin regime by undermining the redistributive power to temper intra-elite power struggles and the social contract in which the population remains silent in exchange for socio-economic development.

SOD is mentioned considerably more often in the 2009 version of the strategy than in the 2003 version (25.41% versus 15.54%; see Table 3-1), reflecting increased attention to SOD after the crisis that placed SOD higher on the securitisation agenda.

When taking a deeper look at the SOD strategies, a clear shift in attention can be discerned. The node 'diversification of export markets', which measures the frequency of statements referring to the need to diversify Russia's export pipelines towards Asia is mentioned twice as frequently in 2009 than before the crisis (see Table 3-1).

In addition to the demand effect of the economic crisis, energy prices plummeted, raising the feeling of dependence on fluctuations in world energy prices. Following the example of the Organization of Petroleum Exporting Countries (OPEC), a Gas Exporting Countries Forum was established to gain more influence on gas price setting. Russia's 2009 energy strategy endorses Russia's cooperation with member countries such as Qatar and Algeria to become less dependent on market forces.

In conjunction with this approach, another strategy to secure higher revenue from energy has emerged. Instead of exporting unrefined energy products, Russia seeks investments in refineries on Russian soil, which would generate higher added value for Russian exports. This diversification of export goods ascertains the same trend as the diversification of export markets. Whereas in 2003 the strategy to produce higher quality goods was only coded 22 times, the pressure of the economic crisis boosted this approach by a factor of 2 (see Table 3-1).

This frequency analysis of coded statements demonstrates heightened attention to SOD since the start of the economic crisis, pressing the need to safeguard Russia's

---

44 Ministerstvo Energetiki Rossiiskoi Federatsii: Energeticheskaya strategiya Rossii na period do 2030 goda. Utverzhdena rasporyazheniem Pravitel'stva Rossiiskoi Federatsii ot 13 noyabrya 2009g. no. 1715-r, 13 November 2009, http://minenergo.gov.ru/aboutminen/energostrategy/.

3. Opening Up the Black Box: Russia's Energy Security Concept 119

revenues by exporting higher finished goods and hedging against drops in export volumes and world market prices.

Table 3-1: Security of Demand Strategies (Energy Strategy 2003, 2009)

Strategy	References		Percentage Coverage	
Year	2003	2009	2003	2009
Security of demand (SOD)	300	352	15.54%	25.41%
Diversification of export markets	19	37	1.80%	4.32%
Diversification of export goods	22	42	1.27%	3.08%
Diversification of economic structure	12	19	0.73%	2.55%

Note: Percentage coverage indicates the share of the source that has been coded on node X. A percentage coverage of 25% on node SOD implies that one-fourth of the document has been coded as addressing issues of security of demand.

3.4.3. Russia's Security of Demand

Based on the discourse analysis, coded statements of the Energy Strategies were inductively aggregated to four categories of Russia's SOD (see Figure 3-2):
- Recovery: sustainable development of the Russian economy
- Rents: secure and stable revenue inflows to the state budget
- Resource nationalism: state control over resources, companies and pipelines
- Reliability: Russia as a reliable partner, integrated in the world economy

The subsequent qualitative analysis discloses several SOD challenges and strategies. The **Recovery** node addresses the link between energy revenues and economic development. Russia has aimed towards economic recovery on the basis of energy wealth since the collapse of the Soviet Union. Stable energy revenues are interpreted as an important part of Russia's broader economic security. As the 2020 strategy reads:

> The strengthening of Russia's position in the world oil and gas markets is of strategic importance to maximise the export potential of our domestic energy sector in the next 20 years and contribute to the economic security of the country.[45]

For almost a decade, the importance of the energy sector was beyond dispute. After the 1998 devaluation of the rouble, one of the driving forces behind Russia's steady average 6% economic growth from 2000 to 2008 was the rising oil price. Because gas

---

[45] Predsedatel' Pravitel'stva Rossiiskoi Federatsii: Energeticheskaya strategiya Rossii na period do 2020 goda. Utverzhdena rasporyazheniem Pravitel'stva Rossiiskoi Federatsii ot 28 avgusta 2003g. no. 1234-r, 28 August 2003, http://www.rg.ru/2003/09/30/energeticheskajastrategija.html, here section 4.7; in the Russian original: Стратегически важным является укрепление позиции России на мировых нефтяном и газовом рынках, с тем чтобы в течение предстоящего двадцатилетия максимально реализовать экспортные возможности отечественного ТЭК и внести вклад в обеспечение экономической безопасности страны.

prices are pegged to oil prices, Russia's most important export product also benefited, increasing the relative importance of the gas share for export revenues.[46]

However, the sustainability of this resource-led development came under pressure as a consequence of the fall in world energy prices in 2008. In contrast to the 2003 document, the 2009 strategy highlights a fundamental tension between the important role of Russia's energy sector to its economy and the dependence on the energy sector as the single driving force behind economic growth. A compromise was found asserting that:

> [...] export of energy sources will remain the most important factor in developing the national economy; however, the level of its influence on the economy will decline.[47]

One means to achieve this goal is raising the share of less energy-intensive sectors, which would lead to innovative economic growth.[48] This call for a diversification of the economic structure—the transformation of an energy exporting country towards a knowledge economy—would insulate the Russian economy from fluctuating energy prices and demand structures (see Table 3-1). According to the 2009 energy strategy, the energy-intensive production of raw materials will develop at a slower pace since market demand will shift to the less energy-intensive high-tech and service sectors. In other words, the structural diversification of Russia's economy is portrayed as an almost natural process, instigated by new demand incentives.

Another strategy attempts to ensure long-term revenue flows by means of active political and diplomatic support for long-term gas export contracts.

> The gas export, thanks to long-term contracts, allows for sufficient Russian supply towards the European market.[49]

Moreover, the take-or-pay provision of such export contracts implies that a consumer must accept the agreed upon amount of gas and, failing to do so, must pay for the contracted volume up to an agreed-upon ceiling. The producer thus enjoys a stable demand for energy.

In addition to these contractual clauses, a diversification policy of export markets has been developed in reaction to the economic crisis (see section 3.4.2). The overall

---

46 Gaddy, Clifford G./ Ickes, Barry W.: Resource rents and the Russian economy, in: Eurasian Geography and Economics, 2005 (vol. 46), no. 8, pp. 559–583.
47 Ministerstvo Energetiki Rossiiskoi Federatsii: Energeticheskaya strategiya Rossii na period do 2030 goda. Utverzhdena rasporyazheniem Pravitel'stva Rossiiskoi Federatsii ot 13 noyabrya 2009g. no. 1715-r, 13 November 2009, http://minenergo.gov.ru/aboutminen/energostrategy/, here section IV 2, in the Russian original: [...] экспорт энергоносителей будет оставаться важнейшим фактором развития национальной экономики, однако степень его влияния на экономику будет сокращаться.
48 Ibid., here section VII 1.
49 Ibid., here section VI 5; in the Russian original: Экспорт газа, осуществляемый преимущественно на основе долгосрочных контрактов, позволит сохранить необходимый объем поставок из России на европейский рынок.

3. Opening Up the Black Box: Russia's Energy Security Concept 121

goal is to increase the proportion of exports towards Asia. By 2030, the share of liquid hydrocarbons supplied to Asia should rise from the current level of 6% to 22–25%, and the export of Russian gas, which currently does not find its way to Asia at all, is expected to reach 19–20%. The recovery node thus indicates the insulation of the national economy against long-term demand and price shocks, thereby improving Russia's SOD.

The **Rents** node is based on the need to secure stable energy rents to fund the state budget, which finances regime survival mechanisms such as social programmes and security services. In contrast to the recovery node, rent-capturing behaviour is focused on short-term gains without wealth creation. The state captures existing wealth for redistribution among elite groups and companies and, to a lesser extent, for social redistribution. The Putin administration has increased the amount of rents collected by the government in the form of corporate taxes.[50] The ownership structure accounts for another mechanism to capture profits. The proposal by the Russian Ministry of Finance to raise the minimum dividend of state-owned companies from 25% to 35% fits this rent-capturing behaviour.[51] Apart from these reported formal taxes, informal taxes disappear into the pockets of regional officials.[52]

This close intertwinement of the energy sector and the state budget is mentioned in the documents under consideration 17 times (see Figure 3-2). The 2003 version of the energy strategy formulates the reciprocal relationship as follows:

> The energy sector, which has complex and manifold relations with the state budget, is the major source of budget income and the major recipient of state spending, thus influencing the income and spending sides of the budget at all levels.[53]

The goal of the revenue side of the budget is to ensure stable income from energy resources. In 2012, the budget derived 50.2% of its revenues from the export of oil and gas.[54] Similar to the observations made within the recovery node, analysis reveals the increasing urgency of budget dependency on energy rents. Whereas the 2003 energy strategy is concerned with a stable inflow of revenues to the state budget, the 2009

---

50 Gaddy, Clifford G./ Ickes, Barry W.: Resource rents and the Russian economy, in: Eurasian Geography and Economics, 2005 (vol. 46), no. 8, pp. 559–583.
51 Topalov, Aleksei: Goskompanii obkladyvayut dividendami, in: Gazeta.ru, 13 May 2013, http://www.gazeta.ru/business/2013/05/13/5319553.shtml.
52 Gaddy, Clifford G./ Ickes, Barry W.: Resource rents and the Russian economy, in: Eurasian Geography and Economics, 2005 (vol. 46), no. 8, pp. 559–583.
53 Predsedatel' Pravitel'stva Rossiiskoi Federatsii: Energeticheskaya strategiya Rossii na period do 2020 goda. Utverzhdena rasporyazheniem Pravitel'stva Rossiiskoi Federatsii ot 28 avgusta 2003g. no. 1234-r, 28 August 2003, http://www.rg.ru/2003/09/30/energeticheskajastrategija.html, here section 4.1.3; in the Russian original: Энергетический сектор связан сложными и разнообразными взаимоотношениями с государственным бюджетом, являясь и основным источником формирования его доходной части, и получателем государственных средств, оказывая влияние на формирование и исполнение бюджетов всех уровней.
54 'Dolyu neftegazovych dokhodov byudzheta RF nuzhno snizit' do 25%—Medvedev', in: Ria Novosti, 26 February 2013, http://ria.ru/economy/20130226/924619759.html.

version urges lowering the high dependency of the state budget on energy rents. On the expenditure side of the budget, the redistribution of oil rents over the ministries, companies and specific projects keeps the elite in power, thereby ensuring regime survival.[55] Most investments in the energy sector are targeted at technical innovation, making the energy sector more energy efficient.[56] However, in 2014, only 14% of the state budget will be invested in the national economy, whereas 23% of the state budget goes to national security and 35% will be directed towards social policy.[57] This budgetary breakdown is typical for semi-authoritarian petro-states in their quest to ensure regime survival: expenditures for national security are meant to deter or repress popular revolts, whereas social spending takes the wind out of the opposition's sails.[58]

Resource nationalism is understood as 'referring to a wide range of strategies that domestic elites employ to increase their control of natural resources'.[59] Resource nationalism historically is closely linked to state sovereignty. The permanent sovereignty over natural resources was seen as a guarantee for economic independence from colonial powers.[60] The Russian energy strategies similarly interpret control over the energy sector as assuring independence from foreign influence.

From an economic perspective, state ownership in the energy sector is a means to capture a larger portion of the financial benefit of hydrocarbon production and exports. However, this economic imperative coexists with a domestic political component. State guarantees of full protection of property rights would only result in an autonomous energy sector with political ambitions.[61] Therefore, resource nationalism is necessary to secure the political power of the current elite.

Another mechanism to build the elite power base by securing control over the energy sector is the overlap between political and management functions. High-level government officials sit on the boards of the two pipeline controlling companies, Transneft and Gazprom.[62] Moreover, Igor Sechin, the chairman of Rosneft, is

---

55 Satpayev, Dosym/ Umbetaliyeva, Tolganai: Sumerechnaya zona ili lovushki perekhodnogo perioda, Almaty: Alyans Analiticheskich Organizaciy, 2013, here p. 127.
56 Ministerstvo Energetiki Rossiiskoi Federatsii: Energeticheskaya strategiya Rossii na period do 2030 goda. Utverzhdena rasporyazheniem Pravitel'stva Rossiiskoi Federatsii ot 13 noyabrya 2009g. no. 1715-r, 13 November 2009, http://minenergo.gov.ru/aboutminen/energostrategy/.
57 Parfent'eva, Irina: Pozitsiya pravitel'stva—zhit' po sredstvam, in: Kommersant.ru, 21 September 2011, http://www.kommersant.ru/doc/1778486.
58 Ross, Michael L.: Does oil hinder democracy?, in: World Politics, 2001 (vol. 53), no. 3, pp. 325–361.
59 Domjan, Paul/ Stone, Matt: A comparative study of resource nationalism in Russia and Kazakhstan, 2004-2008, in: Europe-Asia Studies, 2010 (vol. 62), no. 1, pp. 35–62.
60 Schrijver, Nico J.: Sovereignty over natural resources : balancing rights and duties in an interdependent world, Cambridge: Cambridge University Press, 1997.
61 Gaddy, Clifford G./ Ickes, Barry W.: Resource rents and the Russian economy, in: Eurasian Geography and Economics, 2005 (vol. 46), no. 8, pp. 559–583, here p. 576.
62 Domjan, Paul/ Stone, Matt: A comparative study of resource nationalism in Russia and Kazakhstan, 2004-2008, in: Europe-Asia Studies, 2010 (vol. 62), no. 1, pp. 35–62.

3. Opening Up the Black Box: Russia's Energy Security Concept

a former deputy prime minister in Putin's government who became a member of the Commission on the Strategic Development of the Energy Sector and Ecological Security in 2012. The content of the Energy Strategy reaffirms this privileged relationship. For instance, the 2009 energy strategy recognises the beneficial tax regime for the already over-privileged domestic energy companies as one of the major achievements of the 2003 energy strategy.[63]

This symbiosis of state and energy concerns spills over into foreign policy.

> The strategic goal of the development of the oil sector is [...] to safeguard Russia's political interests in the world; [...].[64]

One instrument of foreign policy influence is the oil and gas pipeline network owned by Transneft and Gazprom. These two guardians not only ensure rents by buying Central Asian gas and oil cheaply and selling it at a much higher price to European countries but also act as political bargaining chips: alternative export pipelines are used to bypass 'unstable' transit countries such as Ukraine and Belarus.

The strengthening of Russia's position in world energy markets is also linked to the presence of national energy companies on foreign soil. The state-controlled energy monopolies must have free access to foreign markets, and the Russian government is willing to actively support this internationalisation policy.

> Laws should be adopted [...] to support national companies in their struggle for resources and energy export markets.[65]

Russian investments in the European downstream sector are seen as a guarantee for Russia's security of demand interests.

The opaque ownership structure and foreign ambitions, in combination with supply interruptions, raised suspicion among European partners about energy as a political weapon. Thus, a growing reputational element exists within the Russian SOD concept. Russia must promote its **Reliability** as an energy supplier. Russia's reputation as a reliable partner that meets its international obligations must convince EU countries of stable energy supplies.

---

63   Ministerstvo Energetiki Rossiiskoi Federatsii: Energeticheskaya strategiya Rossii na period do 2030 goda. Utverzhdena rasporyazheniem Pravitel'stva Rossiiskoi Federatsii ot 13 noyabrya 2009g. no. 1715-r, 13 November 2009, http://minenergo.gov.ru/aboutminen/energostrategy/, here section V 4.
64   Predsedatel' Pravitel'stva Rossiiskoi Federatsii: Energeticheskaya strategiya Rossii na period do 2020 goda. Utverzhdena rasporyazheniem Pravitel'stva Rossiiskoi Federatsii ot 28 avgusta 2003g. no. 1234-r, 28 August 2003, http://www.rg.ru/2003/09/30/energeticheskajastrategija.html, here section VI 2; for the gas sector see also section VI.3; in the Russian original: Стратегическими целями развития нефтяного комплекса являются: [...] обеспечение политических интересов России в мире; [...].
65   Ibid., here section IV.7; Необходимо принять законы [...] поддержки отечественных компаний в борьбе за ресурсы и рынки сбыта энергоносителей.

Russia will act [...] on the principles of predictability, responsibility, mutual trust and consider interests of both suppliers and consumers.[66]

Russia's key challenge is to meet its international and domestic energy commitments. The energy crises between Russia and Ukraine and Belarus delivered a serious blow to this image by raising awareness of the use of energy as a political weapon and scaring away EU consumers. In response, the EU seeks to diversify suppliers.[67]

One strategy to increase reliability is to integrate Russia into the international economy. This approach would create a common ground of rules in which Russia is treated as a full partner. In 2003, WTO accession on favourable conditions and the ratification of the Energy Charter Treaty were set as targets.[68] However, in the 2009 energy strategy, neither WTO accession nor the Energy Charter were mentioned, suggesting that Russia prefers to integrate on its own conditions. When comparing the two versions of the energy strategy within the node 'integration with the world economy,' the shift towards a more confident Russia becomes apparent. Although not yet a WTO member in 2009, Russia already was integrated into the world economy, highlighting that 'Russia already today is in one of the leading positions in the global system of energy trade'. Moreover, Russia is ready to use this leverage: 'The maximum effective use of Russia's energy potential to realise its full integration within the world energy market is a strategic goal of the foreign energy policy'.[69] However, this more assertive stance could harm Russia's image as a reliable partner.

Alternative pipelines, such as the Nord Stream and South Stream, would add to Russia's reliability due to the decline in supply interruptions provoked by unstable transit countries. As a result, Russia would enhance its role as the sole guarantor of a stable energy supply and the only actor responsible for potential interruptions. Transit countries could no longer be blamed when something goes wrong. Reliability thus would become more closely linked to Russia's own capacity to comply with long-term energy contracts.

---

66   Ministerstvo Energetiki Rossiiskoi Federatsii: Energeticheskaya strategiya Rossii na period do 2030 goda. Utverzhdena rasporyazheniem Pravitel'stva Rossiiskoi Federatsii ot 13 noyabrya 2009g. no. 1715-r, 13 November 2009, http://minenergo.gov.ru/aboutminen/energostrategy/, here section V.9; in the Russian original: Политика России в указанной сфере осуществляется [...] на принципах предсказуемости, ответственности, взаимного доверия и учета интересов производителей и потребителей.
67   Smith Stegen, Karen: Deconstructing the 'energy weapon': Russia's threat to Europe as case study, in: Energy Policy, 2011 (vol. 39), no. 10, pp. 6505–6513, here p. 6508.
68   Predsedatel' Pravitel'stva Rossiiskoi Federatsii: Energeticheskaya strategiya Rossii na period do 2020 goda. Utverzhdena rasporyazheniem Pravitel'stva Rossiiskoi Federatsii ot 28 avgusta 2003g. no. 1234-r, 28 August 2003, http://www.rg.ru/2003/09/30/energeticheskajastrategija.html, here section 3.2.
69   Ministerstvo Energetiki Rossiiskoi Federatsii: Energeticheskaya strategiya Rossii na period do 2030 goda. Utverzhdena rasporyazheniem Pravitel'stva Rossiiskoi Federatsii ot 13 noyabrya 2009g. no. 1715-r, 13 November 2009, http://minenergo.gov.ru/aboutminen/energostrategy/.

3.5. Conclusion: Russia's Energy Security Concept

The Russian case indicates that an additional concept of energy security exists beyond SOS. The Russian elite articulated SOS issues in their energy strategy as well as what can be called SOD.

Although it is an energy producer, Russia faces considerable challenges in relation to SOS. Similar to consuming countries, the four A's (i.e., availability, acceptability, accessibility, and affordability) all play an important role in securing stable domestic supplies to the end-consumer. Evidence of both aspects of energy security were coded abundantly, and twice as much attention is still focused on SOS. This finding modifies the statement that producer countries will mainly be concerned about SOD issues. SOS remains the most important concern of the Russian elite.

Although the academic literature is biased towards studying importing countries' SOS, the comparative analysis of the 2003 and 2009 versions of the Russian energy strategy demonstrates that Russia's attention to SOD increased. This reflects the sense of urgency following the price and demand shocks in the wake of the 2008 global financial crisis. The urgency to ensure stable and reliable demand for Russia's energy products rose on the security agenda.

The discourse analysis identified four constituent parts of Russia's SOD: recovery, rents, resource nationalism, and reliability. The recovery aspect of SOD is a self-defence mechanism to insulate the Russian economy from price and demand shocks in the long run. Energy is linked to economic security, which ensures the independence of the country. Several diversification strategies to ensure demand stability in the long run were identified: diversification of export goods, export markets and the economic structure. The rents aspect does not focus on the existence of the state economy as a whole but on the survival of the elite themselves. The direct link between the energy sector and the state budget has been highlighted. Redistribution of existing wealth through social policies and by reinforcing the security apparatus ensures the elite will remain in power. Resource nationalism allows the political elite to control energy companies, thereby keeping the economic elite from becoming overly powerful. In relation to Russia's export markets, the reputation of being a reliable energy partner gains importance. Russia must convince consumer countries of its ability to provide stable supplies.

Internal tensions between these aspects of energy security clearly exist. A stringent resource nationalism policy harms the reliability status and stands in stark contrast with liberalisation policies. These intra-relationships and the inter-relationships between SOD and SOS should be analysed further.

Figure 3-1: Coded references to SOS in Russia's energy strategies, 2003 and 2009 (NVivo 10)

SECURITY OF SUPPLY RUSSIAN FEDERATION	2	1359
ACCEPTABILITY	2	385
Ecologic security	2	90
Energy efficiency and conservation	2	265
social security	2	29
ACCESSIBILITY	2	323
Diversification of domestic energy routes	2	47
Diversification of suppliers (Central Asia)	2	5
Domestic stable supply to end-consumer	2	55
Early warning system for threats ES	2	3
Liberalization of the energy market	2	132
Regional disproportion energy supply (access)	2	81
AFFORDABILITY	2	123
Domestic energy prices	2	88
Production cost energy higher	2	33
AVAILABILITY	2	527
Available reserves	2	54
Diversification energy fields	2	41
Diversification energy mix	2	105
Domestic demand for energy	2	63
Energy Production	2	99
Investment in Energy sector	2	165

3. Opening Up the Black Box: Russia's Energy Security Concept

Figure 3-2: Coded references to SOD in Russia's energy strategies, 2003 and 2009 (NVivo 10)

SECURITY OF DEMAND RUSSIAN FEDERATION	2	651
RECOVERY	**2**	**262**
contractual take or pay	2	4
Diversification of (export) goods	2	64
Diversification of economic structure	2	31
Diversification of export markets	2	56
Energy condition to economic development and growth	2	49
Foreign demand for energy	2	36
Increase Export Level	2	22
RELIABILITY	**2**	**40**
integration in international economy	2	30
Reliable energy exporter to EU	2	10
RENTS	**2**	**53**
Energy Revenue to State Budget	2	17
World Energy Price Fluctuations	2	36
RESOURCE NATIONALISM	**2**	**294**
Control over Transit Pipelines	2	28
Energy instrument foreign policy	2	12
State control on energy market	2	164
Strong energy companies TEK	2	90

Part II. Energy Policy in Central Asia

Farkhod Aminjonov

4. Challenges Along the Way Towards a Maximally Secure Central Asian Gas System

4.1. Introduction

Possessing reserves of approximately 20 trillion cubic meters of natural gas, Central Asia is becoming increasingly attractive to energy-thirsty larger powers surrounding the Central Asian region. While energy importers address the need to ensure steady imports of energy and the security of energy supplies, Central Asian gas exporters aim to secure their ability to constantly export energy to obtain a steady income.

The Central Asian countries can roughly be divided into net consumer (Tajikistan and Kyrgyzstan) and net producer countries (Kazakhstan, Turkmenistan, and Uzbekistan). The Central Asian energy system during the Soviet period was constructed in such a way that the stability and reliability of energy supplies were maintained through a resource-sharing mechanism. The resource-sharing mechanism ensured the stability of energy supplies even after the disintegration of the Soviet Union. The mechanism was quite simple: the consumer countries of Kyrgyzstan and Tajikistan ensured a continuous flow of water and a certain amount of electricity during the summer to the producer countries, Kazakhstan, Turkmenistan, and Uzbekistan, which channelled fuel and gas to them in return.[1] Although there is still a demand for fossil energy by the consumer countries, the current geopolitical and economic realities have challenged the effectiveness of this exchange mechanism.

Having experienced the negative impacts of excessive dependence on Russian pipelines, Central Asian exporters are now pursuing the diversification of energy export routes to obtain access to various energy markets while avoiding Russian territory. However, pipelines are the only cost-efficient way to transport energy from this landlocked region, and Central Asia is surrounded by larger powers (Russia, China, Europe, and South Asia) that often compete for energy resources. Thus, particular consideration is required when pursuing the diversification of energy export routes. This paper discusses the factors that may threaten and that are already affecting the security of energy supplies for countries within the Central Asian energy system (CAES), such as asymmetrical interdependent energy supply relations among energy actors within the CAES, in which energy actors interact and affect each other's security; insufficient volume of natural gas production to meet international demand without compromising

---

1 Laldjebaev, Murodbek: The water-energy puzzle in Central Asia: the Tajikistan perspective, in: Water Resources Development, 2010 (vol. 26), no. 1, pp. 23–36, here p. 24.

internal consumption needs and gas exports to neighbouring consumer countries; the absence of effective enforcement mechanisms to coordinate responses to insecurities of energy supplies; and, in a long-term perspective, the underdevelopment of renewable energy sources in the overall energy balance to secure the availability of clean and sufficient energy supplies for both population and economic needs for the foreseeable future.

4.2. Conceptualising Energy Security

A large amount of work has been conducted on energy security within the framework of national security, economic gains, international politics, energy interdependence, and the diversification of energy sources/energy export routes. Various scholars have examined the concept from different angles, but they have failed to produce a universal definition of energy security.[2]

One group of scholars argues that energy security is constructed around the recognition that oil, gas, and their renewable counterparts need to be considered first and foremost as commodities. Thus, energy export/import relations are to be exercised through market-based transactions.[3] Others emphasise the linkage between energy security, international politics, and national security.[4] State actors are usually guided by a particular type of logic suggesting that they cannot trust energy security to market forces alone. Thus, state actors often interpret threats to energy security as threats to national security.[5] For instance, in choosing routes to export energy, states often consider the political ramifications of various route options.[6]

There have been two major shifts in the literature in defining the concept of energy security. The first shift is characterised by broadening sources vital to providing energy security. The evaluation of energy security is no longer limited to the security of oil and gas supplies and the diversification of energy transporting routes and energy markets. The diversification of the energy mix, by including alternative energy

---

2   Ciuta, Felix: Conceptual notes on energy security: total or banal security?, in: Security Dialogue, 2010 (vol. 41), no. 2, pp. 123–144, here p. 126.
3   Goldthau, Andreas/ Witte, Jan Martin: From energy security to global energy governance, in: Journal of Energy Security, 23 March 2010, http://www.ensec.org/index.php?option=com_content&view=article&id=234:from-energy-security-to-global-energy-governance&catid=103:energysecurityissuecontent&Itemid=358.
4   Allison, Roy/ Jonson, Lena: Central Asian security: the new international context, London: Royal Institute of International Affairs, 2001; Müller-Kraenner, Sascha: Energy security: re-measuring the world, London: Earthscan, 2008; Moran, Daniel/ Russel, James A.: Energy security and global politics: the militarization of resource management, London: Routledge, 2009; Proedrou, Filippos: EU energy security in the gas sector: evolving dynamics, policy dilemmas and prospects, Aldershot: Ashgate, 2012.
5   Moran, Daniel/ Russel, James A.: Energy security and global politics: the militarization of resource management, London: Routledge, 2009, p. 2.
6   Shaffer, Brenda: Energy politics, Philadelphia, PA: University of Pennsylvania Press, 2009.

Challenges Along the Way Towards a Secure Central Asian Gas System 133

sources such as wind, solar, nuclear, and hydropower, has become an important aspect of energy security to ensure the long-term security of energy supplies.[7]

The second shift is about expanding the range of actors whose energy security concerns must be considered. Conventional understanding of energy security emphasises the importance of energy (mainly oil and gas) supplies for importing countries. As a result, energy security has been associated with energy self-sufficiency and the security of energy supplies. When a country begins importing a large amount of oil and gas, it becomes vulnerable to potential energy sanctions. This compromises its energy security, understood as the security of energy supplies.[8] '[C]onventional energy security seeks to assure supply while assuming that demand is given [...]'.[9]

Recent studies on energy security have shifted the definition of energy security from a sole focus on the importers' perspective to the relationship between energy importers and exporters, thus emphasising the vulnerability of the latter in energy export/import relations. The main line of argument is that 'exporters worry about energy demand the way energy importers worry about energy supply'.[10] Energy security, from the exporters' perspective, is defined as the security of demand to generate economic growth and to maintain social stability,[11] which can be threatened by a variety of factors, such as resource wars or economic sanctions imposed by importing states.[12]

The most recent approach to energy security places individuals' energy needs at the centre of interest because an abundance of energy resources in a country does not necessarily mean that individuals are continuously provided with sufficient energy.[13] The human dimension of energy security, or the so-called 'energy services security', places individuals at the centre of interest in designing energy policy and emphasises individuals' access to energy resources in sufficient volume at affordable price.[14]

---

7   Von Hippel, David/ Suzuki, Tatsujiro/ Williams, James H./ Savage, Timothy/ Hayes, Peter: Energy security and sustainability in Northeast Asia, in: Energy Policy, 2011 (vol. 39), no. 11, pp. 6719–6730; Winzer, Christian: Conceptualizing energy security, in: Energy Policy, 2012 (vol. 46), no. 7, pp. 36–48; Sovacool, K. Benjamin: The methodological challenges of creating a comprehensive energy security index, in: Energy Policy, 2012 (vol. 48), pp. 835–840.
8   Wesley, Michael: Energy security in Asia, London: Routledge, 2007, p. 43.
9   Vivoda, Vlado: Evaluating energy security in the Asia-Pacific region: a novel methodological approach, Brisbane: Griffith Asia Institute, p. 2, http://www98.griffith.edu.au/dspace/bitstream/handle/10072/36043/65347_1.pdf?sequ (accessed 3 May 2013).
10 Bahgat, Gawdat: Energy security: an interdisciplinary approach, Chichester: Wiley, 2011, p. 226.
11 Yergin, Daniel: The quest: energy, security, and the remaking of the modern world, New York: Penguin Press, 2011, p. 267.
12 Allison, Roy/ Jonson, Lena: Central Asian security: the new international context, London: Royal Institute of International Affairs, 2001.
13 Shaffer, Brenda: Energy politics, Philadelphia, PA: University of Pennsylvania Press, 2009, p. 91; Buzar, Stefan: Energy poverty in Eastern Europe: hidden geographies of deprivation, Aldershot: Ashgate, 2007.
14 Alhaji, Anas F.: What is energy security? Definitions and concepts (Part 3/5), in: Middle East Economic Survey, 2007 (vol. 50), no. 45, http://archives.mees.com/issues/213/articles/8255.

The list of factors that can threaten the security of supply and demand is quite inclusive:

> internal instability, civil wars, ethnic violence that can disturb energy production; terroristic attacks on energy infrastructure; politically motivated suspension of oil and gas supply; economic sanctions against energy producing countries; war between energy producers; territorial disputes that can significantly slow down cooperation among parties in energy sector and so on.[15]

However, 'energy security is not just about countering the wide variety of threats. It is also about the relations among nations, how they interact with each other, and how every impact their overall security'.[16] Yergin considers energy security a system 'composed of the national policies and international institutions that are designed to respond in a coordinated way to energy supply disruptions'.[17]

Taking into account all of the above-mentioned aspects, energy security can be defined as a condition states enjoy when they are confident that they will have adequate and sustainable energy supplies for the foreseeable future. The CAES is a framework/complex system within which various energy actors interact and affect each other's security. Given this definition of energy security, the security of the CAES is the condition in which all Central Asian states enjoy energy security (for both population and economy needs) simultaneously. Adequate energy supplies basically means that states have enough energy resources to meet their needs. The sustainability of energy supplies implies that present needs are met without compromising energy supplies for future generations. The CAES entails balancing the energy interests of all. Reaching consensus is difficult but necessary if the end goal is to ensure that everyone enjoys energy security.

4.3. The Central Asian Gas System

During the Soviet era, the stability and reliability of energy supply flows were ensured by instructions from a single political centre (Moscow). After the disintegration of the Soviet Union, Russia inherited the infrastructure that the Central Asian states needed to transport energy out of the region, creating excessive dependence on the Russian pipeline network and energy market. In the 1990s, the Central Asian states continued bartering energy with each other and Russia, almost the same way they had in the unified Soviet energy system. However, regional energy exporters' dissatisfaction with the terms of the energy trade dictated by Russia and the willingness of other external customers to invest in the construction of pipeline networks to transport

---

15   Bahgat, Gawdat: Energy security: an interdisciplinary approach, Chichester: Wiley, 2011, p. 15.
16   Yergin, Daniel: The quest: energy, security, and the remaking of the modern world, New York: Penguin Press, 2011, p. 264
17   Yergin, Daniel: The quest: energy, security, and the remaking of the modern world, New York: Penguin Press, 2011, p. 267

energy while avoiding Russia transformed the CAES. As a result, two interlinked levels emerged within the Central Asian energy system: first, energy supply relations within the Central Asian region; second, energy export/import between Central Asian producers and external customers.

The Central Asia-Centre (CAC) gas pipeline network is still the main natural gas transportation system from Central Asia to Russia, where the gas enters the pipeline system operated by the Russian state-owned company Gazprom and is then re-exported to Ukraine and Europe.[18] The planned Pre-Caspian pipeline along the Caspian coast aims to increase the volume of Central Asian gas that is transported to Russia.[19] However, Turkmenistan began exporting natural gas to Iran in 1997 and is now planning to triple its export.[20] Additionally, starting with a small quantity of natural gas imports from Central Asia in 2009, China is rapidly increasing the volume of energy imported from the region. Central Asia is also perceived by Europeans as a good alternative to loosen energy dependence on Russian natural gas. Thus, the EU has initiated several projects to export Central Asian natural resources while bypassing Russia. Of all of the projects planned within the Southern Corridor, the priority in Brussels is given to the Nabucco project, a pipeline to transport up to 31 billion cubic metres (bcm) of gas from Central Asia via Turkey to Europe. Turkmenistan, as a major natural gas producer, is expected to be one of the key suppliers to Europe. In addition to European customers, South Asian countries are very interested in a share of the Central Asian natural gas reserves via the planned TAPI pipeline project from Turkmenistan via Afghanistan to Pakistan and India.

4.4. Diversification of Central Asian Gas Exports

Because there has been no truly effective enforcement mechanism since the end of the Soviet Union to secure reliable energy supplies within the Central Asia energy system, regional exporters are attempting to establish symmetrical interdependent relations with major customers so that there is less incentive on either side to cause energy supply disruptions. Shaffer argues that energy security is provided when suppliers and consumers are interdependent in energy supply relations. The extent to which each side possesses alternative supply or market options, including transport

---

18 International Energy Agency: Optimizing Russian natural gas: reform and climate policy, Paris: IEA, 2006, p. 31, http://www.iea.org/publications/freepublications/publication/russiangas2006.pdf.
19 Gould, Tim/ Murray, Isabel/ Sinton, Jonathan/ Graczyk, Dagmar/ Segar, Christopher: Perspectives on Caspian oil and gas development, Paris: IEA Directorate of Global Energy Dialogue, 2008, p. 62, http://www.asiacentral.es/docs/caspian_perspectives_iea_dec08.pdf.
20 'Turkmenistan, Iran launch gas pipeline', in: Pipeline and Gas Journal, 2011 (vol. 238), no. 1, http://www.pipelineandgasjournal.com/turkmenistan-iran-launch-gas-pipeline.

infrastructure, determines the extent of interdependence in energy relations.[21] On the one hand, the diversification of energy export routes ensures alternative ways of transporting energy for energy consumers, which decreases the impact of technical failures or energy supply disruptions by either energy exporter or transit countries. On the other hand, diversification provides alternative energy markets for producers, which increases exporting countries' bargaining power so that they can sell energy at the highest possible price.[22] In addition to the existing Turkmenistan–Iran gas pipelines, the newly constructed Central Asia–China gas pipeline that runs from Turkmenistan through Uzbekistan and Kazakhstan to China decreased regional gas exporters' dependence on Russia. Although Russia remains the major importer of Central Asian natural gas, with increasing alternative transport capacities, Central Asian exporters are now less vulnerable to unilateral disruptions of gas imports by Russia.

There is clearly a demand for the region's energy resources, both within the region and by external customers. Taking into account the fact that Central Asia (mainly Kazakhstan, Turkmenistan, and Uzbekistan) possesses significant natural gas reserves, providing sufficient energy for both economic and population needs should not be a problem. However, despite its energy reserves, the overall energy supply-demand relations are far from stable and reliable within the Central Asian energy system.

Although, hypothetically, the diversification of energy transporting routes benefits both exporters and importers, many factors determine the success of such diversification:
- geography (the distance between exporting and importing countries; security/vulnerability of transport routes),
- political relations among energy actors,
- availability of sufficient energy resources to meet energy demand and transport infrastructure,
- policy (commitment to implement),
- resources (capacity and willingness to pay the price to secure access to alternative energy).[23]

Central Asian energy exporters' desire to further diversify export routes is understandable. However, policy makers must consider particular characteristics of the region and natural gas supply deals when pursuing the further diversification of energy export routes because this diversification may not necessarily contribute to overall security within the CAES.

---

21 Shaffer, Brenda: Energy politics, Philadelphia, PA: University of Pennsylvania Press, 2009, p. 4.
22 Hubinger, Vaclav: Southern Corridor: a strategy for sustainable energy cooperation with Central Asia, in: EUCAM Watch, no. 4, 2009, p. 4, http://www.fride.org/download/EUCAM_Newletter4_ENG_may09.pdf.
23 Vivoda, Vlado: Diversification of oil import sources and energy security: a key strategy or an elusive objective?, in: Energy Policy, 2009 (vol. 37), no. 11, pp. 4615–4623, here p. 4620.

For a land-locked region such as Central Asia, cross-border gas pipelines are the only cost-efficient way to export energy. The construction of such pipelines requires significant investments from both the producer and customer sides. Because new international gas pipelines need to operate for at least fifteen to twenty years before investments can be recouped, natural gas is often traded within the framework of long-term supply contracts.[24] These characteristics of the natural gas trade and the fact that the region is surrounded by major powers (Russia, China, India, and Europe) that often compete for energy resources establish a particular type of energy supply relations in which energy actors might be willing to use various political, economic, and military tools to force Central Asian exporting countries to fulfil their obligations at any cost. Although there have been partially successful efforts to diversify away from Russia and possibly obtain higher prices for the exported gas, this strategy is constrained by a lack of production capacity and the conflict between export desire and regional consumption needs.

4.5. The Monopoly of the Russian Pipeline System

In an attempt to decrease its dependence on Russia's pipeline grid, some former Soviet republics adopted a 'two-track' or 'offend no one' foreign policy strategy. For instance, Azerbaijan could not afford to offend Russia but wanted to export its oil to the United States and Europe. Thus, Azerbaijan opted for two export routes, one through Russia and another through Georgia. When the Russian route was closed due to the war in Chechnya in 1999, the two-track strategy proved to be very useful because Azerbaijan could still use the pipeline to Georgia.[25] However, Central Asian energy producers did not or could not adopt a 'two-track' strategy and remained significantly dependent on Russian pipeline system for a long time.

The Russian government effectively used Central Asian exporters' dependence on Russian pipelines to promote its economic and political interests. Central Asian gas producers had to sell their energy resources to Moscow, which partially re-exported that energy to Europe at two and sometimes three times the purchase price or supplied its southern regions with the energy. Prior to the 2000s, the price paid by Russia for natural gas did not correspond to the market value of the gas and was often in the form of barter. Since 2006, energy trade relations between Russia and Central Asian producers have taken the form of cash payments. Russia purchased Turkmen gas at the price of US$60 per 1,000 cubic metres (cm) in 2006. During the last quarter of 2006, the price rose to US$100, and it increased to US$150 per 1,000cm in the second

---

24 Shaffer, Brenda: Energy politics, Philadelphia, PA: University of Pennsylvania Press, 2009, p. 38.
25 Yergin, Daniel: The quest: energy, security, and the remaking of the modern world, New York: Penguin Press, 2011, p. 58

half of 2008.[26] However, when Gazprom bought Turkmen gas for less than US$100 per 1,000cm in 2006, it resold that gas through RosUkrEnergy to Europe for US$250 per 1,000cm.[27] The inability to sell its energy resources directly to Europe was one of the reasons Turkmenistan agreed to these terms.

In an attempt to block projects that could challenge its almost complete monopoly over the region's gas exports, Russia agreed to pay a higher price for Central Asian natural gas. Russia signed an agreement with Turkmenistan in 2008, according to which Russia was obligated to purchase Turkmen gas in the amount of 70–80bcm per year for the European price of US$350 per 1,000cm.[28] Due to the Russian–Ukrainian gas crisis in 2009, the European price for natural gas fell to US$280 per 1,000cm, which meant that Russia could no longer profit from re-exports or swaps of Central Asian gas to Europe. Thus, Russia effectively used an explosion that occurred on the CAC pipeline as an excuse to cut gas imports from the region. Consequently, Turkmenistan experienced significant economic losses, which accounted for US$1 billion every month during the period of cut-off.[29] Anderi Grozin, director of the Central Asia Department at the CIS Institute in Moscow, highlighted the seriousness of the nine months of gas export disruption by arguing that Turkmenistan owed its financial survival to a Chinese loan of 2 billion Euros.[30]

Ongoing negotiations between regional energy exporters and potential European and South Asian natural gas importers show that the goal pursued by exporters is not only to diversify their dependence on Russia but also to obtain access to as many energy markets as possible. However, the realisation of the planned pipeline projects is likely to negatively impact the energy security of the Central Asian consumer countries.

4.6. The Pitfalls of Natural Gas Export Diversification

With all of the existing energy transportation projects upgraded and planned ones constructed, there is doubt that the Central Asian producer countries will be able to meet their supply obligations and to sensibly use the additional transport capacities. Russia remains an important customer of the regional energy resources via the CAC

---

26 Gould, Tim/ Murray, Isabel/ Sinton, Jonathan/ Graczyk, Dagmar/ Segar, Christopher: Perspectives on Caspian oil and gas development, Paris: IEA Directorate of Global Energy Dialogue, 2008, p. 11, http://www.asiacentral.es/docs/caspian_perspectives_iea_dec08.pdf.
27 Bahgat, Gawdat: Europe's energy security: challenges and opportunities, in: International Affairs, 2006 (vol. 82), no. 5, pp. 961–975, here p. 961.
28 Henderson, Creelea: Shifting sands in Central Asia: geopolitics of natural gas flows, in: Perspective, 2010 (vol. 20), no. 2, http://www.bu.edu/iscip/Vol20/henderson.html.
29 'Central Asia considers alternatives to Gazprom for exporting gas. Will Silk Road producers actually prefer Europe to Russia?', in: Central Asian Online, 18 November 2011, http://centralasiaonline.com/en_GB/articles/caii/features/main/2011/11/18/feature-01?change_locale=true.
30 Tynan, Deirdre: Gas flows again to Russia, while discontent simmers, in: Gundogar, 19 January 2010, http://www.gundogar.org/?0220048966000000000000011000000.

Challenges Along the Way Towards a Secure Central Asian Gas System 139

gas pipeline network, with the capacity to transport 50bcm per year and, if upgraded, 90bcm per year.[31] Russia is also interested in the construction of the Pre-Caspian pipeline (from Turkmenistan via Kazakhstan to Russia) to significantly increase the overall capacity of the CAC.[32] Turkmenistan is planning to increase its gas supply to Iran from 6–8bcm to 20bcm per year.[33] Another major energy importer that has recently entered the market is China. Turkmenistan was initially obliged to export 30bcm per year of natural gas to China according to the agreement signed in 2006. However, the new agreement that was signed two years later increased the volume to 40bcm per year by 2015. During the Shanghai Cooperation Organisation's summit in Beijing in June 2012, the countries' presidents, Hu Jintao and Gurbanguly Berdimuhammedov, agreed to increase the amount of gas exports even further, to 65bcm per year.[34] China will be receiving an additional 10bcm per year from Uzbekistan according to the agreement signed in 2010.[35] Moreover, China is hoping, and Uzbekistan is showing willingness, to increase the export volume to 25bcm per year by building a third line of the Uzbek section of the Central Asia–China pipeline.[36] Among the major planned pipeline projects, the TAPI pipeline, with a capacity of 33bcm per year,[37] and the Nabucco project, with a capacity up to 31bcm per year,[38] stand out. It is worth mentioning that there are many controversies regarding the viability of these two gas pipeline projects. However, even without TAPI and Nabucco, all other projects with a high possibility of realisation may still result in a mismatch between a region's production capacity (without compromising regional consumption needs) and external demand for Central Asian gas.

In fact, there are already signs that it would be quite challenging for some regional gas exporters to increase their export capacity. For instance, outdated and inefficient natural gas transportation systems, growing internal energy demand, and the fact that no major natural gas reserves have recently been explored are indications of

---

31 International Energy Agency: Optimizing Russian natural gas: reform and climate policy, Paris: IEA, 2006, p. 31 http://www.iea.org/publications/freepublications/publication/russiangas2006.pdf.
32 Socor, Vladimir: Russia resuming gas imports from Turkmenistan on a small scale, in: Eurasia Daily Monitor, 4 January 2010 (vol. 7), Issue 1, http://georgiandaily.com/index.php?option=com_content&task=view&id=16309&Itemid=132.
33 'Turkmenistan, Iran launch gas pipeline', in: Pipeline and Gas Journal, 2011 (vol. 238), no. 1, http://www.pipelineandgasjournal.com/turkmenistan-iran-launch-gas-pipeline.
34 Socor, Vladimir: China to increase Central Asian gas imports through multiple pipelines, in: Eurasia Daily Monitor, 9 August 2012 (vol. 9), Issue 152, http://www.jamestown.org/single/?no_cache=1&tx_ttnews%5Btt_news%5D=39751.
35 'Central Asia-China gas pipeline, Turkmenistan to China', in: Hydrocarbons-technology, http://www.hydrocarbons-technology.com/projects/centralasiachinagasp/ (accessed 1 May 2013).
36 'Uzbekistan i Kitai: vzaimovigodnoe sotrudnichestvo', in: russian.china.org.cn, 6 June 2012, http://russian.china.org.cn/exclusive/txt/2012-06/06/content_25591078.htm.
37 'Turkmenistan–Afghanistan–Pakistan–India gas pipeline: South Asia's key project', in: PetroMin, April–June 2011, pp. 6–12, http://elordenmundial.files.wordpress.com/2013/02/tap-pipeline.pdf.
38 Hasanov, H.: Negotiations on Trans-Caspian gas pipeline held successfully in Brussels, in: Trend, 10 March 2012, http://en.trend.az/capital/energy/2001834.html.

Uzbekistan's physical incapability to increase its exports. The fact that Uzbekistan had to cut its gas export to Russia for 40 days to meet its internal energy needs in 2012[39] and that it supplies less than the agreed amount of gas to Tajikistan (132 instead of 155 million cubic metres of gas)[40] can be considered signs that Uzbekistan will face challenges in supplying approximately 15bcm of gas to Russia[41], 10bcm to China, and 4.5bcm to southern Kazakhstan through the Tashkent–Chimkent–Bishkek–Almaty pipeline system.[42] Given the fact that there has been no natural gas production boom in Central Asia and the region's gas export capacity remains at approximately 55–65bcm per year, even Turkmenistan, with its massive gas reserves, may face technical, economic, and security challenges to keep up with international energy demand.

4.7. Conflicts Between Export and Domestic and Regional Consumption Needs

Despite the fact that there are energy shortages for regional consumption, Central Asian energy producers will most likely continue increasing their export capacity to meet growing international demand. There are two major reasons for regional exporters' willingness to restrict domestic consumption in favour of increasing export capacity: first, regional exporters are attempting to compensate their economic losses from subsidising energy for domestic consumption by generating higher revenues from energy export; second, asymmetrical power relations between Central Asian energy producers and such external customers as Russia, China, and, potentially, Europe will force regional exporters to go along with the system (in which energy export to the external markets is prioritised) rather than challenge it.

4.7.1. Energy Subsidies, Energy Efficiency

One of the reasons for Central Asian natural gas exporters' policies of restricting domestic consumption through rationing is the use of subsidised gas for political purposes domestically. Low prices for natural gas on the domestic market result in over-consumption and lower profits for producers (both private and state-owned companies), making it unattractive to invest in upgrading domestic energy infrastructure that can

---

39 'Uzbekistan vozobnovil eksport gaza Rossiyu posle mesyachnogo pereryva', in: Kazenergy, 29 January 2013, http://kazenergy.com/ru/press/2011-04-21-10-41-35/7764-2013-01-29-11-07-35.html.
40 Akhmadov, Erkin: Uzbekistan introduces new laws on gas supply, in: CACI Analyst, 9 January 2013, http://www.cacianalyst.org/?q=node/5905.
41 Sharip, Farkhad: Uzbekistan's quest for Aral Sea oil may weaken Kazakhstan's position in the Caspian, in: Eurasia Daily Monitor, 2 February 2012 (vol. 9), Issue 23, http://www.jamestown.org/single/?no_cache=1&tx_ttnews%5Btt_news%5D=38962.
42 US Energy Information Administration: Kazakhstan energy data, 18 September 2012, http://www.eia.gov/countries/cab.cfm?fips=KZ.

Challenges Along the Way Towards a Secure Central Asian Gas System 141

significantly increase efficiency to avoid energy waste and expanding pipeline networks to supply distant regions and a larger number of people with energy.[43] In 2011, natural gas subsidies cost US$4.36 billion to the budget of Turkmenistan, which was 14.8% of its gross domestic product (GDP). In Uzbekistan, subsidies amounted to US$9.09 billion (approximately 18.9% of the GDP).[44]

Losses caused by outdated and inefficient energy production and transportation infrastructures cost Uzbekistan approximately 4.5% of its GDP every year.[45] Kazakhstan consumes only half of its overall gas production and exports the other half (Kazakhstan produced 21.2bcm of natural gas in 2012 and consumed only 10.5bcm) because it lacks extensive internal gas supply networks to transport energy from resource-rich regions to distant consumption centres.[46] Increasing the volume of energy production by building large power-generation plants seems to be a priority area for regional energy producers. However, according to some experts, because the level of energy waste at the consumer end is extremely high, investments in energy projects to increase efficiency would most likely cost about half as much as building new plants with almost the same results.[47] For instance, even though Uzbekistan ranks 20th worldwide in terms of gas reserves,[48] poor transportation and distribution infrastructures account for losses of 20bcm of natural gas per year.[49] Another example is the CAC pipeline network. The lack of maintenance and investment over time has almost halved the operational capacity of the system to less than 50bcm per year.[50] According to the International Energy Agency, at least 30bcm per year is wasted. Even Gazprom itself acknowledges that investments in the pipelines would save up to 10bcm of gas per

---

43 Clements, Benedict et al.: Energy subsidy reform: lessons and implications, Washington, DC: International Monetary Fund, 2013, pp. 15–16, http://www.imf.org/external/np/pp/eng/2013/012813.pdf.
44 'Fossil fuel consumption subsidy rates as a proportion of the full cost of supply 2011', in: IEA: World Energy Outlook, 2012, http://www.iea.org/subsidy/index.html; Clements, Benedict et al.: Energy subsidy reform: lessons and implications, Washington, DC: International Monetary Fund, 2013, p. 48, http://www.imf.org/external/np/pp/eng/2013/012813.pdf.
45 World Bank: Uzbekistan: the economics of efficiency. Uzbekistan pushes to reduce energy consumption in industry, 30 April 2013, http://www.worldbank.org/en/results/2013/04/30/uzbekistan-the-economics-of-efficiency.
46 Bisenov, Naubet: Neighbourly negotiations, in: Energy Global, 12 March 2013, http://www.energyglobal.com/news/pipelines/articles/Neighbourly_negotiations_an_analysis_of_central_asian_energy_pipelines.aspx.
47 Mirimanova, Natalia: Interview with Rainer Behnke, 2009, p. 3, http://www.eucentralasia.eu/file admin/PDF/Commentaries/Interview_INOGATE.pdf.
48 BP statistical review of world energy, June 2012, http://www.bp.com/content/dam/bp/pdf/Statistical-Review-2012/statistical_review_of_world_energy_2012.pdf.
49 Townsend, Jacob/ King, Amy: Sino-Japanese competition for Central Asian energy: China's game to win, in: China and Eurasia Forum Quarterly, 2007 (vol. 5), no. 4, pp. 23–45, here p. 27.
50 International Energy Agency: Optimizing Russian natural gas: reform and climate policy, Paris: IEA, 2006, p. 31, http://www.iea.org/publications/freepublications/publication/russiangas2006.pdf.

year,[51] which is one-fifth of the overall natural gas export from the region. Low energy prices in domestic markets make it unattractive to invest in energy efficiency projects and expand internal pipeline networks. Having experienced a significant economic loss due to subsidising natural gas, Central Asian energy producers are trying to compensate that loss by increasing the volume of gas exports, which will most likely negatively impact the availability of already insufficient energy for domestic and regional consumption.

4.7.2. Asymmetrical Power Relations Within the Central Asian Energy System

Central Asian energy producers' attempt to reduce their energy dependence on Russia by establishing interdependent relations with external customers was successful, to a certain extent, especially with China. However, there is still asymmetry in the extent of vulnerability to energy supply/demand disruptions for energy producers within the Central Asian energy system. The higher the cost of the termination or drastic alteration of energy relations for an actor, the more vulnerable this actor is.[52] Currently, most of the Central Asian energy resources are transported through Russian and Chinese pipelines. However, neither China nor Russia considers Central Asian energy a vital source for their economy. Russia needs Central Asian oil and gas mainly to fulfil its supply obligations towards Europe. Central Asian gas constitutes an insignificant portion of the Chinese overall energy balance (out of 130bcm of natural gas that China consumed in 2011, the Central Asian share was less than 15%, which was approximately 0.5% of the overall energy balance).[53] Given the fact that selling natural gas accounts for approximately half of Turkmenistan's[54] budget and significantly contributes to Kazakhstan and Uzbekistan's budgets, Central Asian energy exporters are more vulnerable to energy supply disruptions than either Russia or China.[55]

Due to asymmetrical power relations within the CAES, some countries hold potential power advantages over others, in which the latter are very limited in their foreign policy choices. In this sense, more vulnerable states are likely to go along with the

---

51 Fredholm. Michael: Natural gas: an expensive trickle, in: Transitions Online, 10 April 2008, http://www.tol.org/client/article/19509-an-expensive-trickle.html?print.
52 Keohane, Robert O./ Nye, Joseph S.: Power and interdependence. 3rd edition, New York: Longman, 2001.
53 BP Statistical Review of World Energy, June 2012, http://www.bp.com/content/dam/bp/pdf/Statistical-Review-2012/statistical_review_of_world_energy_2012.pdf; Batkar, Mamta: The ultimate guide to China's voracious energy use, in: Business Insider, 17 August 2012, http://www.businessinsider.com/china-energy-use-2012-8?op=1.
54 'Difficulties loom for a Turkmen–China energy deal', in: Forbes, 8 March 2011, http://www.forbes.com/sites/energysource/2011/03/08/difficulties-loom-for-a-turkmen-china-energy-deal/.
55 Townsend, Jacob/ King, Amy: Sino-Japanese competition for Central Asian energy: China's game to win, in: China and Eurasia Forum Quarterly, 2007 (vol. 5), no. 4, pp. 23–45, here p. 23

Challenges Along the Way Towards a Secure Central Asian Gas System 143

system rather than attempt to change it.[56] However, in an attempt to increase bargaining power and secure energy demand, regional natural gas exporters support various energy export diversification projects. However, regional exporters' physical inability to produce sufficient energy to meet international demand may lead to competition among external customers for the region's energy resources. The Shanghai Cooperation Organisation is the only (and quite vague) institutional framework to regulate energy export/import relations between regional gas producers and major customers (China and Russia).[57] In the absence of an effective enforcement mechanism to coordinate energy supply relations, the consequences of such competition are unpredictable.

If Central Asian gas exporters are relatively weaker players in energy export/import relations with China and Russia, they are in a far better position to dictate terms in gas supply relations with Central Asian consumer countries. Tajikistan and Kyrgyzstan's gas reserves are insignificant, and these countries mainly rely on imports from neighbouring Uzbekistan and Turkmenistan. Tajikistan and Kyrgyzstan's excessive dependence on hydrocarbon imports from neighbouring countries makes them vulnerable to energy supply disruptions. For instance, gas supply disruptions from Uzbekistan and Turkmenistan to Tajikistan in 2008 caused countrywide electricity blackouts throughout the entire winter period.[58] Even though Kyrgyzstan has developed seven oil fields and two gas fields, difficult geological conditions keep the recovery rate very low.[59] Increasing the volume of energy exports to external customers is likely to negatively impact the Central Asian energy balance, making even less energy available for export to neighbouring consumer countries.

Having experienced severe energy shortages in the past, Kyrgyzstan and Tajikistan are already speeding up the construction of the Rogun and Kambarata hydropower plants. For Tajikistan and Kyrgyzstan, increasing hydropower potential to meet their energy needs has become a national security priority. Taking into account the huge hydropower potential of these countries, the construction of the Kambarata I (with 1.9 gigawatt) and Rogun (3.6 gigawatt) hydropower plants will increase their bargaining power vis-à-vis Uzbekistan and provide greater energy security.

During the summer of 2008, the outlet of water to produce electricity increased, which significantly decreased the level of water in the reservoirs for the winter to

---

56 Garrison, Jean A./ Abdurahmonov, Ahad: Explaining the Central Asian energy game: complex interdependence and how small states influence their big neighbors, in: Asian Perspective, 2011 (vol. 35), no. 3, pp. 381–405, here p. 387.
57 Ogutcu, Mehmet/ Ma, Xin: Geopolitics of energy: China and the Central Asia, in: Insight Turkey, 2007 (vol. 9), no. 4, pp. 44–62, here p. 55.
58 Laldjebaev, Murodbek: The water-energy puzzle in Central Asia: the Tajikistan perspective, in: Water Resources Development, 2010 (vol. 26), no. 1, pp. 23–36.
59 Dorian, James P.: Central Asia: A major emerging energy player in the 21st century, in: Energy Policy 2006 (vol. 34), no. 5, pp. 544–555, here p. 548.

produce electricity to meet at least the minimal internal consumption needs.[60] The price hike for Uzbek gas worsened Tajikistan's energy security conditions. Just before the winter, Uzbekistan doubled the price of natural gas exported to Tajikistan. Although the Tajikistani government succeeded in renegotiating the gas price from US$300 to US$240 per 1,000cm (US$90 higher than the previous price)[61], it was still higher than the price Tajikistan was willing to pay. Uzbekistan's desire to generate higher revenues by selling energy to external customers (China and Russia) is understandable. However, Uzbekistan's energy policy to increase the volume of natural gas exports to Russia and China at the expense of supplies to Central Asian consumers forced Kyrgyzstan and Tajikistan to take counter measures to meet their energy needs by constructing the Rogun and Kambarata hydropower plants, which impact the water-energy balance within the region.[62] The Almaty Agreement of 1992[63] was supposed to keep the Soviet-era resource-sharing mechanism functioning 'until the Central Asian states could reach a solution amenable to all parties'.[64] However, frequent energy supply disruptions to Kyrgyzstan and Tajikistan by producer countries prove that the Almaty Agreement is no longer effective.

4.8. Sustainable Energy Supplies

State actors' perceptions of energy security in Central Asia are limited to ensuring the highest possible price for exporters' gas through the diversification of energy supply routes and securing continuous supplies of hydrocarbons for importing countries. The reason for such a limited understanding of energy security is that although all elements of energy security are considered important, states are sometimes interested in promoting one aspect of energy security more than others.[65] Energy export security (mainly hydrocarbons) to generate higher revenues to the budget and energy import security to satisfy customers' economic needs justify state actors' particular interests in energy supply security within the CAES. However, it is impossible to address the complex of energy security by considering just one part of it.[66] Equating energy security

---

60 Laldjebaev, Murodbek: The water-energy puzzle in Central Asia: the Tajikistan perspective, in: Water Resources Development, 2010 (vol. 26), no. 1, pp. 23–36, here p. 29.
61 Ibid., p. 30.
62 Kasymov, Shavkat: Dammed or damned: Tajikistan and Uzbekistan wrestle over water-energy nexus, in: World Policy, 2 April 2013, http://www.worldpolicy.org/blog/2013/04/02/dammed-or-damned-tajikistan-and-uzbekistan-wrestle-over-water-energy-nexus.
63 In February 1992, the five Central Asian republics signed in Almaty the 'Agreement on cooperation in joint management, use and protection of interstate sources of water resources'.
64 Dinar, Ariel/ Dinar, Shlomi/ McCaffrey, Stephen/ Mckinney, Daene: Bridges over water: understanding transboundary water conflict, negotiation and cooperation, Singapore: World Scientific Publishing, 2007, p. 294.
65 Shaffer, Brenda: Energy politics, Philadelphia, PA: University of Pennsylvania Press, 2009, p. 93.
66 Von Hippel, David/ Suzuki, Tatsujiro/ Williams, James H./ Savage, Timothy/ Hayes, Peter: Energy security and sustainability in Northeast Asia, in: Energy Policy, 2011 (vol. 39), no. 11, pp. 6719–6730.

Challenges Along the Way Towards a Secure Central Asian Gas System 145

to the stability of energy supplies would mean that in the best-case scenario, oil and gas supply security would be ensured (which is not the case in the CAES). However, a lack of attention by state actors, which are key players in the energy sector, towards the development of alternative energy sources and increased efficiency of energy production and transportation by employing new technologies impact the availability of clean energy resources for the foreseeable future.

Overemphasising the importance of fossil fuels in the energy sector explains the lack of attention by governmental agencies towards the development of the renewable energy sector to ensure the sustainability of energy supplies. Despite the fact that Central Asian countries enjoy an abundance of clean energy sources (5.5% of the world's economically efficient hydro potential is mainly in Tajikistan and Kyrgyzstan and an availability of solar energy of an average of 8–10 hours of sunshine per day),[67] they represent only a tiny proportion of the overall energy balance. In 2009, renewables, mainly hydropower, accounted for only 1% of Kazakhstan's, 2% of Uzbekistan's, and 0.001% of Turkmenistan's energy sectors.[68] Uzbekistan and Kazakhstan have recently shown interest in developing clean energy sources. Kazakhstan's state agencies received special instructions from President Nursultan Nazarbaev to turn the capital, Astana, into a 'green city', which will be supplied by 100% renewable energy by 2017.[69] Kazakhstan is showing interest in developing its nuclear, wind, and solar energy sources.[70] According to the strategy 'Kazakhstan 2050', 'alternative and renewable energy sources must account for at least a half of country's total energy consumption by 2050'.[71] The president of Uzbekistan has also signed a decree on 'measures to further develop alternative sources of energy' in the country.[72] However, these countries still lack the capacity to increase the share of renewables in the overall energy balance.

---

67 Zakhidov, R.A.: Central Asian countries energy system and role of renewable energy sources, in: Applied Solar Energy, 2008 (vol. 44), no. 3, pp. 218–223.
68 'Renewable Energy Country Profile—Kazakhstan', IRENA, http://www.irena.org/REmaps/countryprofiles/asia/Kazakhstan.pdf; 'Renewable Energy Country Profile—Uzbekistan', IRENA, http://www.irena.org/REmaps/countryprofiles/asia/Uzbekistan.pdf; 'Renewable Energy Country Profile—Turkmenistan', IRENA, http://www.irena.org/REmaps/countryprofiles/asia/Turkmenistan.pdf (all accessed 12 May 2013).
69 'Astana k EXPO-2017 dolzhna stat' energoeffektivnym "zelenym" gorodom—Nazabaev', in: Kazenergy, 23 January 2013, http://kazenergy.com/ru/press/2011-04-21-10-24-20/7685---expo-2017----qq--.html.
70 Vermenichev, Andrei: Potentsial yadernoi energetiki Kazakhstana sostavlyaet 15–20 protsentov ot generiruemoi v strane energii, in: Kazenergy, 30 January 2013, http://kazenergy.com/ru/press/2011-04-21-10-24-20/7767--15-20-.html.
71 'Strategy "Kazakhstan-2050": new political course of the established state', address by the President of the Republic of Kazakhstan, Nursultan Nazarbaev, 14 December 2012, http://www.mod.gov.kz/mod-en/index.php/address-by-the-president-of-the-republic-of-kazakhstan-leader-of-the-nation-nnazarbayev-strategy-kazakhstan-2050-new-political-course-of-the-established-state.
72 'Prezident Islam Karimov podpisal Ukaz o razvitii al'ternativnykh istochnikov energii', in: anons.uz, 1 March 2013, http://www.anons.uz/article/economics/8711/.

Disputes over the water-energy nexus between Uzbekistan and Central Asian consumer countries pose a serious challenge to the realisation of the region's full hydropower potential. Hydropower provides approximately 98% of Tajikistan's electricity, and only 10% of its total hydropower potential is currently being tapped.[73] Further development of consumer countries' hydropower potential (i.e., the construction of the Kambarata and Rogun hydropower plants) will enable a surplus of electricity for internal consumption, which will most likely result in increasing the volume of clean energy exports to neighbouring producer countries as well as external energy consumers.[74] A surplus of electricity will not entirely replace gas and fuel imports, but it will allow Tajikistan and Kyrgyzstan to engage in energy trade negotiations with Uzbekistan on a more equal footing.

4.9. Conclusion

The security of the CAES is the condition in which all Central Asian states enjoy sufficient and sustainable energy supplies for both population and economy needs simultaneously. Due to asymmetrical power relations within the CAES, larger powers attempt to maintain their influence over energy policy choices, and weaker powers are constantly searching for ways to increase their leverage. The dependence of Central Asian exporters on the Russian pipeline network to export their natural gas puts them in a vulnerable position vis-à-vis Russia. Russia effectively used this dependence to promote its economic and political interests. However, the Russian monopoly was challenged by Iranian and Chinese energy pipeline infrastructures that transformed the CAES. This situation constitutes partially successful efforts to diversify away from Russia and possibly to obtain a higher price for the exported gas.

There are several factors that determine the stability and reliability of energy supplies between energy state actors within the CAES. These include natural gas trade deals, which are usually signed on a long-term basis with long term obligations; pipelines, the only cost-efficient way to transport natural gas, which require significant investments from both producers and customers; and the fact that Central Asia is surrounded by larger powers that often compete for energy resources. Taking into account the fact that there has been no natural gas production boom in Central Asia, with all existing pipelines upgraded and new ones constructed, there is doubt that regional producers will be able to extract the necessary amounts of gas to keep up

---

73 Wilby, Robert/ Friedhoff, Michael/ Connell, Richenda/ Rabb, Ben/ Minikulov, Nasridin/ Homidov, Anvar/ Shodmanov, Muzaffar/ Leonidova, Nadezhda: Tajikistan pilot programme for climate resilience (PPCR) project A4: improving the climate resilience of Tajikistan's hydropower sector, Dushanbe: PPCR, 2011, http://www.ppcr.tj/IP/Phase1/Component4/ppcr_a4_-draft_final_report_13oct11%20(Final%20REport).pdf.

74 Tomberg, Igor: Energy industry in Central Asia—challenges and prospects, in: Russian International Affairs Council, 26 April 2012, http://russiancouncil.ru/en/inner/?id_4=350#top.

with international energy demand and to use all of the export capacity. An attempt to fulfil obligations to supply the agreed amount of natural gas to external customers may result in an unstable regional energy market in which less gas is available for Central Asian consumer countries. Having experienced a shortage of energy, especially during the winter period, Tajikistan and Kyrgyzstan are trying to develop their hydropower potential, which will further escalate tension between consumer and producer countries in Central Asia over the energy-water nexus. As a result, producer countries will continue to use Tajikistan and Kyrgyzstan's dependence on Uzbek gas as leverage to block the construction of giant hydropower plants, and both parties' energy security will be compromised.

It is important that energy security policies are designed in such a way that increasing energy export capacity does not compromise the availability of sufficient energy supplies for the regional needs in Central Asia. Subsidizing energy for the domestic market makes it economically inefficient for regional exporters to invest in increasing the volume of energy for internal consumption. In contrast, to compensate for economic loss due to subsidising energy for domestic consumers, regional exporters are likely to continue increasing their export capacity. One of the possibilities to increase the amount of gas for domestic and regional consumption without compromising states' economic and political interests is to invest in energy efficiency projects aimed at saving energy that is currently being wasted due to poor transportation and distribution infrastructure. In the long run, however, the sustainability of energy supplies cannot be achieved without increasing the share of renewable energy sources in the overall energy balance of the Central Asian countries.

Boris Barkanov

5. The Geo-Economics of Eurasian Gas: the Evolution of Russian–Turkmen Relations in Natural Gas (1992–2010)

5.1. Introduction

This chapter examines how Russian relations with Turkmenistan with respect to natural gas have developed during the post-Soviet period (1992–2010). To measure bilateral relations, I use the concept of 'conditions of trade', which relates to gas price, forms of payment, contract duration, and key aspects of bargaining dynamics. Overall, the price Russia pays for Turkmen gas has increased gradually and significantly. Mixed cash/barter payments eventually gave way to full cash payment, while contract duration shifted from short-term deals to a long-term contract concluded in 2003. Finally, bargaining dynamics changed in that Russian state actors have come to play a more central role in negotiations related to the gas trade, while intermediary private actors have been gradually displaced.[1] This has had important consequences for the bilateral relationship: Russia has become a more reasonable partner, adopting a more diplomatic public tone, and showing greater ability to compromise. At the same time, economic issues could be linked to broader political questions related to bilateral ties.

Additionally, this work tests the explanatory potential of several hypotheses regarding these developments. Two of these hypotheses concern the economic structure of the bilateral relationship: supply-demand dynamics related to Russian gas and the availability of alternative pipeline routes for the export of Turkmen gas. A third hypothesis focuses on politics: relations between the Russian state and Gazprom, the partially state-owned Russian gas monopoly. The main findings are: (1) that supply-demand dynamics related to Russian gas were not sufficient to produce changes in the relationship; (2) the Central Asia–China gas pipeline was sufficient to produce the largest price increases after 2006, while the Trans-Caspian gas pipeline was necessary but not sufficient to produce the earlier changes; and (3) political factors associated with state-firm relations in Russia were necessary for changes in price beginning in 1999, form of payment, the time frame of the relationship as well as bargaining dynamics.

---

1   Øverland, Indra: Natural gas and Russia–Turkmenistan relations, in: Russian Analytical Digest, no. 56, 3 March 2009, pp. 9–12, http://www.css.ethz.ch/publications/pdfs/RAD-56.pdf; Pomfret, Richard: Turkmenistan after Turkmenbashi, in: Ahrens, Joachim/ Hoen, Herman W. (eds): Institutional reform in Central Asia: politico-economic challenges, New York: Routledge, 2013, pp. 63–88, here p. 80; 'Gazprom, Naftogaz sign long-term cooperation deal—paper', in: RIA Novosti, 23 October 2008, http://en.ria.ru/russia/20081023/117920413.html.

Why study Russian–Turkmen gas relations? Studying this dyad offers a window into understanding Russia's relations with Central Asia's most important gas exporter and how they might shape the future prospects for Russia's dominant position in the Eurasian market.[2] Specifically, it helps us understand the tools at Russia's disposal to control the flow of Turkmen gas to Europe as well as the trade-offs and potential limitations Russia faces in their successful deployment.

Russia has the largest conventional gas reserves in the world, and as the largest exporter of gas to Europe, it has a dominant position in that market. Although there is some disagreement regarding the actual volumes of gas that are available in Turkmenistan, there is no question that Turkmenistan has the largest conventional gas reserves in Central Asia.[3] Because of its relatively small population, limited domestic demand entails that Turkmenistan has the potential to export significant volumes. Even if available reserves do not justify early post-Soviet hopes of turning Turkmenistan into the 'new Middle East',[4] volumes are potentially sufficiently large to shape market dynamics in consumer countries (see Table A-3 on p. 311 in the annex).

Russia is interested in controlling Turkmen gas for two main reasons. First, during times of declining domestic production and/or increasing domestic and international demand, Turkmen gas provides an important supplement to Russia's overall fuel balance. Second, Russia seeks control to prevent competition with its own gas sales downstream in the EU consumer market. Preventing Central Asian gas from reaching European markets has been an important objective of Gazprom, Russia's monopoly exporter, and of Russian strategy during the post-Soviet period.[5] Additional volumes of gas put downward pressure on price and jeopardise Gazprom's revenues and market share, undermining the financial well-being of the Russian state, which holds a majority stake in and receives tax payments from the gas exporter. Thus, Russia has strong economic incentives to stifle competition and prevent other exporters from undercutting its position in its most valuable consumer market.[6]

Because natural gas is a major input into industrial production, electricity generation, and household heating, it also has strategic political significance. Thus, protection

---

2   'Eurasian' in this article refers to the trade between the former Soviet states and Europe. However, this market is evolving and now includes sales to China. This article focuses on Russian control of gas flows to Europe, and an examination of the Chinese market is largely outside its scope.
3   Williams, Selina: BP cuts Russia, Turkmenistan natural gas reserves estimates, in: Wall Street Journal, 12 June 2013, http://online.wsj.com/article/BT-CO-20130612-706046.html.
4   Olcott, Martha B.: Pipelines and pipe dreams: energy development and Caspian society, Journal of International Affairs, 1999 (vol. 53), no. 1, pp. 305–323.
5   Locatelli, Catherine/ Rossiaud, Sylvain: Russia/CIS gas, Dundee: POLINARES, 2012 (POLINARES Working Paper 71), p. 10, http://www.polinares.eu/docs/d5-1/polinares_wp5_chapter5_1.pdf.
6   Bochkarev, Danila: 'European' gas prices: implications of Gazprom's strategic engagement with Central Asia, in: Pipeline and Gas Journal, 2009 (vol. 236), no. 6, http://pipelineandgasjournal.com/%E2%80%9Ceuropean%E2%80%9D-gas-prices-implications-gazprom%E2%80%99s-strategic-engagement-central-asia?page=show.

5. The Geo-Economics of Eurasian Gas

of its dominant market position enhances Russia's potential political power. The interests of European consumers are exactly the opposite: to ensure relatively low prices by encouraging competition and to diversify sources of gas and transit routes to minimise political dependency. The latter concerns became particularly salient after Europe suffered gas shortages, becoming collateral damage of bilateral gas conflicts between Russia and Ukraine in 2006 and 2009.[7]

In addressing Russia's dominant position in EU gas markets, most academic and policy related literature has focused on the possibility of making available alternative sources of supply. With respect to Central Asia, several authors have emphasised the importance of developing export pipelines independent of Russian control, so that Central Asian, and especially Turkmen, gas can flow freely to consumer markets, notably in Europe.[8] Other research has identified challenges faced by bypass projects, including, among other things, a Russian strategy to undermine them.[9] However, as

---

7   'Ukraine–Russia gas dispute—call for stronger EU energy policy', in: European Parliament, 12 January 2006, http://www.europarl.europa.eu/sides/getDoc.do?language=EN&type=IM-PRESS&reference=20060112STO04233&secondRef=0; Ratner, Michael/ Belkin, Paul/ Nichol, Jim/ Woehrel, Steve: Europe's energy security: options and challenges to natural gas supply diversification, Washington, DC: Congressional Research Service, 2013, p. 1, http://www.fas.org/sgp/crs/row/R42405.pdf.

8   Adams, Jan S.: Pipelines and pipedreams: can Russia continue to dominate Caspian Basis energry?, in: Problems of Post-Communism, 1998 (vol. 45), no. 5, pp. 26–36, here pp. 30–31; Odom, William: The Caspian Sea littoral states: the object of a new 'Great Game'?, in: Dettke, Dieter (ed.): A Great Game no more: oil, gas, and stability in the Caspian Sea region, Washington, DC: Friedrich-Ebert-Stiftung, 1999, pp. 77–84, here p. 83; Lubin, Nancy: Pipe dreams: potential impacts of energy exploitation, in: Harvard International Review, 2000 (vol. 22), no. 1, pp. 66–69, here p. 67; Kalicki, Jan H.: Caspian energy at the crossroads, Foreign Affairs, 2001 (vol. 80), no. 5, pp. 120–134; Adams, Jan S.: Russia's gas diplomacy, in: Problems of Post-Communism, 2002 (vol. 49), no. 3, pp. 14–22; Cutler, Robert M.: The Caspian energy conundrum, in: Journal of International Affairs, 2003 (vol. 56), no. 2, pp. 89–102; Hansen, Sander: Pipeline politics: the struggle for control of the Eurasian energy resources, The Hague: Clingendael Institute for International Relations, 2003; Haghayeghi, Mehrdad: The coming of conflict to the Caspian Sea, in: Problems of Post-Communism, 2003 (vol. 50), no. 3, pp. 32–41; Dorian, James P.: Central Asia: a major emerging energy player in the 21st century, in: Energy Policy, 2006 (vol. 34), no. 5, pp. 544–555; Kandiyoti, Rafael: What price access to the open seas? The geopolitics of oil and gas transmission from the Trans-Caspian republics, in: Central Asian Survey, 2008 (vol. 27), no. 1, pp. 75–93; Sír, Jan/ Horák, Slavomír: China as an emerging superpower in Central Asia: the view from Ashkhabad, in: China and Eurasia Forum Quarterly, 2008 (vol. 6), no. 2, pp. 75–88; Marketos, Thrassy: Eastern Caspian Sea energy geopolitics: a litmus test for the U.S.–Russia–China struggle for the geostrategic control of Eurasia, in: Caucasian Review of International Affairs, 2009 (vol. 3), no. 1, pp. 2–19, http://www.cria-online.org/6_2.html; Peyrouse, Sébastien: Berdymukhammedov's Turkmenistan: a modest shift in domestic and social politics, in: China and Eurasia Forum Quarterly, 2010 (vol. 8), no. 3, pp. 47–66; Akiner, Shirin: Silk roads, great games, and Central Asia, in: Asian Affairs, 2011 (vol. 42), no. 3, pp. 391–402.

9   Heslin, Sheila N.: Key constraints to Caspian pipeline development: status, significance and outlook, Houston, TX: Rice University, James A. Baker III Institute for Public Policy, 1998, http://bakerinstitute.org/media/files/Research/33b24a64/key-constraints-to-caspian-energy-development-status-significance-and-outlook.pdf; Olcott, Martha B.: Pipelines and pipe dreams: energy development and Caspian society, in: Journal of International Affairs, 1999 (vol. 53),

some observers have noted, Russian policies undertaken to implement this strategy have varied widely, with important consequences for bilateral relations.[10] This variation suggests that Russia has multiple strategies for achieving its objective of continued dominance, which in turn implies that resistance to Russia's dominance in the downstream market depends on numerous factors, of which building new pipelines that source alternative gas resources is only one.

This work thus contributes to our understanding of the geo-economics of Eurasian gas by examining how and why Russian relations with Turkmenistan have changed and the implications of these relations for Russia's dominant position in Eurasian gas markets and control of Central Asian gas flows, notably to Europe. The main focus of analytical attention is the drivers of Russian policy. However, as relations cannot be reduced to Russian policy alone, the chapter also examines Turkmen bargaining insofar as it has shaped the conditions of the bilateral gas trade.

The discussion begins by conceptualising how Russia and Turkmenistan might interact as competitors in the gas market and the consequences of this interaction for the division of wealth. The section that follows defines what I call the 'conditions

---

no. 1, pp. 305–323; Finon, Dominique/ Locatelli, Catherine/ Mima, Silvana: Long-term competition between gas infrastructures developments in Asia. The construction on the Siberia and Caspian export development by cross-border pipelines, Grenoble: Institut d'Économie et de Politique de l'Énergie, 2000; Kubicek, Paul: Russian energy policy in the Caspian basin, in: World Affairs, 2004 (vol. 166), no. 4, pp. 207–217; Kazantsev, Andrei: Russian policy in Central Asia and the Caspian Sea region, in: Europe-Asia Studies, 2008 (vol. 60), no. 6, pp. 1073–1088; Cornell, Svante E.: Trans-Caspian pipelines and Europe's energy security, in: Cornell, Svante E./ Nilsson, Niklas (eds): Europe's energy security: Gazprom's dominance and Caspian supply alternatives, Stockholm/ Washington, DC: Central Asia–Caucasus Institute & Silk Road Studies Program, 2008, pp. 141–153, http://www.silkroadstudies.org/new/docs/publications/EnergySecurity.pdf; Liuhto, Kari: The EU needs a common energy policy—not separate solutions by its member states, in: Liuhto, Kari (ed.): EU–Russia gas connection: pipes, politics and problems, Turku: Publications of Pan-European Institute, 2009, pp. 109–140, http://www.utu.fi/fi/yksikot/tse/yksikot/PEI/rap ortit-ja-tietopaketit/Documents/Liuhto%200809%20web.pdf; Locatelli, Catherine: Russian and Caspian hydrocarbons: energy supply stakes for the European Union, in: Europe-Asia Studies, 2010 (vol. 62), no. 6, pp. 959–971; Attanasi, Emil D./ Freeman, Philip A.: Role of stranded gas from Central Asia and Russia in meeting Europe's future import demand for gas, in: Natural Resources Research, 2012 (vol. 21), no. 2, pp. 193–220; Horák, Slavomír: Turkmenistan's shifting energy geopolitics in 2009–2011, in: Problems of Post-Communism, 2012 (vol. 59), no. 2, pp. 18–30; Kubicek, Paul: Energy politics and geopolitical competition in the Caspian Basin, in: Journal of Eurasian Studies, 2013 (vol. 4), no. 2, pp. 171–180.

10 Bahgat, Gawdat: Pipeline diplomacy: the geopolitics of the Caspian Sea region, in: International Studies Perspective, 2002 (vol. 3), no. 3, pp. 310–327; Stulberg, Adam N.: Moving beyond the great game: the geoeconomics of Russia's influence in the Caspian energy bonanza, in: Geopolitics, 2005 (vol. 10), no. 1, pp. 1–25; Makhmudov, Rustam: The growing role of natural gas in the Eurasian energy games, in: Central Asia and the Caucasus, 2007 (vol. 48), no. 6, pp. 51–68, http://www.ca-c.org/journal/2007-06-eng/05.shtml; Pirani, Simon: Central Asian and Caspian gas production and the constraints on export, Oxford: Oxford Institute for Energy Studies, 2012, http://www.oxfordenergy.org/wpcms/wp-content/uploads/2012/12/NG_69.pdf; Ericson, Richard E.: Eurasian natural gas: significance and recent developments, in: Eurasian Geography and Economics, 2012 (vol. 53), no. 5, pp. 615–648.

of trade' and demonstrates how they changed during the period under study. Next, three explanations for this transformation—changes in the supply-demand dynamics of Russian gas, the emergence of new, non-Russian export routes from Turkmenistan, and political relations between the Russian state and Gazprom—are evaluated. The conclusion returns to the question of Russian control of Turkmen gas flows to Europe in light of the drivers of its policies, establishing the circumstances under which Russia is likely to succeed as well as the limitations of its Central Asian strategy for continued dominance in the European gas markets.

5.2. Pipeline Power and Russian Strategy

To understand the character of exporter relations in the Central Asian context, it is necessary to consider an important source of economic power in the gas trade: the pipeline network for the transport of natural gas. After the collapse of the Soviet Union, Turkmenistan became an independent state no longer subordinate to Moscow. However, because the Soviet pipeline system from Central Asia passed through Russia, it depended entirely on Gazprom, which controlled the network, to export gas to Europe.[11] As a state offering transit services to the European market, Russia held a monopoly. This created a situation of asymmetric power that favoured Russia, as Russia controlled the possibilities for export of Turkmen gas.

How this power is used, however, is not pre-determined. For the sake of simplicity, we might identify two strategies for Russia relative to Turkmenistan: 'winner-takes-all' and 'rent-sharing'. Under a 'winner-takes-all' strategy, Russia abuses its power, acts like a monopolist, and shuts Turkmenistan out of the lucrative European trade. This undermines competition and maintains prices at higher levels downstream. In the short term, all the wealth from Europe accrues to Russia; Turkmenistan is excluded from the market and receives nothing. Although attractive in the short term, this is a risky strategy over the long term because it creates incentives for Turkmenistan to try to break the Russian transit monopoly by building alternative pipelines. If Turkmenistan succeeds, this can lead to competition downstream, lower prices, lower revenues for Russia (albeit higher revenues for Turkmenistan, which previously earned nothing), and a transfer of wealth to consumers. Under a 'rent-sharing' strategy, by contrast, Russia agrees to share part of the revenues from the European market. Everything else equal, if the distribution of wealth between exporters is satisfactory, this is an economically stable outcome in that there is no financial incentive for Turkmenistan to try to access the EU market on its own. In the short term, 'rent-sharing' implies lower revenues from the European market for Russia than 'winner-takes-all'. However, 'rent-sharing'

---

11   Stern, Jonathan P.: The future of Russian gas and Gazprom, New York: Oxford University Press, 2005, p. 66.

also entails that competition is limited over the long term and that the wealth available to exporters as a whole is greater. Because collusion maintains prices at higher levels, there is no transfer of wealth to consumers. In other words, 'rent-sharing' allows exporters to maximise their long-term wealth at the expense of consumers. Overall, the data suggest that during the 1990s, Gazprom adopted a 'winner-takes-all' strategy toward Turkmenistan. After 1999, a 'rent-sharing' strategy began to emerge but was only fully implemented a decade later.

5.3. Evolution of Bilateral Relations

This section examines how changes in Russian policy have affected bilateral relations. To measure these relations, I focus on the 'conditions of trade' (COT). In economics, terms of trade (TOT) most commonly refers to '[…] the relative price, on world markets, of a country's exports compared to its imports'.[12] Although important, TOT excludes many factors that are also crucial to understanding relations between natural gas exporters. By contrast, COT includes TOT (pricing) but also takes into account: (1) forms of payment, (2) contract duration, and (3) bargaining dynamics. To develop a comprehensive view of the bilateral relationship, all four aspects, which changed significantly between the Yeltsin (1992–1999) and Putin (2000–2010) periods, will be considered.

5.3.1. Pricing (Terms of Trade)

With the collapse of the Soviet Union, gas prices within the former Soviet republics did not immediately converge to external (European) prices but remained low. In other words, not all consumer markets were the same. Most notably, European markets were far more attractive to gas producers than markets in the former Soviet Union (FSU).

During the Soviet period, Turkmenistan received a hard currency 'quota' from the central state for ostensible gas exports to Europe.[13] According to several sources, the Soviet system of sharing the European export market persisted through 1992–1993.[14] In 1993, Russia exported 92.7 billion cubic meters (bcm) of gas to Europe, while Turkmenistan exported 8.2bcm to the same market. However, Gazprom blocked export access in October 1993, and in the following month, Russia annulled its natural gas

---

12   Alan Deardorff's 'Glossary of International Economics', http://www-personal.umich.edu/~alandear/glossary/t.html (accessed 26 September 2013).
13   Stern, Jonathan P.: Soviet and Russian gas: the origins and evolution of Gazprom's export strategy, in: Mabro, Robert/ Wybrew-Bond, Ian (eds): Gas to Europe: the strategies of four major suppliers, Oxford: Oxford University Press, 1999, pp. 135–199, here p. 160.
14   Sagers, Matthew J.: The Russian natural gas industry in the mid-1990s, in: Post-Soviet Geography, 1995, (vol. 36), no. 9, pp. 521–564, here p. 557; Cutler, Robert M.: Turkey and the geopolitics of Turkmenistan's natural gas, in: Review of International Affairs, 2003 (vol. 1), no. 2, pp. 20–33; 'Turkmenistan raises revenue for gas but may switch exports', FT Energy Newsletters—East European Energy Report, 19 February 1992, p. 7.

5. The Geo-Economics of Eurasian Gas

export intergovernmental agreement with Turkmenistan.[15] Turkmenistan was also not paid for two months of European sales, an amount equal to US$185 million.[16] In 1994, Russian exports composed the entirety of gas exports to Europe from the former Soviet republics.[17]

Eventually, Turkmen exports became exclusively destined for markets within the FSU,[18] where prices were lower than those in Europe. Because Ukraine is the largest non-Russian market among former Soviet states, the present discussion focuses on the trade between Turkmenistan and Ukraine. Conflicts emerged almost immediately, as Turkmenistan demanded hard currency payment at 'world prices',[19] which was hardly attractive to (or perhaps even feasible for) Ukraine. When negotiations broke down, and Turkmen exports to Ukraine stopped, Gazprom had to make up the difference. Because a very large proportion (90% at the time)[20] of Russian exports to Europe transited through Ukraine, Gazprom had few options. As a near-monopolist transit state, Ukraine could refuse to transit gas absent supplies from Russia. More immediately, Ukraine could simply siphon transit gas to supply its own consumers with gas that was intended for Gazprom's customers in Europe. As a result, Gazprom also sold gas to Ukraine at depressed prices in exchange for accumulated debt. This became a central aspect of Russia's relations with Ukraine early on when, in September 1993, President Yeltsin offered debt cancellation in exchange for Russian control of the Black Sea fleet and Ukraine's nuclear warheads.[21]

In short, after the dust of the Soviet collapse settled, Turkmen gas exports became linked to the consumer markets of the FSU, where prices and consumer capacity to

---

15 Sagers, Matthew J.: The Russian natural gas industry in the mid-1990s, in: Post-Soviet Geography, 1995, (vol. 36), no. 9, pp. 521–564, here p. 557; Cutler, Robert M.: Turkey and the geopolitics of Turkmenistan's natural gas, in: Review of International Affairs, 2003 (vol. 1), no. 2, pp. 20–33.
16 'Turkmenistan threatens Ukraine with gas cut, in: International Gas Report, 18 February 1994 (cited in Hancock, Kathleen: Bringing economics back in: how relation-specific assets shed light on Russia–Turkmenistan economic relations. Paper presented at the annual meeting of the International Studies Association, San Diego, CA, USA, 22 March 2006, p. 27, http://citation.allacademic.com//meta/p_mla_apa_research_citation/0/9/9/2/3/pages99238/p99238-27.php.
17 Sagers, Matthew J.: The Russian natural gas industry in the mid-1990s, in: Post-Soviet Geography, 1995, (vol. 36), no. 9, pp. 521–564, here p. 557.
18 Adams, Jan S.: Pipelines and pipedreams: can Russia continue to dominate Caspian Basis energy?, in: Problems of Post-Communism, 1998 (vol. 45), no. 5, pp. 26–36, here p. 30.
19 Stern, Jonathan P.: Soviet and Russian gas: the origins and evolution of Gazprom's export strategy, in: Mabro, Robert/ Wybrew-Bond, Ian (eds): Gas to Europe: the strategies of four major suppliers, Oxford: Oxford University Press, 1999, pp. 135–199, here p. 160.
20 Victor, David/ Victor, Nadejda: The Belarus connection: exporting Russian gas to Germany and Poland: geopolitics of natural gas study, Stanford, CA: Stanford Institute for International Studies, 2004, p. 13 (Working Paper 26), http://iis-db.stanford.edu/pubs/20603/Yamal_final.pdf.
21 Fredholm, Michael: Natural gas trade between Russia, Turkmenistan, and Ukraine: agreements and disputes, Stockholm: Stockholm University, Department of Oriental Languages, 2008, p. 12 (Asian Cultures and Modernity Research Report no. 15), http://gpf-europe.com/upload/iblock/2fa/fredholm.ukraine.russia.gas.rr15.pdf.

pay were significantly lower than in Europe. Gazprom also sold gas to these markets. However, in contrast to Turkmenistan, Gazprom had a lucrative portfolio of consumers in Europe. Overall pricing for Gazprom was thus higher because it arose from a mix of high prices in Europe and low prices in the former Soviet republics.

Obviously, linking Turkmen exports to markets in the FSU had enormous implications for the distribution of wealth. By preventing Turkmenistan from selling gas to the lucrative European market and channelling it exclusively to low value markets in the FSU, Gazprom was essentially taking the entire European pie for itself.[22] From this perspective, we can say that Gazprom adopted the 'winner-takes-all' strategy discussed above.

Even before coming to power, Vladimir Putin had developed a special interest in the gas business,[23] and his arrival foreshadowed important changes in the relationship between the Russian state and Gazprom. In late 1999, Gazprom agreed to begin purchasing 20bcm of Turkmen gas for US$36 per 1,000 cubic metres (cm). According to Stern, the chairman of Gazprom, Rem Vyakhirev, '[…] had been under severe pressure from the Russian government—then headed by prime minister (soon to be President) Vladimir Putin—to mend fences with the Turkmen President […]'.[24] On his first foreign trip as president in May 2000, Putin stated, after meeting with the President of Turkmenistan, Saparmurat Niyazov, that although gas pricing was 'a question for commercial talks […] it is clear that it must correspond to international rates and must be beneficial to both Russia and Turkmenistan'.[25]

In this context, Putin's allusion to international prices must be understood as a reference to prices in Europe, which at the time was the only available consumer market other than those in the FSU. It was widely understood that Europe had the most valuable market, with the highest prices. Thus, the president was suggesting that prices of Turkmen gas could be raised closer to European levels. In light of Gazprom's strategy of paying Turkmenistan as little as possible for its gas during the 1990s, the pres-

---

22 Aris, Ben: Weaning off the barrel, in: The Moscow Times, 3 June 2002.
23 Nemtsov, Boris/ Milov, Vladimir: The Nemtsov White Paper, Part II: Gazprom, 28 September 2008, http://larussophobe.wordpress.com/2008/09/28/the-nemtsov-white-paper-part-ii-gazprom-the-full-text/ (Russian version at: http://www.nemtsov.ru/?id=705498&PHPSESSID=b2cb c6117a64fb501dc28fb3d1f2a7cb); Harley Balzer: The Putin thesis and Russian energy policy, in: Post-Soviet Affairs, 2005 (vol. 21), no. 3, pp. 210–225, here p. 211; Olcott, Martha B.: The energy dimension in Russian global strategy: Vladimir Putin and the geopolitics of oil, Houston, TX: Rice University, The James A. Baker III Institute for Public Policy, 2004, http://carnegieendow ment.org/files/wp-2005-01_olcott_english1.pdf.
24 Stern, Jonathan P.: The future of Russian gas and Gazprom, New York: Oxford University Press, 2005, p. 74.
25 Stern, David/ Sasin, Igor: Putin upbeat on boosting Turkmen gas deliveries, but no deal cut, in: Agence France Press—English, 19 May 2000, http://lists.topica.com/lists/eurasia/read/message. html?sort=a&mid=1302342793; 'Turkmen, Russian heads talk bilateral relations in Turkmen capital', in: in BBC Monitoring Central Asia Unit, 20 May 2000.

5. The Geo-Economics of Eurasian Gas   157

ident's explicit statement about mutual benefit is best interpreted as a signal that, going forward, higher prices would lead to a more equitable distribution of wealth. Although an actual price was not decided upon at the time, Putin announced that, together with his counterpart Niyazov, '[...] [we] have agreed to increase Turkmen gas shipments to the Russian Federation by 10 BCM a year. [...] next year the amount will be 30 BCM and then 40 BCM, and then 50 BCM'.[26] At the time, the press reported that both parties were pleased with the negotiations.[27] Table 5-1 presents the volumes purchased by Gazprom during the 2000–2010 period.[28]

Table 5-1: Volumes of Turkmen Gas Purchased Annually by Gazprom

Year	Volume (in bcm)
2000	20
2004	5–6
2005	6–7
2006	42
2007	42.6
2008	42.3
2009	11.8
2010	10.7

With time, price not only began to rise but became formally linked to prices in Europe. Figure 5-1 presents the evolution of price (US$/1,000cm) paid by Gazprom for Turkmen gas as well as the price of Russian gas at the German border.[29]

---

26 'Russian, Turkmen heads meet in Ashkhabad', in: BBC Summary of World Broadcasts, 22 May 2000.
27 Rotar, Igor: '"Sredneaziatskoe turne" zavershilos"', in: Nezavisimaya gazeta, 23 May 2000.
28 Data on volumes were taken from a variety of public sources: 2000: Cutler, Robert M.: Turkey and the geopolitics of Turkmenistan's natural gas, in: Review of International Affairs, 2003 (vol. 1), no. 2, pp. 20–33, here p. 23; Gorst, Isabel/ Zaman, Amberin: Partners halt work on Caspian gas line, in: Platt's Oilgram News, 23 May 2000 (vol. 78), no. 99, p. 1; 'Gazprom concluded a long-term gas purchase contract with Turkmenistan', Gazprom press release, 10 April 2003, http://www.gazprom.com/press/news/2003/april/article62404/; 'Gazprom and Turkmenistan update Turkmen gas supply conditions for 2006', in: RuStocks.com, 29 December 2005, http://www.rustocks.com/index.phtml/Pressreleases/0/1/8735?filter=2005; 'Gazprom delegation visits Turkmenistan', in: RuStocks.com, 5 September 2006, http://www.rustocks.com/index.phtml/Pressreleases/GAZP/12/10091?filter=2006; Gazprom: Gas purchases, http://www.gazprom.com/about/production/central-asia/ (accessed 27 September 2013).
29 Data on prices for Turkmen gas were taken from various public sources: December 1999: Cutler, Robert M.: Turkey and the geopolitics of Turkmenistan's natural gas, in: Review of International Affairs, 2003 (vol. 1), no. 2, pp. 20–33, here p. 23; Gorst, Isabel/ Zaman, Amberin: Partners halt work on Caspian gas line, in: Platt's Oilgram News, 23 May 2000 (vol. 78), no. 99, p. 1; 'Gazprom concluded a long-term gas purchase contract with Turkmenistan', Gazprom press release, 10 April 2003, http://www.gazprom.com/press/news/2003/april/article62404/; 'Russia to drop barter for cash only payment for Turkmen gas', in: ICIS, 29 April 2005, http://www.icis.com/heren/articles/2005/04/29/9283639/russia-to-drop-barter-for-cash-only-payment-for-turkmen-gas.html; 'Gazprom and Turkmenistan update Turkmen gas supply conditions for 2006', in: RuStocks.com,

Figure 5-1: Evolution of Price Paid by Gazprom for Turkmen Gas Versus the Price of Russian Gas at the German Border (US$/1,000cm)

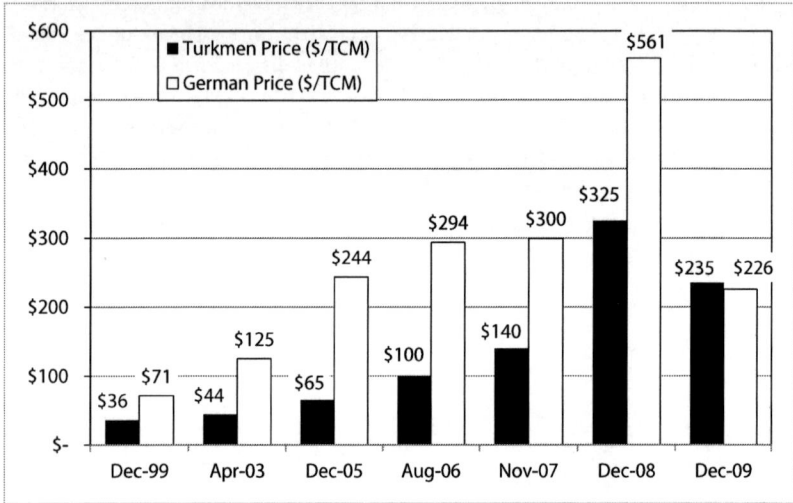

Figure 5-2 depicts the change in the Turkmen price as a proportion of the German border price.[30]

In 1999, Turkmen gas fetched half the German price, a relative level that would only be surpassed in 2009. After nearly three years of negotiations, the parties agreed in 2003 to a higher price. However, as a ratio of the German price, the value actually fell, reflecting the fact that prices in Germany were growing rapidly, nearly doubling since the last agreement. In December 2004, Turkmenistan unsuccessfully attempted to raise its price to US$60.[31] The following April (2005), both parties reaffirmed the

---

29 December 2005, http://www.rustocks.com/index.phtml/Pressreleases/0/1/8735?filter=2005; 'Gazprom delegation visits Turkmenistan', in: RuStocks.com, 5 September 2006, http://www.rustocks.com/index.phtml/Pressreleases/GAZP/12/10091?filter=2006; 'Gazprom delegation visits Turkmenistan', in: RuStocks.com, 27 November 2007, http://www.rustocks.com/index.phtml/Pressreleases/GAZP/1/13279?filter=2007-11; 'Turkmenistan, Russia agree to resume gas supplies in 2010, ending impasse', in: IHS Global Insight, 23 December 2009, http://www.ihs.com/products/global-insight/industry-economic-report.aspx?id=106594749; Socor, Vladimir: Russia resumes gas imports from Turkmenistan, in: Asia Times online, 6 January 2010, http://www.atimes.com/atimes/Central_Asia/LA06Ag02.html.

30 Own calculations based on International Monetary Fund: Russian Natural Gas border price in Germany, http://www.imf.org/external/np/res/commod/External_Data.xls (accessed 12 September 2013).

31 Pirani, Simon: Central Asian and Caspian gas production and the constraints on export, Oxford: Oxford Institute for Energy Studies, 2012, http://www.oxfordenergy.org/wpcms/wp-content/uploads/2012/12/NG_69.pdf; Hancock, Kathleen J.: Escaping Russia, looking to China: Turkmenistan pins hopes on China's thirst for natural gas, in: China and Eurasia Forum Quarterly,

previous price, which was supposed to remain in place during 2005–2006. Confronted with continuous Turkmen demands for a higher price, however, Gazprom subsequently agreed on four occasions to increase its purchase price: in December 2005, it agreed to pay US$65 per 1,000cm for the following year. In August 2006, the price was raised to US$100 per 1,000cm during 2007–2009, to be renegotiated in mid-2009. In November 2007, Gazprom yielded again and agreed to a price of US$130 per 1,000cm for the first half of 2008, to be raised to US$150 during the second half of the year.[32]

Figure 5-2: Change in the Turkmen Gas Price as a Proportion of the German Border Price

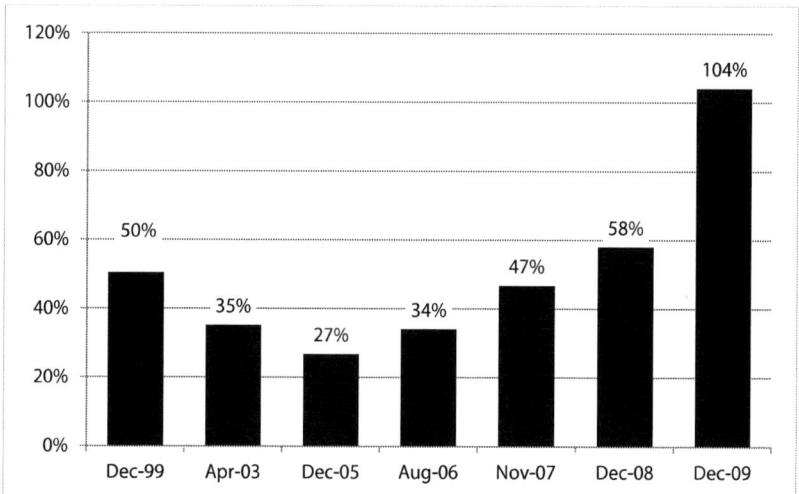

The x axis denotes months of negotiated agreements. The ratio includes German prices for the same month.

At the time of the November 2007 agreement, Gazprom announced in a press release that beginning in 2009, the price would be '[...] set according to the price formula based on the [sic] market principles'.[33] The following January, the Minister for Industry and Energy, Viktor Khristenko, announced that Russia was 'considering using a gas price formula in purchase contracts with all Central Asian countries, similar to that

---

2006 (vol. 4), no. 3, pp. 67–87, here p. 73; Fredholm, Michael: Natural gas trade between Russia, Turkmenistan, and Ukraine: agreements and disputes, Stockholm: Stockholm University, Department of Oriental Languages, 2008, pp. 9–10 (Asian Cultures and Modernity Research Report no. 15), http://gpf-europe.com/upload/iblock/2fa/fredholm.ukraine.russia.gas.rr15.pdf.
32 Socor, Vladimir: Russia surging farther ahead in race for Central Asian gas, in: Eurasia Daily Monitor, 16 May 2007 (vol. 4), no. 96, http://www.jamestown.org/programs/edm/single/?tx_ttnews[tt_news]=32747&tx_ttnews[backPid]=171&no_cache=1.
33 'Gazprom delegation visits Turkmenistan', in: RuStocks.com, 27 November 2007, http://www.rustocks.com/index.phtml/Pressreleases/GAZP/1/13279?filter=2007-11.

used in gas sales on European markets'.[34] This appears to have been a response to Turkmen demands for higher prices, perhaps instigated by comments of American and European Commission officials that 'the price of Turkmen gas under current export contracts is very low against the backdrop of the current natural gas market situation in Europe, and might be higher'.[35] At the time, Gazprom chief executive officer (CEO), Alexei Miller, noted that part of the difficulty in raising the price of Turkmen gas was that 'traditionally [the] key market for Turkmen gas is Ukraine [...]. It is clear that it will take time for the Ukrainian economy to be adapted to European prices'.[36] Thus, the pricing of Turkmen gas remained linked to FSU markets.

The following March, after a meeting between Gazprom CEO Miller and his counterparts from KazMunaiGaz (Kazakhstan), Uzbekneftgas (Uzbekistan), and Turkmengaz, Gazprom announced that, 'based upon the interests of the national economies and considering the international commitments with regard to the [sic] energy supply reliability and continuity, starting from 2009 natural gas will be sold at European prices'.[37] It was also noted that these prices would be 'considerably higher than their current prices'.[38] Before the year started, the parties agreed to a fixed price upwards of US$300–350 per 1,000cm.[39] It took nearly a decade, but the 'winner-takes-all' strategy had finally given way to 'rent-sharing'.

In 2009, the question of price re-emerged as result of the global economic crisis (discussed below). After an extended disruption to trade following a pipeline explo-

---

34 Rodova, Nadia: Central Asian states to hike gas price in 2009: Turkmenistan, Uzbekistan, Kazakhstan plan European prices, in: Platts Oilgram News, 12 March 2008 (vol. 86), no. 51, p. 4.
35 Gazprom's payments for Turkmen gas result from negotiations in which the question of how much Gazprom can afford to pay figures importantly. However, owing to the nature of Gazprom's contracts, Turkmenistan cannot be certain of Gazprom's European revenues and overall financial situation. The practical result of comments of external actors that Gazprom is understating its European revenues (and thus overstating the financial hardship of higher Turkmen prices) during negotiations is that Turkmenistan has grounds to demand a higher price. Rodova, Nadia: Turkmenistan wants 30% gas price increase for Russia, in: Platts Oilgram News, 26 November 2007, (vol. 85), no. 233, p. 1.
36 Ibid.
37 'On working meeting of Alexey Miller and heads of gas companies from Kazakhstan, Uzbekistan and Turkmenistan', Gazprom press release, 11 March 2008, http://www.Gazprom.com/press/news/2008/march/article64101.
38 Rodova, Nadia: Central Asian states to hike gas price in 2009: Turkmenistan, Uzbekistan, Kazakhstan plan European prices, in: Platts Oilgram News, 12 March 2008 (vol. 86), no. 51, p. 4; 'Turkmenistan, Russia agree to resume gas supplies in 2010, ending impasse', in: IHS Global Insight, 23 December 2009, http://www.ihs.com/products/global-insight/industry-economic-report.aspx?id=106594749; Fredholm, Michael: Natural gas trade between Russia, Turkmenistan, and Ukraine: agreements and disputes, Stockholm: Stockholm University, Department of Oriental Languages, 2008 (Asian Cultures and Modernity Research Report no. 15), http://gpf-europe.com/upload/iblock/2fa/fredholm.ukraine.russia.gas.rr15.pdf.
39 'Turkmenistan, Russia agree to resume gas supplies in 2010, ending impasse', in: IHS Global Insight, 23 December 2009, http://www.ihs.com/products/global-insight/industry-economic-report.aspx?id=106594749.

5. The Geo-Economics of Eurasian Gas 161

sion on 22 December, the parties amended their supply agreement, linking the sale price to 'prices to be based on a fluctuating European price formula'.[40] This marked the formal introduction of a European 'netback' pricing formula,[41] an idea that had already been floated in March 2008.[42] The netback price is the European contract price that Gazprom receives minus the costs of transit through Russia.[43] According to one source, Gazprom paid US$220 to US$250 for Turkmen gas in 2010.[44] They also agreed on lower volumes; beginning in January 2010, Gazprom would purchase up to 30bcm,[45] below the 42.3bcm that the company purchased before the crisis.[46] Indeed, Gazprom only purchased 10.7bcm from Turkmenistan in 2010 (see Table 5-1).[47]

5.3.2. Form of Payment

Neo-classical economics pays little attention to form of payment because payment in cash or cash equivalents is so common in the advanced economies that it is easily taken for granted. However, in addition to using cash or its equivalent, payment can be settled through barter (with goods). Usually, this type of settlement is less desirable because it is less liquid and fungible than money and is associated with higher transaction costs. During the 1990s, it was nearly ubiquitous in the FSU.

Informal payment practices emerged during the Yeltsin period for a variety of reasons.[48] As Russia's largest company and the supplier of the most common fuel in Russia's primary energy balance, Gazprom was at the centre of many of the non-

---

40 Ibid.
41 Socor, Vladimir: Russia resumes gas imports from Turkmenistan, in: Asia Times online, 6 January 2010, http://www.atimes.com/atimes/Central_Asia/LA06Ag02.html.
42 'On working meeting of Alexey Miller and heads of gas companies from Kazakhstan, Uzbekistan and Turkmenistan', Gazprom press release, 11 March 2008, http://www.Gazprom.com/press/news/2008/march/article64101; also see International Energy agency: Perspectives on Caspian oil and gas development, Paris: IEA, 2008, p. 11, http://www.asiacentral.es/docs/caspian_perspectives_iea_dec08.pdf.
43 In fact, Gazprom charges different prices to different European consumers. It is unclear how these different prices relate in practice to the price derived from the new formula.
44 Socor, Vladimir: Russia resumes gas imports from Turkmenistan, in: Asia Times online, 6 January 2010, http://www.atimes.com/atimes/Central_Asia/LA06Ag02.html.
45 'Turkmenistan, Russia agree to resume gas supplies in 2010, ending impasse', in: IHS Global Insight, 23 December 2009, http://www.ihs.com/products/global-insight/industry-economic-report.aspx?id=106594749.
46 Cf. Gazprom, Gas purchases, http://www.Gazprom.com/about/production/central-asia/ (accessed 12 September 2013).
47 Ibid.
48 Barkanov, Boris: Mercantilist development in Russia: the legitimacy of state power, state identity, and the Energy Charter regime (1990–2010), University of California at Berkeley, unpublished dissertation, 2011; Woodruff, David: Money unmade: barter and the fate of Russian capitalism, Ithaca, NY: Cornell University Press, 2000; Gaddy, Clifford G./ Gale, William G.: Demythologizing the Russian flat tax, in: Tax Notes International, 14 March 2005 (vol. 43), pp. 983–988, http://www.brookings.edu/~/media/research/files/articles/2005/3/14russia%20gaddy/20050314gaddygale.pdf.

monetary schemes.[49] Instead of settlement in cash, non-payment and barter became widespread in the former Soviet republics' gas trade. This was particularly important, given that Turkmen exports were channelled exclusively to FSU markets.

Gazprom did not cause the non-payments crisis. Many of its accounts payable in Russia were also in arrears. Owing to its exports, however, it still had very large cash revenues. Thus, in compelling Turkmenistan to export only within the FSU, Gazprom was essentially tying Turkmenistan to the least liquid market while keeping the most liquid market (Europe) for itself.[50]

When Russia began purchasing Turkmen gas in 1999, the parties agreed that 40% of payments would be in cash and that the remainder would be in barter. Under Putin, cash payment was gradually restored. The April 2003 framework foresaw that during the first three years, settlement would be half in cash and half in barter.[51] In April 2005, the parties agreed that payment would be entirely in cash.[52]

5.3.3. Contract Duration

Contracts vary in duration. At one extreme are one-time instantaneous transactions where payment is exchanged for a particular good (spot market). At the other extreme are long-term contracts that stipulate exchange over much longer periods, frequently 10, 15, 20, or even 30 years. Historically, the sale of natural gas in Europe by European, Soviet, and (later) Russian producers, was conducted using long-term contracts of up to 30 years.[53]

After the collapse of the Soviet Union, short-term commercial deals with FSU countries, concluded within the framework of long-term intergovernmental agreements, became the norm for Gazprom.[54] Under the December 1999 agreement with Turkmenistan, for example, Gazprom would purchase 20bcm for the following year.[55]

---

49 Woodruff, David: Money unmade: barter and the fate of Russian capitalism, Ithaca, NY: Cornell University Press, 2000.
50 Olcott, Martha B.: International gas trade in Central Asia: Turkmenistan, Iran, Russia and Afghanistan, Stanford, CA: Stanford Institute for International Studies, 2004 (Working Paper 28), http://iis-db.stanford.edu/pubs/20605/Turkmenistan_final.pdf.
51 Global Witness: It's a gas: funny business in the Turkmen–Ukraine gas trade, Washington, DC: Global Witness, 2006, p. 26, http://www.globalwitness.org/library/its-gas-funny-business-turk men-ukraine-gas-trade.
52 'Russia to drop barter for cash only payment for Turkmen gas', in: ICIS, 29 April 2005, http://www.icis.com/heren/articles/2005/04/29/9283639/russia-to-drop-barter-for-cash-only-payment-for-turkmen-gas.html.
53 Energy Charter Secretariat: Putting a price on energy: international pricing mechanisms for oil and gas, Brussels: Energy Charter Secretariat, 2007, http://www.encharter.org/fileadmin/user_upload/document/Oil_and_Gas_Pricing_2007_ENG.pdf.
54 Stern, Jonathan P.: The future of Russian gas and Gazprom, New York: Oxford University Press, 2005, pp. 67–68.
55 Pereplecnin, Mikhail/ Petrov, Eduard: Rossiya vozvrashchaetsya v Turkmeniyu, in: Nezavisimaya gazeta, 18 May 2000, p. 2.

5. The Geo-Economics of Eurasian Gas 163

After the May 2000 meeting between the heads of the Russian and Turkmen states, the press reported that a 30-year deal with a maximum annual volume of 50bcm had been discussed.[56] On 10 April 2003, the '25-year Cooperation Agreement in the gas industry,' an intergovernmental agreement between Russia and Turkmenistan for the 2004–2028 period, was signed.[57] Within this framework, Gazprom Export and Turkmenneftegaz concluded a long-term contract[58] for the purchase of 80–90bcm of natural gas annually beginning on 1 January 2004 (for a total of 1.8 trillion cubic meters over the term of the agreement).[59]

5.3.4. Bargaining Dynamics

Over the course of two decades, power has been an important factor in gas relations between Russia and Turkmenistan. Both sides have acted in self-interested ways, and power has played a particularly important role during negotiations over price and payment. On several occasions, each side has interrupted gas flows, apparently to affect how wealth from the gas trade is distributed between the various parties. From the perspective of realist IR theory, none of this is surprising.

As we have seen, by the late 1990s, Turkmen exports were being channelled exclusively to the low price, illiquid consumer markets of the FSU. Because of payment problems, intermediary companies became involved in marketing gas, collecting payments, and facilitating barter and/or monetising the trade. In November 1995, Turkmenrosgaz, a joint venture between Gazprom, Itera, and Turkmenneftegaz was created to fulfil the latter function.[60] By mid-1997, significant unpaid debt to Turkmenistan had been

---

56 'Turkmen, Russian heads talk bilateral relations in Turkmen capital', in: BBC Monitoring Central Asia Unit, 20 May 2000.
57 Gazprom in questions and answers, international projects, http://eng.gazpromquestions.ru/?id=2#c685 (accessed 12 September 2013); Tsygankova, Marina: Netback pricing as a remedy for the Russian gas deficit, Kongsvinger: Statistics Norway, 2008 (Discussion Papers no. 554), http://www.ssb.no/a/publikasjoner/pdf/DP/dp554.pdf.
58 Gazprom in questions and answers, international projects, http://eng.gazpromquestions.ru/?id=2#c685 (accessed 12 September 2013).
59 'Turkmenistan, Russia agree to resume gas supplies in 2010, ending impasse', in: IHS Global Insight, 23 December 2009, http://www.ihs.com/products/global-insight/industry-economic-report.aspx?id=106594749; Neff, Andrew: Russia sign accords on gas supply, CAC pipeline modernisation, in: World Markets Research Centre Daily Analysis, 19 August 2003; 'Gazprom approves opening representative office in Turkmenistan', in: Prime-TASS Energy Service, 21 August 2003.
60 In the joint venture, Turkmenrosgaz, Gazprom controlled a 45% stake, Itera held a 4% stake, and Turkmenneftegaz held the remaining 51%. Olcott, Martha B.: International gas trade in Central Asia: Turkmenistan, Iran, Russia and Afghanistan, Stanford, CA: Stanford Institute for International Studies, 2004 (Working Paper 28), http://iis-db.stanford.edu/pubs/20605/Turkmenistan_final.pdf; Fredholm, Michael: Natural gas trade between Russia, Turkmenistan, and Ukraine: agreements and disputes, Stockholm: Stockholm University, Department of Oriental Languages, 2008, p. 37 (Asian Cultures and Modernity Research Report no. 15), http://gpf-europe.com/upload/iblock/2fa/fredholm.ukraine.russia.gas.rr15.pdf; Global Witness: It's a gas: funny business in the Turkmen–Ukraine gas trade, Washington, DC: Global Witness, 2006, http://www.

accumulated by Itera for gas sold to Ukraine. This meant that, in addition to obtaining the lowest available prices for its gas, Turkmenistan was in fact not receiving payment. As a result, Turkmenistan unilaterally dissolved Turkmenrozgas, and the export of Turkmen gas outside of Central Asia came to a halt.[61]

After negotiations in August 1997, Gazprom CEO Rem Vyakhirev 'made a decision that it was impossible to find an acceptable long-term basis for supplying Turkmen gas to former Soviet republics (principally Ukraine) and that these markets would henceforth need to be supplied with Russian gas'[62]. Gazprom's inflexibility constituted a de facto embargo, which lasted nearly two years (February 1997 to December 1998).[63] At the time, representatives of Gazprom commented that Turkmenistan would 'come begging on its knees'.[64] During the negotiations between Gazprom, Itera, and Turkmenistan, Vyakhirev stated publicly that 'Turkmenistan "would be forced to eat sand" if it would not agree to sell gas to Russia under his conditions'.[65] Due to a lack of transport alternatives, Turkmenistan was unable to export gas for nearly two years, with devastating consequences for the Turkmen economy, in which natural gas exports compose 85% of annual revenues.[66]

After the new president came to power and intervened in gas relations directly, Russian policy began to change. During an interview, one Turkmen state official familiar with post-Soviet gas negotiations between Russia and Turkmenistan stated that, with Putin's arrival, the Russian position became more 'reasonable'.[67] As already noted, Putin acknowledged that Turkmenistan should receive 'international' prices and that the trade should be mutually beneficial. In December 1999, Vyakhirev apologised to

---

globalwitness.org/library/its-gas-funny-business-turkmen-ukraine-gas-trade; Stern, Jonathan P.: Soviet and Russian gas: the origins and evolution of Gazprom's export strategy, in: Mabro, Robert/ Wybrew-Bond, Ian (eds): Gas to Europe: the strategies of four major suppliers, Oxford: Oxford University Press, 1999, pp. 135–199, here p. 161.
61 Stern, Jonathan P.: Soviet and Russian gas: the origins and evolution of Gazprom's export strategy, in: Mabro, Robert/ Wybrew-Bond, Ian (eds): Gas to Europe: the strategies of four major suppliers, Oxford: Oxford University Press, 1999, pp. 135–199, here p. 161.
62 Ibid.
63 Adams, Jan S.: Pipelines and pipedreams: can Russia continue to dominate Caspian Basis energy?, in: Problems of Post-Communism, 1998 (vol. 45), no. 5, pp. 26–36.
64 '"Blue Stream" takes shape', in: Oil & Gas of Kazakhstan, 19 April 2000.
65 'Gazprom co-founder Rem Vyakhirev dead at 78', in: The Moscow Times, 13 February 2013. See also: Adams, Jan S.: Russia's gas diplomacy, in: Problems of Post-Communism, 2002 (vol. 49), no. 3, pp. 14–22; Kandiyoti, Rafael: What price access to the open seas? The geopolitics of oil and gas transmission from the Trans-Caspian republics, in: Central Asian Survey, 2008 (vol. 27), no. 1, pp. 75–93; Adams, Jan S.: Pipelines and pipedreams: can Russia continue to dominate Caspian Basis energy?, in: Problems of Post-Communism, 1998 (vol. 45), no. 5, pp. 26–36, here p. 32.
66 Boucek, Christopher: Energy security implications of post-Niyazov Turkmenistan, in: CACI Analyst, 10 January 2007, http://old.cacianalyst.org/?q=node/4378.
67 Interview with the author, spring 2013. For characterization of Gazprom's approach, see for example: Adams, Jan S.: Pipelines and pipedreams: can Russia continue to dominate Caspian Basis energy?, in: Problems of Post-Communism, 1998 (vol. 45), no. 5, pp. 26–36; Adams, Jan S.: Russia's gas diplomacy, in: Problems of Post-Communism, 2002 (vol. 49), no. 3, pp. 14–22.

5. The Geo-Economics of Eurasian Gas

President Niyazov and the Turkmen people for the 'anti-Turkmen' sentiments that had been attributed to him in the press as the embargo unfolded.[68] As bilateral relations appeared to improve, compromise became possible, though certainly not easy.[69]

Under pressure from the new prime minister, Gazprom compromised on price when Russia began purchasing Turkmen gas at the end of 1999. Turkmenistan had demanded a price between US$40–42 per 1,000cm, US$10 more than Gazprom was prepared to pay. The partners agreed to roughly split the difference with an agreed price of US$36.[70] The proportion between cash and barter was also a compromise; Gazprom demanded that payment be 70% in barter and 30% in cash, while Turkmenistan insisted on 50% payment in cash. The final agreement stipulated that payment would be 40% in cash, with the remainder in barter.

During the winter of 2004–2005, Turkmenistan demanded a price increase in response to rising prices in Europe. After Gazprom refused, Ashgabat demonstrated its willingness to exercise power by cutting off supplies entirely.[71] Although Russia prevailed and price remained at the same level for another year, Gazprom nevertheless agreed to shift to full cash payment (from 50%) earlier than previously negotiated.[72]

Perhaps the most interesting example of Russia's new bargaining posture can be seen in its response to the gas crisis of 2009. The 2008 global economic crisis was a double blow for Russian gas exports: European gas prices (linked to dramatically falling oil prices) and demand both fell. As a result, Gazprom found itself in the uncomfortable position of buying Turkmen gas that it no longer needed (due to the fall in European, Russian, and FSU demand)[73] at prices that exceeded those it received from European

---

68 Stern, Jonathan P.: The future of Russian gas and Gazprom, New York: Oxford University Press, 2005, p. 74.
69 'Vladimir Putin ends working visits to Uzbekistan and Turkmenia', in: RIA Novosti, 20 May 2000.
70 Gorst, Isabel/ Zaman, Amberin: Partners halt work on Caspian gas line, in: Platt's Oilgram News, 23 May 2000 (vol. 78), no. 99, p. 1; Cutler, Robert M.: Turkey and the geopolitics of Turkmenistan's natural gas, in: Review of International Affairs, 2003 (vol. 1), no. 2, pp. 20–33.
71 Hancock, Kathleen J.: Escaping Russia, looking to China: Turkmenistan pins hopes on China's thirst for natural gas, in: China and Eurasia Forum Quarterly, 2006 (vol. 4), no. 3, pp. 67–87, here p. 73; Fredholm, Michael: Natural gas trade between Russia, Turkmenistan, and Ukraine: agreements and disputes, Stockholm: Stockholm University, Department of Oriental Languages, 2008, pp. 9–10 (Asian Cultures and Modernity Research Report no. 15), http://gpf-europe.com/upload/iblock/2fa/fredholm.ukraine.russia.gas.rr15.pdf.
72 'Russia to drop barter for cash only payment for Turkmen gas', in: ICIS, 29 April 2005, http://www.icis.com/heren/articles/2005/04/29/9283639/russia-to-drop-barter-for-cash-only-payment-for-turkmen-gas.html.
73 'Turkmenistan, Russia agree to resume gas supplies in 2010, ending impasse', in: IHS Global Insight, 23 December 2009, http://www.ihs.com/products/global-insight/industry-economic-report.aspx?id=106594749; Pannier, Bruce: Pipeline explosion raises tensions between Turkmenistan, Russia, in: RFE/RL, 14 April 2009, http://www.rferl.org/content/Pipeline_Explosion_Stokes_Tensions_Between_Turkmenistan_Russia/1608633.html; According to Henderson, netback pricing (discussed below) was introduced at this time. Henderson, James: The pricing debate over Russian gas exports to China, Oxford: Oxford Institute for Energy Studies, 2011, p. 24, www.oxfordenergy.org/wpcms/wp-content/uploads/2011/10/NG-561.pdf. For a similar interpretation,

consumers. Some observers speculated that Gazprom lost 'more than $1 billion' during the first quarter of 2009, purchasing Central Asian gas, and that the agreement to pay 'European prices' was a 'mistake'.[74]

Gazprom's attempt to renegotiate turned ugly on April 9, 2009 when an explosion occurred on the Central Asia–Centre pipeline (near the Uzbek–Turkmen border) that connects Turkmenistan to Russia, stopping the flow of gas entirely.[75] Given each side's propensity to play hardball, it is perhaps not surprising that the downturn in European markets produced a major conflict in bilateral relations. Each party blamed the other for the malfunction and resulting cessation of trade. According to Turkmenistan, Gazprom did not give sufficient notice of its intention to reduce gas purchases. Ashgabat claimed that more time would have been necessary to prepare production wells and transport infrastructure for the change to lower volumes while avoiding technical malfunction.[76] According to one press report, 'Russian officials have remained diplomatic about the issue, leaving Russian media to take the lead in directing blame toward Turkmenistan.'[77] Thus, on the day of the accident, Anatoly Dmitrievsky, the Director of the Oil and Gas Research Institute under the Russian Academy of Sciences,[78] stated on TV that the explosion was due to disrepair of the old Turkmen pipeline system (built in the late 1960s and early 1970s). He also suggested that Turkmen dispatchers monitoring the system were at fault.[79]

---

see 'Turkmenistan–China natural gas pipeline: transforming dream into reality', in: Global Gas Transport, 1 July 2010, http://www.globalgastransport.info/archive.php?id=733.

74 Pannier, Bruce: Pipeline explosion raises tensions between Turkmenistan, Russia, in: RFE/RL, 14 April 2009, http://www.rferl.org/content/Pipeline_Explosion_Stokes_Tensions_Between_Turkmenistan_Russia/1608633.html; Blagov, Sergei: Russia struggles to revive energy ties with Turkmenistan, in: Eurasia Daily Monitor, 2009 (vol. 6), no. 230, http://www.jamestown.org/programs/edm/single/?tx_ttnews[tt_news]=35839&tx_ttnews[backPid]=485&no_cache=1.

75 Pannier, Bruce: Pipeline explosion raises tensions between Turkmenistan, Russia, in: RFE/RL, 14 April 2009, http://www.rferl.org/content/Pipeline_Explosion_Stokes_Tensions_Between_Turkmenistan_Russia/1608633.html.

76 Vasánczkii, Luça Z.: Gas exports in Turkmenistan, Paris: Institut français des relations internationales, 2011, www.ifri.org/downloads/noteenergielvasanczki.pdf; 'Turkmenistan: gas blast ignites Turkmen–Russian row, in: Eurasianet.org, 9 April 2009, http://www.eurasianet.org/departments/insightb/articles/eav041009b.shtml.

77 Pannier, Bruce: Pipeline explosion raises tensions between Turkmenistan, Russia, in: RFE/RL, 14 April 2009, http://www.rferl.org/content/Pipeline_Explosion_Stokes_Tensions_Between_Turkmenistan_Russia/1608633.html.

78 'Russian, European and Ukrainian experts consider European energy sector pressing challenges at Gazprom's headquarters', Gazprom press release, 21 February 2011, http://www.Gazprom.com/press/news/2011/february/article109580/.

79 Pannier, Bruce: Pipeline explosion raises tensions between Turkmenistan, Russia, in: RFE/RL, 14 April 2009, http://www.rferl.org/content/Pipeline_Explosion_Stokes_Tensions_Between_Turkmenistan_Russia/1608633.html; 'Turkmenistan: gas blast ignites Turkmen–Russian row, in: Eurasianet.org, 9 April 2009, http://www.eurasianet.org/departments/insightb/articles/eav041009b.shtml.

5. The Geo-Economics of Eurasian Gas

How then is this episode different from relations in the 1990s and the embargo at the end of the decade? To begin with, Gazprom adopted a relatively conciliatory tone. According to one press report, on the day of the accident, Gazprom CEO Miller only addressed the incident indirectly but was diplomatic nevertheless: '[Turkmenistan] is our strategic partner in the gas sector, with whom we are linked by many years of fruitful cooperation', Miller said in a clear move to ease the looming stand-off. 'We are sure that in future [sic] this cooperation will only strengthen'.[80] In fact, since the turn of the century, Gazprom had repeatedly emphasised the importance of its 'strategic' relationship with the Central Asian gas exporter.

Second, the dispute was settled not by Gazprom but at the highest political level in Russia.[81] The Russian Foreign Ministry immediately became actively involved in the dispute,[82] reflecting the fact that for Russia, gas revenues were only one part of a larger portfolio of important issues regarding which Moscow wanted cooperative relations with Turkmenistan.[83] During the 1990s, Gazprom was primarily responsible for the gas trade in Central Asia, and it was able to keep state officials at arm's length to protect its autonomy. In fact, decision making was dominated by company management and its allies among intermediary firms. In consequence, economic relations were not embedded in broader political relations. This is important. When economic relations are embedded in cooperative political relations, economic conflicts can be tempered as economic interests become subordinated to political ones.[84]

Negotiations lasted nearly nine months. Although the pipeline was quickly repaired, no Turkmen gas was acquired by Gazprom during this period, producing losses equivalent to 25% of GDP for the Central Asian exporter.[85] The company proposed two options: either the volumes it purchased could be reduced (by 80%), or the purchase

---

80 'Pipeline blast halts Turkmen gas exports to Russia', in: Reuters, 9 April 2009, http://www.Reuters.com/article/2009/04/09/turkmenistan-russia-gas-idUSL959981620090409.
81 Fedynsky, Peter: Russia and Turkmenistan vying for Central Asian gas, in: Voice of America, 22 December 2009, http://www.voanews.com/content/russia-and-turkmenistan-vying-for-central-asian-gas-79905322/369383.html; Blagov, Sergei: Russia struggles to revive energy ties with Turkmenistan, in: Eurasia Daily Monitor, 2009 (vol. 6), no. 230, http://www.jamestown.org/programs/edm/single/?tx_ttnews[tt_news]=35839&tx_ttnews[backPid]=485&no_cache=1.
82 Gurt, Marat: Turkmenistan accuses Russia of violating gas agreement, in: Reuters, 10 April 2009, http://uk.Reuters.com/article/2009/04/10/turkmenistan-russia-idUKLA39873720090410; 'Russia blamed for pipeline blast', in: BBC News, 10 April 2009, http://news.bbc.co.uk/2/hi/asia-pacific/7993625.stm; Peyrouse, Sébastien: Berdymukhammedov's Turkmenistan: a modest shift in domestic and social politics, in: China and Eurasia Forum Quarterly, 2010 (vol. 8), no. 3, pp. 47–66.
83 Cutler, Robert M.: Turkey and the geopolitics of Turkmenistan's natural gas, in: Review of International Affairs, 2003 (vol. 1), no. 2, pp. 20–33, here p. 30.
84 Bhadrakumar, M.K.: Russia takes control of Turkmen (world?) gas, in: Asia Times Online, 30 July 2008, http://www.atimes.com/atimes/Central_Asia/JG30Ag01.html.
85 Peyrouse, Sébastien: Berdymukhammedov's Turkmenistan: a modest shift in domestic and social politics, in: China and Eurasia Forum Quarterly, 2010 (vol. 8), no. 3, pp. 47–66.

price could be lowered (by 40%).[86] Eventually, on 22 December, the parties amended their supply agreement. According to one source, Gazprom would pay US$220–250 for Turkmen gas in 2010.[87] They also agreed on lower volumes beginning in January 2010; Gazprom only purchased 10.7bcm from Turkmenistan in 2010 instead of the contracted 30bcm.[88] Furthermore, not everyone at Gazprom agreed that the company should pay European market prices.[89] Although the values used to calculate Gazprom's post-2009 margins are only speculative and should be interpreted carefully, my estimations suggest that Gazprom's re-export of Turkmen gas is not nearly as profitable as it was earlier and may even be a loss-making enterprise.

Perhaps most importantly, the sales price would be determined 'based on a fluctuating European price formula.'[90] This marked the formal introduction of European 'netback' pricing.[91] By contractually linking price to the European market through a formula, Russia was essentially limiting bargaining over the most controversial question. Going forward, the price was set to change automatically as a function of prices in Europe. To 'resolve the problems of [...] the price-setting process' was a goal set by Putin in May 2000.[92] However, it took nearly a decade to find a solution.

To summarise, the change was not that Russia had abandoned the use of power to coerce its partner. Rather, the bargaining dynamics had changed, as evidenced by the fact that Russia had become more reasonable in negotiations, had altered its public tone, and was more willing to compromise. More importantly, the negotiation participants had changed. Although Gazprom still played a central role, the key decision makers were high-level state officials, while intermediary agents were gradually eliminated. This meant that economic issues could be linked to broader political issues related to the bilateral relationship.

---

86 Dan, Feng: Analysis on natural gas geo-politics in Central Asia-Russia region, Paper submitted to the 21st World Energy Congress 2010, p. 8, http://www.worldenergy.org/documents/congresspapers/140.pdf.
87 Socor, Vladimir: Russia resumes gas imports from Turkmenistan, in: Asia Times online, 6 January 2010, http://www.atimes.com/atimes/Central_Asia/LA06Ag02.html.
88 Gazprom: Gas purchases, http://www.gazprom.com/about/production/central-asia/ (accessed 27 September 2013).
89 Vasánczkii, Luça Z.: Gas exports in Turkmenistan, Paris: Institut français des relations internationales, 2011, p. 9, www.ifri.org/downloads/noteenergielvasanczki.pdf.
90 'Turkmenistan, Russia agree to resume gas supplies in 2010, ending impasse', in: IHS Global Insight, 23 December 2009, http://www.ihs.com/products/global-insight/industry-economic-report.aspx?id=106594749.
91 Socor, Vladimir: Russia resumes gas imports from Turkmenistan, in: Asia Times online, 6 January 2010, http://www.atimes.com/atimes/Central_Asia/LA06Ag02.html.
92 'Turkmen, Russian heads talk bilateral relations in Turkmen capital', in: BBC Monitoring Central Asia Unit, 20 May 2000.

5.4. Explaining Evolution in the Conditions of Trade

5.4.1. Changing Supply-Demand

The preceding discussion has shown how the conditions of trade changed during the post-Soviet period. How did this happen? One explanation focuses on gas supply-demand dynamics in Russia.[93] During the 1990s, Gazprom had significant surplus gas. On the supply side, its major deposits were still relatively young and promised ample future volumes in the medium term. At the same time, domestic demand was relatively limited because of Russia's economic depression during the 1990s. These surpluses shrank toward the end of the decade as Gazprom's deposits grew older and more depleted and the Russian economy began to recover, giving rise to growth in demand for gas after the 1998 August default.[94] From this perspective, Gazprom became more cooperative because it had a greater need for Turkmen gas to fill its growing commitments to consumers (domestic and international) amid dwindling reserves.[95] This explanation highlights how economic structure shapes bargaining power; specifically, Gazprom's need for Turkmen gas shifted the balance of power in the latter's favour.[96]

At the end of the 20th century, it was becoming increasingly clear that Gazprom would need Turkmen gas to supplement the dwindling indigenous gas component in its fuel balance. However, under Vyakhirev's management, the company only came to an agreement with Turkmenistan in 1999, after being pressured by Vladimir Putin, then soon to be President. Indeed, earlier in the month, Vyakhirev was still playing hard ball, announcing publicly that '[…] if no solution is found by 1 January 2000, the parties shall have to postpone the conversations for another year'.[97] Thus, supply and demand factors were not sufficient.

In addition, this account cannot explain why it took so long to raise prices to European levels and leaves unexplained the delay in concluding long-term contracts,

---

93 Blank, Stephen: Turkmenistan and Central Asia after Niyazov, Carlisle, PA: Strategic Studies Institute of the US Army War College, 2007, www.strategicstudiesinstitute.army.mil/pdffiles/pub791.pdf (alternative URL: http://www.isn.ethz.ch/Digital-Library/Publications/Detail/?ots591=0c54e3b3-1e9c-be1e-2c24-a6a8c7060233&lng=en&id=46890).
94 One consequence of the default was that the Russian ruble was dramatically devalued. By making imports more expensive and exports more competitive, the economy was stimulated through increased exports and import substitution.
95 Rosner, Kevin: Gazprom and the Russian state, London: GMB Publishing, 2006, p. 20.
96 In a situation of interdependence between two states, changes in the level of dependence of one state have consequences regarding how power is distributed between the parties. An increase in actor A's dependence on B creates power for actor B, ceteris paribus. For a general discussion of power under interdependence, see Keohane, Robert/ Nye, Joseph: Power and interdependence, New York: Addison Wesley, 2001, pp. 9–17.
97 'Russia draws a glowing gas export picture for Turkmenistan, but President is being sceptical', in: Azer Press, 18 December 1999.

the elimination of intermediaries as the Russian state took a more central role in the gas trade, and how economic issues became embedded in the broader political relationship. These shortcomings reveal a more general problem with an explanation that focuses on economic structure, power, and outcome but tells us very little about how processes, in this case bargaining dynamics, change.

5.4.2. Alternative Export Pipelines

A second explanation focuses on the importance of additional export outlets for Turkmenistan.[98] This explanation is logically similar to the previous one in focusing on the effects of economic structure on the relative distribution of power between actors in producing outcomes. Specifically, more outlets meant that Turkmenistan had more options and thus greater leverage over Russia. Three pipeline projects are relevant: the Central Asia–China gas pipeline from Turkmenistan to China via Uzbekistan and Kazakhstan; the Korpedzhe–Kurt-Kui gas pipeline to Iran; and the Trans-Caspian gas pipeline. A map depicting these pipelines is found in the annex.

The completion of the Central Asia-China pipeline was a significant development because it opened an export route not involving Russian transit to a major market. A project to export gas to China, initiated by Turkmenistan in 1993,[99] finally developed momentum in 2006 when then Presidents Niyazov and Hu Jintao concluded an intergovernmental agreement to build the pipeline and begin gas sales.[100] Thus, the concrete plan to develop a new export route, which became the foundation for aggressive Turkmen bargaining,[101] appears to have been a sufficient condition for some of the later (after 2006) price increases, which were the most significant price increases in terms of their impact on rent distribution. By contrast, the Korpedzhe–Kurt-Kui pipe-

---

98 Blank, Stephen: Turkmenistan and Central Asia after Niyazov, Carlisle, PA: Strategic Studies Institute of the US Army War College, 2007, www.strategicstudiesinstitute.army.mil/pdffiles/pub791.pdf (alternative URL: http://www.isn.ethz.ch/Digital-Library/Publications/Detail/?ots591=0c54e3b3-1e9c-be1e-2c24-a6a8c7060233&lng=en&id=46890); Blum, Doug/ Saivetz, Carol: Fishing in troubled waters: Putin's Caspian policy (Transcript), Cambridge, MA: Harvard University, Belfer Center for Science and International Affairs, 2 May 2001.
99 Dorian, James P.: Turkmenistan's future in gas and oil hinges on certainty for export options, in: Oil and Gas Journal, 2002, (vol. 100), no. 417, http://www.ogj.com/articles/print/volume-100/issue-41/general-interest/focus-on-turkmenistan-turkmenistans-future-in-gas-and-oil-hinges-on-certainty-for-export-options.html.
100 Turkmenistan committed to supplying China with 30bcm annually for 20 years beginning in 2009. Tomberg, Igor: Energy policy and energy projects in Central Eurasia, in: Central Asia and the Caucasus, 2007, no. 6(48), pp. 38–50, http://www.ca-c.org/journal/2007-06-eng/04.shtml.
101 Cornell, Svante E.: Trans-Caspian pipelines and Europe's energy security, in: Cornell, Svante E./ Nilsson, Niklas (eds): Europe's energy security: Gazprom's dominance and Caspian supply alternatives, Stockholm/ Washington, DC: Central Asia–Caucasus Institute & Silk Road Studies Program, 2008, pp. 141–153, http://www.silkroadstudies.org/new/docs/publications/EnergySecurity.pdf; Kandiyoti, Rafael: What price access to the open seas? The geopolitics of oil and gas transmission from the Trans-Caspian republics, in: Central Asian Survey, 2008 (vol. 27), no. 1, pp. 75–93.

5. The Geo-Economics of Eurasian Gas   171

line to Iran explains very little about the conditions of trade and bilateral relations, as the 1997 Russian gas embargo continued for nearly two years after the pipeline's completion.

The Trans-Caspian pipeline project was launched in 1996 at the behest of the US.[102] It began to move forward in a serious way in response Gazprom's transit embargo of Turkmenistan, and documents related to its implementation were signed in November 1999 during a meeting of the OSCE in Istanbul.[103] The timing suggests that the project may have played a role in the 1999 agreement. Stern has also connected Russia's position at the time to developments regarding the Trans-Caspian pipeline.[104] That former Russian minister of Fuel and Energy, Viktor Kalyuzhny, explicitly drew a connection between Gazprom's abuse of Turkmenistan, the fate of the underwater Caspian pipeline, and the geopolitical implications for Russia in late 1999 provides compelling direct evidence in support of this interpretation. According to one media report,

> [...] the Russian government [...] continues its efforts to try and minimize US influence in Russia's backyard—particularly the planned PSG Trans-Caspian gas project [...]. Viktor Kalyuzhny [...] added that Turkmenistan should be seen 'as an equal partner in the gas market' and that [...] 'if someone in Russia now lays claim to a monopoly, that will force Turkmenistan to look for other options, including the Trans Caspian project'.[105]

However, as discussed above, Gazprom had to be pressured to come to an agreement, suggesting that developments regarding the Trans-Caspian pipeline were by themselves not sufficient to produce the 1999 agreement.

5.4.3. State Relations with Gazprom

A final explanation, focusing on the changes in relations between the Russian state and Gazprom that occurred under Vladimir Putin, is necessary to develop a comprehensive view of why relations changed. In particular, these political changes appear necessary to account for the resumption of trade in 1999, the emergence of a long-term perspective reflected in long-term contracts, and new bargaining dynamics involving compromise, a more prominent state role in the gas trade at the expense of private intermediaries, and the linkage between the gas trade and a broader set of political issues. The remainder of this section relates changes in state-firm relations to variation in these components of the conditions of trade during the post-Soviet period.

---

102 Tomberg, Igor: Energy policy and energy projects in Central Eurasia, in: Central Asia and the Caucasus, 2007, no. 6(48), pp. 38–50, http://www.ca-c.org/journal/2007-06-eng/04.shtml.
103 '"Blue Stream" takes shape', in: Oil & Gas of Kazakhstan, 19 April 2000.
104 Stern, Jonathan P.: The future of Russian gas and Gazprom, New York: Oxford University Press, 2005, p. 74.
105 'Russia ratifies Blue Stream tax protocol, Turks follow suit', in: European Gas Markets, 15 December 1999.

When Russia became an independent state in 1992, economic reform, and destatisation in particular, was a top priority under Boris Yeltsin.[106] In the same year, Gazprom was transformed from a state-owned concern into a 'Russian' joint stock company (RAO).[107] A year later, a trust agreement was concluded with CEO Vyakhirev, whereby the latter would hold a majority of shares owned by the state and represent its interests in managing the company. This was part of a much broader process to reduce state authority over business decision-making in the Russian economy. The result was that gas policy, both domestic and foreign, was largely vested in the company management.[108]

Although Gazprom managers assumed responsibility for providing important public services (for example, selling gas to domestic consumers at a large discount and using gas sales to Ukraine as leverage in attempting to broker a deal concerning the continued presence of Russia's Black Sea fleet in Ukraine), they also had significant autonomy to pursue their own interests, which often conflicted with those of the state.[109] This included maximising short-term revenues from the gas trade by adopting a 'winner-take-all' strategy toward its Central Asian competitors, most notably Turkmenistan. Thus, the timing of the final transfer of authority over gas policy to Gazprom through the trust agreement corresponds closely to Russia's refusal to export Turkmen gas in October 1993 and its final abrogation of the natural gas export intergovernmental agreement with Turkmenistan a month later. The 1990s saw a proliferation of intermediaries in the gas trade, which facilitated barter but were also a vehicle for illegal private extraction from the gas business.[110]

For Vladimir Putin, closer political relations with Central Asia were a top priority[111], and this necessitated both a long-term perspective on the gas trade and substantially altered bargaining dynamics. Early changes in 1999 were a product of Putin's new pri-

---

106 Barkanov, Boris: Mercantilist development in Russia: the legitimacy of state power, state identity, and the Energy Charter regime (1990–2010), University of California at Berkeley, unpublished dissertation, 2011.
107 Radygin, Alexander: State-owned holding companies in Russia, paper prepared for the conference 'Corporate governance of state-owned enterprises in China', Beijing, 18–19 January 2000, p. 16, http://www.oecd.org/corporate/ca/corporategovernanceofstate-ownedenterprises/1923693.pdf. Only two other companies were accorded the status of Russian joint stock company: the electricity monopoly RAO UES (Unified Energy Systems) and Norilsk Nickel.
108 Rutland, Peter: Battle rages over Russia's natural monopolies, in: Transition, 1997 (vol. 8), no. 3, pp. 13, 15.
109 Barkanov, Boris: Mercantilist development in Russia: the legitimacy of state power, state identity, and the Energy Charter regime (1990–2010), University of California at Berkeley, unpublished dissertation, 2011.
110 Ibid.; Rutland, Peter: Lost opportunities: energy and politics in Russia, NBR Analysis, 1997 (vol. 8), no. 5, pp. 5–31, here pp. 20–21, http://www.nbar.org/publications/analysis/pdf/vol8no5.pdf.
111 Kubicek, Paul: Russian energy policy in the Caspian basin, in: World Affairs, 2004 (vol. 166), no. 4, pp. 207–217; Stulberg, Adam N.: Moving beyond the great game: the geoeconomics of Russia's influence in the Caspian energy bonanza, in: Geopolitics, 2005 (vol. 10), no. 1, pp. 1–25.

orities. He also articulated his vision during his first month as president in a visit to Ashgabat in May 2000. At the time, discussions began concerning a long-term contract. Over time, the cast of characters changed as the state began to play a more important role in negotiations and intermediaries were gradually eliminated. As a result, the gas trade became embedded in a broader set of questions related to the interests of the Russian state. Finally, Russia also became more reasonable, willing to compromise, and diplomatic, repeatedly emphasising the strategic nature of the relationship.

5.5. Conclusion: Implication for Russian Gas Dominance in Europe

Given how and why Russia's relationship with Turkmenistan has evolved, what can we conclude about Russia's prospects for controlling Turkmen gas flows to Europe to protect its dominant market position? The emergence of a long-term perspective reflected in the conclusion of 25-year contracts makes it more difficult for other parties to acquire gas and thus difficult to build pipelines on commercial terms. Private investors are not likely to expend the considerable capital necessary to finance such projects, given uncertainty about resources. Even if pipelines were built, Russia has shown that it is prepared to pay higher prices to secure this gas. This means that it may not be economically feasible to ship the gas and sell it in Europe, as the price necessary to outbid Gazprom would likely make it too expensive in the downstream European market (see the chapter by Kustova). Although this lowers Gazprom and state revenues through lower (potentially zero) margins on the resale of Turkmen gas to European customers, it also prevents gas-on-gas competition downstream, which would undermine Russia's dominant position in Europe and have important economic and political consequences over the long term.

The change in bargaining dynamics suggests that Russia is also willing to expend significant political capital through the efforts of state officials at the highest levels to pursue its goal of controlling Turkmen gas. It is also likely to leverage other political factors by linking them to the gas trade. That Russia has become 'more reasonable' suggests that these linkages could be advantageous to Turkmenistan, although one should not discount the potential for Russia to also use coercion.

However, there are important limitations to Russia's approach. First, it appears to be most successful in a high demand environment or at least one in which demand outstrips Russian supply by a large margin. When demand falls, Russia must decide whether to substitute its own supplies for Turkmen gas to prevent Turkmen gas from being marketed independently of Russian control. Obviously, the supply available in Turkmenistan is important. In any case, there are important technical issues to be considered: to what extent can Gazprom curtail production, and is there sufficient capacity in the pipeline network to reroute gas in the volumes necessary? There are also economic tradeoffs: what is the effect of substitution on revenues? Insofar as Turkmen gas is more expensive than Russian gas, substitution will be difficult because it impacts Gazprom's bottom line and the financial well-being of the Russian state. This effect is magnified as the volume of Turkmen gas in question increases.

The 2009 renegotiation demonstrated that Russia is willing to import only limited volumes when its own supplies are sufficient to meet its commitments and Turkmen gas is relatively expensive. However, Turkmenistan has demonstrated that it is prepared to sell gas to China at much lower prices than Gazprom; thus, not substituting raises the spectre of Turkmen sales to EU traders at potentially much lower prices. It is possible that Gazprom would be willing to take greater volumes if an alternative marketing and transport arrangement were in fact possible rather than merely hypothetical. Ultimately, Gazprom started sharing the wealth only after a pipeline to China was in the works and it was clear that Turkmenistan would have alternatives.

Finally, there are also important non-economic considerations. Despite Russia's turn to reason and more diplomatic tone, Turkmenistan remains quite suspicious of Russian power overall, especially in the gas trade.[112] This suspicion was reinforced by Gazprom's strategy and tactics during the 1990s but also during the more recent 2009 crisis.[113] Matching EU prices to acquire Turkmen gas when other buyers are available may not be sufficient, as Turkmenistan is eager to diversify export routes and minimise Russian influence overall. There may very well be a political premium that Russia would have to pay to prevent Turkmen gas from being sold to European traders, as Ashgabat values having alternatives that increase its own power relative to that of Moscow.

---

112 Jackson, Alex: Analyzing Turkmenistan's gas exports after the election, in: Natural Gas Europe, 6 March 2012, http://www.naturalgaseurope.com/analyzing-turkmenistans-gas-exports-after-the-election-5242. For a discussion concerning the general scepticism of some CIS states, see for example Kazantsev, Andrei: Russian policy in Central Asia and the Caspian Sea region, in: Europe-Asia Studies, 2008 (vol. 60), no. 6, pp. 1073–1088.
113 Pirani, Simon: Central Asian and Caspian gas production and the constraints on export, Oxford: Oxford Institute for Energy Studies, 2012, http://www.oxfordenergy.org/wpcms/wp-content/uploads/2012/12/NG_69.pdf; Horák, Slavomír: Turkmenistan's shifting energy geopolitics in 2009–2011, in: Problems of Post-Communism, 2012 (vol. 59), no. 2, pp. 18–30.

Paolo Sorbello

6. Pipelines and Hegemonies in the Caspian: a Gramscian Appraisal

6.1. Introduction

The broad purpose of this work is to understand the dynamic relationship between energy and foreign policy, focusing specifically on the case study of Kazakhstan and the circumstances that led to the construction of two oil pipelines, the Caspian Pipeline Consortium (CPC) and the Kazakhstan–China Pipeline (KCP). Did the change in the material conditions of the Kazakh energy sector produce a shift in foreign policy? Or, more specifically, was the new configuration in the oil industry the trigger for the realisation of the multi-vector foreign policy designed in the 1990s? Can 'energy' be considered the catalyst of Kazakh foreign policy?

In the present work, the word 'energy' is regarded as a package term: it encompasses economic, technical, legal, and political characteristics peculiar to the energy sector in order to analyse the formation of foreign policy from the standpoint of a holistic and rigorous understanding of the energy sector. In the narrower terms of this particular case study, 'energy' itself is limited to oil, but the overarching argument could apply to a variety of energy sources.

The scope of the chapter is limited to the case study of two oil pipelines that were built in different historical periods (the CPC, conceived in 1992 and built in 2001, and the KCP, conceived in 1997 and completed in 2009). The choice of cases rests on the fact that these are not 'legacy pipelines'[1] inherited by Kazakhstan from the Soviet Union. Consequently, the time frame for this research is the post-Soviet period.

6.2. Brief Literature Review of the Role of Energy from the Domestic Arena to Foreign Policy[2]

There is an absence of substantial literature on the roles of energy in foreign policy decision-making that are unrelated to the issue of 'energy security' (i.e., dependence on energy supplies from abroad). Ever since the oil crises in the 1970s, academic literature

---

1 Stulberg, Adam N.: Eurasia's pipeline tangle: practical lessons from cross-border pipeline operations, in: Russia in Global Affairs, 24 September 2011, http://eng.globalaffairs.ru/number/Eurasias-Pipeline-Tangle--15337.
2 This work lacks both a thorough review of the literature and an analysis of Kazakh foreign policy, due to space and scope limitations. Nevertheless, a comprehensive study has been previously performed by the author and is currently being revised and updated.

has devoted significant effort to elucidating the role of energy in foreign policy. Given the status of energy dependency of these countries, however, the main focus was how natural resources are vital for independent conduct in world politics and the economic and political impact of supply disruptions. For example, Krasner addressed energy and foreign policy in the United States.[3] His argument is based on the use of raw materials as a feature in decision-making, and he contends that the final goal of the state is to pursue the national interest. In his realist account, Krasner characterises American decision-making in international commodity markets as a way to pursue broad foreign policy goals. Following a similar line, Ziegler recently argued that 'China's rapidly growing energy demands mesh closely with broader foreign policy goals'.[4]

6.2.1. Curses, Diseases and Rents

Looking at supplier states, the literature, especially the economics literature, has focused on the role of energy in the structure of the national economies of hydrocarbon-exporting states. Terms such as 'rentierism', 'resource curse', and 'Dutch disease' were applied in turn to Middle Eastern, South American, and post-Soviet countries alike. 'Rentierism' refers to 'a government's ability to provide popular social services and patronage while taxing populations lightly or not at all'.[5] The problem is that reliance on external revenues can undermine internal stability and hinder democratic practices, as the country can avoid taxing its citizens. Rentierism, however, does not fully apply to the post-Soviet experience and therefore seldom appears in the literature on the region.[6]

Another branch of the economics scholarship focuses on energy prices, arguing that the price of a commodity, especially when it is a matter of dependency, influences states' behaviour in the international arena. Russia and the other post-Soviet countries, however, are price-takers in terms of oil. Furthermore, the price of exported gas is indexed to that of oil (especially for the contracts stipulated by the Russian state-owned gas company Gazprom and Central Asian exporters). Some studies have con-

---

3   Krasner, Stephen: Defending the national interest: raw materials investments and U.S. foreign policy, Princeton, NJ: Princeton University Press, 1978.
4   Ziegler, Charles E.: The energy factor in China's foreign policy, in: Journal of Chinese Political Science, 2006 (vol. 11), no. 1, pp. 1–24, here p. 4.
5   Ross, Michael: Does oil hinder democracy?, in: World Politics, 2006 (vol. 53), no. 3, pp. 325–361. For a critique, see: Heinrich, Andreas/ Pleines, Heiko (eds): Challenges of the Caspian resource boom: domestic elites and policy-making, Basingstoke: Palgrave Macmillan, 2012.
6   Cf. e.g., Franke, Anja/ Gawrich, Andrea/ Alakbarov, Gurban: Kazakhstan and Azerbaijan as post-Soviet rentier states: resource incomes and autocracy as a double 'curse' in post-Soviet regimes, in: Europe-Asia Studies, 2009 (vol. 61), no. 1, pp. 109–140.

6. Pipelines and Hegemonies in the Caspian: a Gramscian Appraisal 177

sidered energy prices as the relevant variable for explaining Russian foreign policy; they ignore the fact that the inherent correlation is descriptive more than explanatory.[7] The application of the concepts of the 'resource curse' or 'Dutch disease' is more popular for the post-Soviet world.[8] These terms define the possible risks connected to mismanaging the windfall from energy revenues, especially when the country under scrutiny has an unbalanced economic structure and a dependency on energy exports.

6.2.2. Energy and International Relations

In the discipline of International Relations (IR), the focus on energy is generally centred on policy aspects and research is generally based on journalistic and policy reports. The aim is to describe and give recommendations. As an early draft of a recently published article outlines, 'the theoretical ambition remains low' and the emphasis on 'energy diplomacy' causes 'a strongly agent-centric picture of dependencies and interdependencies between energy producers and buyers'.[9]

In contrast, 'energy geopolitics'[10] focuses on crises surrounding dependence[11] (pipeline politics, shortages, cut-offs) and interprets energy choices as the materiali-

---

7   Hill, Fiona: Energy empire: oil, gas, and Russia's revival, London: The Foreign Policy Centre, 2004; Tabata, Sinichiro (ed.): Dependent on oil and gas, Sapporo: Slavic Research Center, 2006. Nanay, Julia: Russia's role in the Eurasian energy market, in: Perovic, Jeronim/ Orttung, Robert/ Wenger, Andreas (eds): Russian energy power and foreign relations: implications for conflict and cooperation, New York: Routledge, 2009, pp. 109–131.
8   Spruyt, Hendrik/ Ruseckas, Laurent: Economics and energy in the South, in: Menon, Rajan/ Fedorov, Yuri E./ Nodia, Ghia (eds.): Russia, the Caucasus and Central Asia: the 21st century security environment, Armonk, NY: M.E. Sharpe, 1999; Karl, Terry L.: Crude calculations: OPEC lessons for the Caspian region, in: Ebel, Robert/ Menon, Rajan (eds.): Energy and conflict in Central Asia and the Caucasus, Lanham, MD: Rowman & Littlefield, 2000; Kalyuzhnova, Yelena: Economies and energy, in: Kalyuzhnova, Yelena/ Jaffe, Amy M./ Lynch, Dov/ Sickels, Robin C. (eds.): Energy in the Caspian region: present and future, Basingstoke: Palgrave Macmillan, 2002, pp. 58–85; Oomes, Nienke/ Kalcheva, Katerina: Diagnosing Dutch disease: does Russia have the symptoms?, Washington, DC: IMF, 2007 (Working Paper, WP/07/102); Goldman, Marshall I.: Petrostate: Putin, power, and the new Russia, Oxford: Oxford University Press, 2008; Baev, Pavel K.: Russian energy policy and military power: Putin's quest for greatness, London: Routledge, 2009; Hannes Meissner: The resource curse and rentier states in the Caspian region, Hamburg: German Institute of Global and Area Studies, 2010 (Working Paper no. 133).
9   Aalto, Pami/ Dusseault, David/ Kennedy, Michael D./ Kivinen, Markku: Russia's energy relations in Europe and the Far East: towards a social structurationist approach to energy policy formation, in: Journal of International Relations and Development (forthcoming, advance online publication,18 January 2013)—here I used an earlier unpublished draft of the paper, which is substantially different from the published one.
10   Mitchell, John V./ Beck, Peter/ Grubb, Michael: The new geopolitics of energy, London: Royal Institute of International Affairs, 1996.
11   Hadfield, Amelia: Energy and foreign policy: EU-Russia energy dynamics, in: Smith, Steve/ Hadfield, Amelia/ Dunne, Tim (eds.): Foreign policy: theories, actors, cases, Oxford: Oxford University Press, 2007, pp. 441–462; Balmaceda, Margarita M.: Energy dependency, politics and corruption in the former Soviet Union: Russia's power, oligarchs' profits and Ukraine's missing energy policy, 1995–2006, London: Routledge, 2008.

sation of actions directed to the realisation of the national interest.[12] The theoretical lens for this interpretation is of course neo-realist and several authors cannot conceal their western bias.[13] An older review of the literature reveals the numerous strands of scholarship dealing with 'energy policy'.[14] Some authors regard the structure/agent divide in the literature as the reason for the lack of theoretical reach.[15]

'In energy policy analysis, it is crucial to acknowledge the diversity of actors'[16] and abandon the state-centric world view that confines all non-state and transnational actors to inconsequential roles. Particularly in the case of Russia and other post-Soviet states, however, where the energy sector is usually controlled by the state, several studies focusing on energy-exporting countries have highlighted (and condemned) the use of energy as a weapon against consumer countries.[17] Following this view, the 'energy tool' is used as a means to reach foreign policy objectives. This trend in the academic and policy world, widespread in American and EU circles, was triggered by Vladimir Putin's rise to power in Russia,[18] higher energy prices, and Russo–Ukrainian (and, to a lesser extent, Russo–Lithuanian and Russo–Belarusian) energy cut-offs. By the same contingency-related token, the 'energy weapon' discourse waned once oil prices were stabilised, Putin allowed Dmitri Medvedev to succeed him, rather than seiz-

---

12 Rutland, Peter: Oil, politics and foreign policy, in: Lane, David (ed.): The political economy of Russian oil, Lanham, MD: Rowman & Littlefield, 1999, pp. 163–188.
13 Kalicki, Jan H./ Goldwyn, Dan L. (eds.): Energy and security: a new foreign policy strategy, Baltimore, MD: Johns Hopkins University Press, 2005; Bahgat, Gawdat: Europe's energy security: challenges and opportunities, in: International Affairs, 2006 (vol. 82), no. 5, pp. 961–975.
14 Aalto, Pami/ Westphal Kirsten: Introduction, in: Aalto, Pami (ed.): The EU–Russia energy dialogue: securing Europe's future energy supplies?, Aldershot: Ashgate, 2007, pp. 1–21, here pp. 4–5.
15 Aalto, Pami/ Dusseault, David/ Kennedy, Michael D./ Kivinen, Markku: Russia's energy relations in Europe and the Far East: towards a social structurationist approach to energy policy formation, in: Journal of International Relations and Development (forthcoming, advance online publication,18 January 2013)—here I used an earlier unpublished draft of the paper, which is substantially different from the published one.
16 Aalto, Pami/ Dusseault, David/ Kennedy, Michael D./ Kivinen, Markku: Russia's energy relations in Europe and the Far East: towards a social structurationist approach to energy policy formation, in: Journal of International Relations and Development (forthcoming, advance online publication,18 January 2013), pp. 1–29, here p. 5.
17 These studies are especially popular in the policy world of American think-tanks: Saivetz, Carol: Russia: an energy superpower?, Audit, MIT Center for International Studies, December 2007, http://web.mit.edu/cis/editorspick_saivetz07_audit.html; Mankoff, Jeffrey: Eurasian energy security, New York: Council on Foreign Relations, 2009 (Council Special Report no. 43), http://www.cfr.org/world/eurasian-energy-security/p18418; Woehrel, Steven: Russian energy policy toward neighboring countries, Washington, DC: Congressional Research Service, 2009 (no. RL34261), https://www.fas.org/sgp/crs/row/RL34261.pdf; Smith, Keith C.: Managing the challenge of Russian energy policies: recommendation for U.S. and EU leadership, Washington, DC: Center for Strategic and International Studies, 2010, http://csis.org/files/publication/101123_Smith_ManagingChallenge_Web.pdf.
18 Milov, Vladimir: The use of energy as a political tool, in: The EU–Russia Review, 2006, no. 1, pp. 12–21, http://www.isn.ethz.ch/Digital-Library/Publications/Detail/?ots591=0c54e3b3-1e9c-be1e-2c24-a6a8c7060233&lng=en&id=48824.

ing all power through the amendment of the Constitution, and Gazprom lost its behemoth's disguise as independent gas producers conquered larger and larger shares of the Russian domestic market and Europe applied stricter regulations to imports and the distribution of natural gas (the Third Energy Package).

In spite of evidence that the 'energy weapon' concept has only narrow and casual explanatory power, scholars and opinion-makers have revived its use, with special reference to the EU–Russia relation. It is telling that the argument in such publications is sustained on the weak legs of newspaper articles.[19]

6.2.3. Pipelines and Foreign Policy

As aptly noted by Sovacool:

> Traditional public policy experts and energy analysts often view pipelines simply as energy delivery mechanisms that transport fuels from one point to another. When more 'critical' assessments occur, these are mostly about improving pipeline performance and minimising the risks of failure, accident, and leakage, and they rarely discuss issues relating to national development, equity, and the environment.[20]

In IR, research on pipelines is carried out as research on the means for either resource nationalism or energy interdependence. Both strands of scholarship see pipelines as fixtures of the international environment. Economists, in turn, look at issues of economies of scale, life-cycles, upfront investment, inflexibility, natural monopolies, and 'the tyranny of distance'[21] that surround pipelines. Pipeline (geo)politics and pipeline (geo)economics nevertheless have failed to analyse the true political consequences the infrastructure carries in terms of foreign policy decision-making. Stulberg makes a compelling argument in his assessment of 'pipeline politics' as more than just an instrument of resource nationalism, given that supply disruptions are exceptional cases and rather expensive weapons that cannot be realistically wielded at no cost. His critique of the state-centric departure point allows for a greater and more comprehensive analysis of the 'multidimensional strategic interaction'.[22] I argue that the playing field where these interactions take place can be interpreted as a 'historical bloc', as conceived by Gramsci (see below).

---

19 Smith Stegen, Karen: Deconstructing the 'energy weapon': Russia's threat to Europe as case study, in: Energy Policy, 2011 (vol. 39), no. 10, pp. 6505–6513, here 6505. It is notable that Smith Stegen references an article by Andrew Kramer on the New York Times as literature 'adherent [to] the energy weapon thesis' (ibid., p. 6506).
20 Sovacool, Benjamin K.: The interpretive flexibility of oil and gas pipelines: case studies from Southeast Asia and the Caspian Sea, in: Technological Forecasting & Social Change, 2011 (vol. 78), no. 4, pp. 610–620, here p. 618.
21 Stulberg, Adam N.: Strategic bargaining and pipeline politics: confronting the credible commitment problem in Eurasian energy transit, in: Review of International Political Economy, 2012 (vol. 19), no. 5, pp. 808–836.
22 Ibid., p. 830.

In this realm, the issues of who builds, who owns, and who feeds the pipeline are the most relevant. In the cases to be analysed here, the ownership of the CPC is predominantly private, whereas the KCP is fully in the hands of the Kazakh and Chinese governments. Private funds, aided by government lenders, were at the origin of the construction of the CPC, while the KCP's construction was the outcome of an intergovernmental framework agreement in a large-scale co-operation in the oil and gas sector. The source of crude oil for both pipelines is the northeastern Caspian region in Kazakhstan. The history of the changing ownership of the CPC is useful to detect larger foreign policy manoeuvres, as is the mutated equilibrium of government and firms' participation in the upstream oil projects of Tengiz and Kashagan.

Business ventures and pipelines become politically relevant to the host country's foreign ministry when their structures, construction and operations are likely to affect relationships with investment partners and consumers abroad. Internally, they become important when considering the socio-economic externalities of both employment and rents from the export of oil. In a country such as Kazakhstan, state-owned companies KazMunaiGaz (oil and gas company), KazTransOil (oil pipeline operator) and Samruk-Kazyna (state-holding company and national welfare fund) are all actors that are strongly connected to the government, although they might pursue strategies at odds with Astana. Lastly, failing to consider technical and environmental issues allows us to paint only a partial image of pipelines' true roles in the hegemonic configurations both within the country and among the actors in the region.

6.2.4. The Gramscian Contribution

The writings of Antonio Gramsci were produced during a very tense period for Italian society. His political writings preceded and followed the Russian Revolution of 1917, the birth of fascism, and the split within the Italian Socialist Party. His theoretical works were developed during his imprisonment under the threat of fascist censorship in the 1920s and 1930s.

In this work, I concentrate on two important concepts, both products of Gramsci's historical materialist view of world politics. First, the concept of the 'historical bloc' is useful in its analysis of the material conditions determining the economic, political and social position of all actors in a determinate society. Gramsci understood that the capitalist system existed in the context of its historical specificity, which has allowed and allows for systemic transformations. In the case of Kazakhstan, the considered actors are the government, national companies and transnational companies.[23] The legal

---

23 Considering society at large, or 'civil society' in the Gramscian sense, is both beyond the possibility of a study of this scale and possibly pointless, given the weak role of the population in the government's decision-making process. An examination of the relationship between oil workers and the rest of the actors could be the subject of a future work with a broader scope.

system, as well as the terms of the contracts, the economic agreements, the allocation of rent, the specificity of Kazakh oil, the board structure of companies and the ministerial roster are among the factors considered to determine the historical bloc. The configuration of material forces informs the background before which social relations between the actors occur. According to Gramsci, these relations are always hegemonic and imply the attempt of one coalition of actors to impose their worldview and social customs on society at-large.[24] Of course, this work transcends the national boundaries within which Gramsci focused his analysis.[25]

Second, the Gramscian concept of hegemony departs from a non-positivist, 'interpretative' approach.[26] Instead of focusing on coercion, Gramscian scholars emphasise the role of consent in building the necessary legitimacy for the ideas of the ruling class to be hegemonic: 'A hegemonial structure of world order is one in which power takes a primarily consensual form'.[27] The considered social force is found both within and above states. The positivist, state-centred approach to hegemony in International Political Economy (IPE) presented by Hirschman, Keohane, Gilpin, and Lake, among others, is dismissed here as partial and incapable of an analytical reach that goes beyond what is easily 'operationalised' and calculated.[28] IPE scholars, however, have managed to refine the theory of hegemonic stability and retain more explanatory power than realist and neo-realist scholars, although they continue to over-concentrate on

---

24 Gramsci, Antonio: Quaderni del carcere, Torino: Einaudi, 1977. See also: Cox, Robert: Social forces, states and world orders: beyond international relations theory, in: Millennium, 1981 (vol. 10), no. 2, pp.126–155; Cox, Robert: Gramsci, hegemony and international relations: an essay in method, in: Millennium, 1983 (vol. 12), no. 2, pp. 162–175; Gill, Stephen: American hegemony and the trilateral commission, Cambridge: Cambridge University Press, 1990; Overbeek, Henk: Restructuring hegemony in the global political economy: the rise of transnational neoliberalism in the 1980s, London: Routledge, 2003; Robinson, William: Gramsci and globalisation: from nation-state to transnational hegemony, in: Bieler, Andreas/ Morton, Adam (eds): Images of Gramsci: connections and contentions in political theory and international relations, London: Routledge, 2006, pp. 165–180; Morton, Adam D.: Unravelling Gramsci: hegemony and passive revolution in the global political economy, London: Pluto Press, 2007; Thomas, Peter D.: The Gramscian moment: philosophy, hegemony and Marxism, Leiden: Brill, 2009; Dale, Gareth (ed.): First the transition, then the crash: Eastern Europe in the 2000s, London: Pluto Press, 2011.
25 Buci-Glucksmann, Catherine: Gramsci and the state, London: Lawrence and Wishart, 1980; Showstack Sassoon, Anne: Gramsci's politics, Minneapolis, MN: University of Minnesota Press, 1987.
26 Lake, David: Leadership, hegemony, and the international economy: naked emperor or tattered monarch with potential?, in: International Studies Quarterly, 1993 (vol. 37), no. 4, pp. 459–489.
27 Cox, Robert: Social forces, states and world orders: beyond international relations theory, in: Millennium, 1981 (vol. 10), no. 2, pp.126–155, here p. 153.
28 Hirschman, Albert O.: National power and the structure of foreign trade, Berkeley, CA: University of California Press, 1945; Keohane, Robert: After hegemony: cooperation and discord in the world political economy, Princeton, NJ: Princeton University Press, 1984; Gilpin, Robert: The political economy of international relations, Princeton, NJ: Princeton University Press, 1987; Lake, David: Leadership, hegemony, and the international economy: naked emperor or tattered monarch with potential?, in: International Studies Quarterly, 1993 (vol. 37), no. 4, pp. 459–489.

the ontological identity between hegemony and domination.[29] In Gramsci, 'hegemony' and 'the hegemon' are considered as two separate entities. Hegemony ought to be considered 'as a type of order that includes the different actors and social groups within the system under examination'.[30] As Morton noted, one of the key differences between the positivist and Gramscian approaches can be found in Gramsci's concept of the state: 'the state is the entire complex of practical and theoretical activities with which the ruling class not only justifies and maintains its dominance, but manages to win the active consent of those over whom it rules'.[31] In addition, Gramsci specified that a hegemonic dynamic is realised when

> the dominant group is in concrete coordination with the general interests of the subordinate groups and life within the state is conceived as a continuous formation and overcoming of instable equilibriums, between the interests of the fundamental group and those of the subordinate groups.[32]

I argue that it is possible to consider social interactions that constitute hegemony through both a national and a transnational framework.

In the case of post-Soviet Russia, this contested concept has been applied by realist[33] and Gramscian[34] scholars alike. 'Organised consent' is what Gramsci deems to be decisive in the establishment of a hegemonic relationship in a particular society. The basis of this conclusion is that neither coercion nor even 'potential imperiality'[35] are sufficient to win over the other actors due to the lack of legitimacy.

This chapter argues that there were two historical blocs in post-Soviet Kazakhstan. The first was formed after independence with the welcoming of foreign firms into the national oil sector. The second emerged with the discovery of oil off the Caspian shores in 2000, matured with the 2004 agreement with China on the construction of the eastward pipeline, and solidified with the modification of the Production Sharing Agreement legislation in 2005.

29　A notable exception is Joseph, Jonathan: Hegemony: a realist analysis, London: Routledge, 2002.
30　Burges, Sean W.: Consensual hegemony: theorizing Brazilian foreign policy after the Cold War, in: International Relations, 2008 (vol. 22), no. 1, pp. 65–84, p. 71.
31　Gramsci, Antonio: Quaderni del carcere, Torino: Einaudi, 1977 (Quaderno 15, par. 10), p. 1765 (quoted in: Morton, Adam D.: Unravelling Gramsci: hegemony and passive revolution in the global political economy, London: Pluto Press, 2007, p. 120).
32　Ibid., p. 1584, author's translation.
33　Donaldson, Robert/ Nogee, Joseph: The foreign policy of Russia: changing systems, enduring interests, Armonk, NY: M.E. Sharpe, 2005; Zevelev, Igor: NATO's enlargement and Russian perception of Eurasian political frontiers, Final Report to NATO Academic Forum, 2000, http://www.nato.int/acad/fellow/98-00/zevelev.pdf.
34　Worth, Owen: Hegemony, international political economy and postcommunist Russia, Aldershot: Ashgate, 2005.
35　Münkler, Herfried: Empires: the logic of world domination from ancient Rome to the United States, Malden, MA: Polity, 2001, p. 41.

6. Pipelines and Hegemonies in the Caspian: a Gramscian Appraisal 183

Gramsci argues that the formation of a historical bloc necessarily produces a hegemonic relationship between the actors involved. The actors he had in mind in the 1920s and 1930s were the state and the civil society, which became concrete actors in their own right, resembling the Marxian dichotomy of the structure–superstructure. I argue that it is possible to shift the focus towards the relationship between the state, state-owned enterprises (National Oil Companies, or NOCs, in my case) and Transnational Companies (TNCs).

6.3. Methodology

Research for this chapter benefits from both qualitative and quantitative data collected in academic and political circles in the West and early investigations in Kazakhstan. As this is an ongoing project, both the investigation and the findings should be considered preliminary. Here, I present an overview of the practical issues surrounding the Kazakh export pipelines and attempt to frame my theoretical approach in relation to the investigative goals.

The rationale for the choice of the case study is explained by two main reasons. First, my academic interest on the interplay between energy and foreign policy has already channelled my focus on the post-Soviet region, and in earlier graduate work I researched the role of energy in Russian foreign policy towards Kazakhstan.[36] Moved by the same goal of finding a theoretical framework that could fit the analysis of multiple different cases while retaining its validity, I switched the focus to Kazakhstan itself and attempted to examine the issue from a different perspective. As in my previous work, the cases I chose allow for a relatively simple problematisation of the issue, which can then be extended by adding variables and unexplored questions. The proximity of Kazakhstan to its export markets, the relative independence of its customers from Kazakh energy supplies, the blocked political system, the common history with Russia and the shared security threats with China allow for an analysis that is more 'relaxed', less politically charged on the East–West divide, and more closely related to the actual technicalities of energy, from the sulphur content to the return on investment for an oil field. The second reason behind the case choices was my current fieldwork in Kazakhstan, which will undoubtedly enhance my ability to test the hypotheses and gain further insight on the causal events that have led to the construction of the pipelines beyond the official press releases.

A qualitative approach is employed to study hegemonic relationships and establish the historical bloc configurations on which these are established. However, mate-

---

36 Sorbello, Paolo: The role of energy in Russian foreign policy towards Kazakhstan, Saarbrücken: Lambert Academic Publishing, 2011.

rial conditions such as market structures and geological data are quantitative factors that require a thorough analysis before embarking on the qualitative assessment.

6.4. The Caspian, Oil and Foreign Policy

The study of the material conditions upon which hegemonic relationships are formed must start from the legal framework concerning the Caspian waters. This is both a technical question that concerns the use of offshore subsoil resources and a political case explaining Kazakh foreign policy. Throughout this work, I purposefully avoid calling the Caspian a 'sea' because of the unresolved legal question around its status. It is a case of *nomen est omen*: in fact, the difference between defining it as a 'sea' or a 'lake' (or even making an *ad hoc* arrangement) is decisive on the issue of subsoil ownership and use, as well as right of navigation and construction of infrastructures. The question became imminent after the break-up of the Soviet Union, as the number of littoral states increased from two to five and the significance of oil and gas exploration and trade became more relevant for that body of water.

Russia and Iran initially rejected any negotiation that failed to consider the 1921 Russo–Persian Treaty of Friendship and the 1940 Treaty on Commerce and Navigation as starting points. Kazakhstan was eager to subdivide the seabed and surface according to the international convention on the Law of the Sea. Kazakhstan was not invited to the 1993 Astrakhan summit, where Russia, Azerbaijan and Turkmenistan gathered to discuss the status of the Caspian. In response, a consortium of the most active Western companies in Kazakhstan was invited to conduct explorations in the 'Kazakh portion' of the Caspian. This move upset Moscow, which sent a note to the British Embassy[37] stating:

> The Caspian Sea is an enclosed water reservoir with a single ecosystem and represents an object of joint use, within whose boundaries all issues or activities, including resources development, have to be resolved with participation of all Caspian states.[38]

Kazakhstan and Russia had never been more distant on the issue. As negotiations continued, in 1995, Vyacheslav Gizzatov, Deputy Foreign Minister of Kazakhstan at the time, agreed on the consideration of the Caspian as a lake, provided that natural resources were split among the legitimate owners (according to the 1970 Soviet delimitation). Later, Russian Foreign Minister Yevgenii Primakov proposed the establishment

---

37 It is relevant to note that the cable was sent to the British Embassy because of the prominent role of BP in Azerbaijan. However, the timing also suggests that the Kazakh move concurred in the diplomatic row. Alexandrov, Mikhail: Russian–Kazakh contradictions on the Caspian Sea legal status, in: Russian and Euro-Asian Bulletin, February 1998 (old website: http://www.cerc.unimelb.edu.au/bulletin/bulfeb98.htm, accessed 26 September 2010; cached: http://archive.is/www.cerc.unimelb.edu.au).
38 Alexandrov, Mikhail: Uneasy alliance: relations between Russia and Kazakhstan in the post-Soviet era, 1992–1997, Westport, CT: Greenwood Press, 1999, p. 284.

6. Pipelines and Hegemonies in the Caspian: a Gramscian Appraisal 185

of a modified median line for the delimitation of the exclusive economic zone, while Russia retained the right to participate in any new business venture. It was not until 1998, however, that Russia and Kazakhstan (unlike their neighbours) reached an agreement that would partition the northeastern Caspian region into national sectors.[39] In 2002, Presidents Vladimir Putin and Nursultan Nazarbaev reached a further accord on the exploration and exploitation of oil and gas fields offshore, located near the bilaterally agreed-upon border. The 'energy pact' gave all rights to Rosneft' and KazMunaiGaz (KMG), the national oil companies; however, the legal question cannot be considered resolved, as sudden and unexpected changes can occur at any time.

Studying the complexity of the juridical status of the Caspian allows the researcher to isolate the legal variable in the interpretation and analysis of the negotiations between governments and between governments and TNCs. The Caspian region holds significant oil reserves, and although they are not comparable to Middle Eastern reserves, they are judged to be on the same level as the North Sea reserves. Kazakhstan holds the vast majority of these reserves.[40] The lack of a comparable domestic or regional market for the oil extracted provides an 'international' dimension to energy issues. Oil had to be exported either via tanker across the Caspian or via the Soviet-era pipeline until 2001. The new millennium saw the realisation of new infrastructure projects, both private and state-owned, which provided new export avenues via Russia to the West and to China. The international diplomatic situation rendered it less acceptable to sell oil (or gas) to Iran, which was viewed as a favourable option both for Central Asian exporters in terms of diversification and for Iran, whose energy-thirsty centres are located in the north, far from the southern energy-producing regions. The system of 'swaps' (Iran paid for imports with the export of its own oil) was suspended in the mid-1990s, when American opposition to any country entertaining trade relations with Iran increased. In this instance, the positive relationship between the North Atlantic Treaty Organization (NATO) and Kazakhstan through the Partnership for Peace programme played an important role in deterring Kazakhstan from a stronger partnership with Iran.[41]

---

39 For further details, see Gizzatov, Vyatcheslav: Negotiations on the legal status of the Caspian Sea 1992–1996: view form Kazakhstan, in: Akiner, Shirin (ed.): The Caspian: politics, energy and security, New York: Routledge, 2004, pp. 48–60; Sorbello, Paolo: The role of energy in Russian foreign policy towards Kazakhstan, Saarbrücken: Lambert Academic Publishing, 2011, Chapter 5.
40 The Kazakh government forecasts production to reach 120–150mt per year in 2015. IHS Global Insight, CERA: Country Intelligence Report Kazakhstan, 1 October 2012, p. 22.
41 The Kazakh–Iranian oil trade has continued inconstantly, as demonstrated by various independent and academic reports. Energy Charter Secretariat: Oil flows and export capacity in the Caspian Sea and Black Sea regions, Brussels: Energy Charter Secretariat, 2008, p. 13, http://www.encharter.org/fileadmin/user_upload/document/Oil_Flows.pdf; Babali, Tuncay: Prospects of export routes for Kashagan oil, in: Energy Policy, 2009 (vol. 37), no. 4, pp. 1298–1308, here p. 1302.

The 'geopolitics' of energy in the region are exceptionally complex, given the variety in terms of objective factors (geological, technical) and subjective ones (economic capability, business attitude, political objectives, legal frameworks, culture and history). In this chapter, pipelines are at the centre of the analytical focus as a setting where these factors interact. The peculiarity of the Caspian region needs to be ascertained and singled out, although the analysis could prove valuable and applicable to other situations. Several works on 'pipeline politics' or 'pipeline diplomacy' have demonstrated how these particular transport infrastructures have an influence on and are influenced by a country's foreign policy decision-making.[42] The focus of these works is generally the post-Soviet space, especially Russia, if the author wants to prove the direct impact of politics in pipeline construction, or Azerbaijan and Kazakhstan (with Baku–Tbilisi–Ceyhan and CPC pipeline, respectively), if the goal is to demonstrate how the participation of TNCs allows for a less politicised treatment of the pipeline business. Nevertheless, this is a partial and narrow interpretation of the role of pipelines, especially because they are treated as the mere outcome of political strategies (or *coups de maître* made by able businessmen). Here, in contrast, pipelines are treated as both cause and consequence of the dynamic relationship between governments, NOCs and TNCs, through an analysis that takes into account the historical context wherein this interplay takes place.

This work focuses on Kazakhstan because it offers the possibility to analyse the emergence of two different pipelines serving two different consumer markets. Their role in Kazakh foreign policy has been particularly relevant, and I especially consider Astana–Moscow and Astana–Beijing relations. Additionally, Kazakh foreign policy towards TNCs, both Western and Chinese, had to come to terms with the issue of oil transport, given these companies' involvement in several upstream projects. Furthermore, the domestic and internal power dynamic between the Kazakh government and KMG has been shaped, albeit to a lesser extent, by pipeline-related issues.

---

42 Stulberg, Adam N.: Strategic bargaining and pipeline politics: confronting the credible commitment problem in Eurasian energy transit, in: Review of International Political Economy, 2012 (vol. 19), no. 5, pp. 808–836; Ziegler, Charles E.: The energy factor in China's foreign policy, in: Journal of Chinese Political Science, 2006 (vol. 11), no. 1, pp. 1–24; Chow, Edward C./ Hendrix, Leigh E.: Central Asia pipelines: field of dreams and reality, in: NBR: Pipeline politics in Asia: the intersection of demand, energy markets, and supply routes, Seattle, WA: The National Bureau of Asian Research, 2010, pp. 29–42 (Special Report no. 23), http://nbr.org/publications/specialreport/pdf/Free/123113/SR23_Pipeline_Politics.pdf; Dellecker, Adrian: Caspian Pipeline Consortium: bellwether of Russia's investment climate?, Paris: Institut français des relations internationales, 2008, http://www.ifri.org/downloads/ifrirnvdelleckercpcengjuin2008.pdf; Kandiyoti, Rafael: Pipelines: oil flows and crude politics, London: IB Tauris, 2008; Omonbude, Ekpen J.: The economics of transit oil and gas pipelines: a review of the fundamentals, in: OPEC Energy Review, 2009 (vol. 33), no. 2, pp. 125–139.

The scope and space of this work limits the extent of my research, which is focused mainly on the first two relationships.[43]

6.5. Case Studies

To better understand the relevance of oil pipelines for Kazakhstan's foreign policy, the two cases below explore export pipelines in detail. First, the CPC is examined as a ground-breaking infrastructure that allowed Kazakhstan to avoid exporting its oil through the Transneft network in Russia or via shipments to Baku. The fact that the pipeline is privately owned, however, should not divert the attention from the important role played by the Russian government and Russian companies. Then, the attention is shifted eastward to the construction of the KCP. Being a government-sponsored project, its purpose is linked closely to the exporter's energy strategy and the importer's needs in terms of energy mix and geographic diversification.

6.5.1. The CPC

The CPC was a project conceived immediately after the collapse of the Soviet Union. The discovery of oil on the northeastern Caspian shore in 1974 had led to slow and complicated preparation for the extraction and production of oil, given the region's distance from consumption centres and technical difficulties.[44] The slow start to operations and the break-up of the Union were fortunate coincidences for the Western oil companies, which became interested in developing energy projects in the newly independent states. At the same time, the Russian government had not modified its Soviet-style control of its energy enterprises and claimed stakes in the oil field on which Soviet engineers had worked.

From a private-sector perspective, Western TNCs viewed the former Soviet republics in the Caspian region as a new energy 'El Dorado'[45] where they could establish new ventures and revamp their business at a time when the power balance in the

---

[43] That being said, it is the intention of the author to gain more knowledge of this particular relationship once the fieldwork occurs in Kazakhstan. KazMunaiGaz, for example, only owns 30% of Kazakh oil and gas production and is the designated company to deal with new ventures. This makes it a significant actor in the balance between the government and TNCs. Kennedy, Ryan/ Nurmakov, Adilzhan: Resource nationalism trends in Kazakhstan, 2004–2009, Lysaker: Fridtjof Nansen Institute, 2010, p. 10 (RUSSCASP Working Paper), http://www.fni.no/russcasp/wp-nurmakov-kennedy-kazakhstan.pdf; Olcott, Martha B.: Kazmunaigaz: Kazakhstan's national oil and gas company, Houston, TX: Rice University, James A. Baker III Institute for Public Policy, 2007, http://carnegieendowment.org/files/Kaz_Olcott.pdf.
[44] Yessenova, Saulesh: Tengiz crude: a view from below, in: Najman, Boris/ Pomfret, Richard/ Ralaband, Gaël (eds.): The economics and politics of oil in the Caspian basin, London: Routledge, 2008, pp. 176–198, here p. 180.
[45] Yergin, Daniel: The quest: energy, security, and the remaking of the modern world, New York: Penguin, 2011.

energy global market was tilting towards state-owned majors. The American company Chevron demonstrated interest in the Tengiz oil field after having been involved in the initial stages of exploration in the early 1980s,[46] although it ultimately had to confront the Russian competition before being able to secure a contract with the Kazakh government. In this period, Kazakhstan's immature foreign policy and its desire to enter the post-1991 global market economy moulded its decision-making to the interests of Western oil companies. The deal was signed in April 1993 in Almaty by the Kazakh president Nazarbaev and Chevron's Chairman Kenneth Derr, establishing a 50-50 joint venture. Production began shortly afterwards and the extracted oil was transported to the West via Russian pipelines (joining the Russian Transneft system through the Atyrau-Samara pipeline) or via railroad to the Black Sea, from which point it was shipped, mostly to the Baltic region and to China.[47] In 1996, the ownership of the field was modified; the Kazakh government sold half of its shares in the TengizChevrOil consortium, which operates the Tengiz field until 2033, to ExxonMobil and, later in 2000, a smaller portion to LukARCO, formerly a subsidiary of Russian LUKOIL in partnership with American ARCO.[48]

Moreover, TNCs have sought to gain shares of upstream operations in the fields of Karachaganak and Kashagan. After a few years of negotiations, British Gas (BG) together with Italian ENI obtained a lucrative partnership in the Karachaganak gas field in 1995. The final Production Sharing Agreement (PSA) was signed in 1997, bringing Chevron and LUKOIL into the partnership. In 2012, the Kazakh national oil and gas company, KMG, joined the venture with a 10% stake.[49] Today, Karachaganak produces approximately 15% of the total production of liquids (oil and gas condensate) in Kazakhstan.[50] In 2003, a pipeline link between Karachaganak and the CPC was completed.[51]

In July 2000, geological exploration led to the discovery of the largest field in the previous 30 years in an area of the Caspian. Production at the Kashagan field has yet to start, but reserves are estimated at 4.9 billion tonnes of oil.[52] The hardships

---

46 Olcott, Martha B.: Kazmunaigaz: Kazakhstan's national oil and gas company, Houston, TX: Rice University, James A. Baker III Institute for Public Policy, 2007, p. 11, http://carnegieendowment.org/files/Kaz_Olcott.pdf.
47 Kaiser, Mark/ Pulsipher, Allan: A review of the oil and gas sector in Kazakhstan, in: Energy Policy, 2007 (vol. 35), no. 2, pp. 1300–1314, here p. 1303.
48 LukARCO is now a 100% owned subsidiary of LUKOIL. ARCO sold its shares to BP, which in 2009 sold its 46% share to LUKOIL. LukARCO owns 5% of the Tengiz field and 12.5% of the Caspian Pipeline Consortium.
49 The current shareholders are BG and ENI (29.25% each), Chevron (18%), LUKOIL (13.5%), and KMG (10%).
50 Cf. the KPO consortium website, http://www.kpo.kz/production-metrics.html?&L=0 (accessed 30 May 2013).
51 Kaiser, Mark/ Pulsipher, Allan: A review of the oil and gas sector in Kazakhstan, in: Energy Policy, 2007 (vol. 35), no. 2, pp. 1300–1314, here p. 1304.
52 Cf. the KazMunaiGaz website, http://www.kmg.kz/en/manufacturing/upstream/kashagan/ (accessed 30 May 2013).

6. Pipelines and Hegemonies in the Caspian: a Gramscian Appraisal

connected to the project are several: the depths at which the reserves are located, the high-pressure and high 'sour gas' content, and the seasonal ice that freezes the waters around the offshore platforms for about five months every year are all factors that have slowed the development of the field and raised the total cost of the project accordingly, which is now expected to reach US$150–180 billion. The PSA was signed in 1997, when Kazakhstan allowed exploration in the northeastern offshore section of the Caspian. The Offshore Kazakhstan International Operating Company was composed of ENI-Agip, BG, ExxonMobil, TotalFinaElf, and Royal Dutch Shell (each with a share of 16.67%), Inpex and ConocoPhillips (with 8.33% each). BG sought to exit the partnership as early as 2003, but its shares were only sold in 2005 after the passage of a new tax law in Kazakhstan that made it easier for the government to renegotiate and take possession of shares in existing and future contracts. With this move, KMG was able to enter the new agreement with a 8.33% share, which was later increased to 16.81% to match the ownership of the chief partners (ENI, ExxonMobil, Total, and Shell, whereas ConocoPhillips (8.40%) and Inpex (7.56%) slightly modified their stakes in the project after lengthy negotiations.[53] The name of the venture was changed to the North Caspian Operating Company. The start of the production has been delayed several times, bringing into question the role of ENI as an operating company and significantly inflating the costs of the investment.[54]

To link these oilfields to export markets, Kazakhstan, KMG, and the TNCs involved in upstream operation have sought to foster projects other than simply channelling oil to the Atyrau–Samara pipeline and sending it all through the Russian network, owned and operated by Transneft. Negotiations for the CPC started in 1992. Russo–Kazakh negotiations stalled in 1993 over a technical and environmental issue, which proved to be more of a pretext,[55] brought to the fore by Transneft.

The Caspian Pipeline Consortium was established as a structure that divided ownership and operation. The ownership structure was divided between government shares (Russia 24%, Kazakhstan 19% and Oman 7%) and company shares (Chevron 15%, LukARCO 12.5%, Rosneft'[56]-Shell 7.5%, Mobil 7.5%, BG 2%, Agip 2%, Kazakhstan

---

53 Kennedy, Ryan/ Nurmakov, Adilzhan: Resource nationalism trends in Kazakhstan, 2004–2009, Lysaker: Fridtjof Nansen Institute, 2010, (RUSSCASP Working Paper), http://www.fni.no/russcasp/wp-nurmakov-kennedy-kazakhstan.pdf. Babali, Tuncay: Prospects of export routes for Kashagan oil, in: Energy Policy, 2009 (vol. 37), no. 4, pp. 1298–1308.
54 More details are available on the Offshore Technology website on Kashagan: website: http://www.offshore-technology.com/projects/kashagan/ (accessed 30 May 2013).
55 Sorbello, Paolo: The role of energy in Russian foreign policy towards Kazakhstan, Saarbrücken: Lambert Academic Publishing, 2011, p. 43.
56 The Russian government owned Rosneft' in its entirety before establishing itself at 75% in 2006 after an IPO.

Pipeline Ventures[57] 1.75% and Oryx 1.75%). The operation was divided across the border between the Kazakh section (CPC-K) and the Russian section (CPC-R).

The pipeline capacity was initially approximately 28 million tonnes (mt) per year; in 2012, the capacity was increased from 28 to 35mt per year. Total exports have almost reached the capacity in 2007–2011 and have remained somewhat stable since (30.6mt in 2012).[58] Most of the capacity is filled by Kazakh oil. The final export oil is a mix of crudes coming from Tengiz, Karachaganak and other smaller Russian and Kazakh producers that are connected to the pipeline. The mix is called CPC Blend and it pioneered the practice of a quality bank to compensate for the characteristics of the oil that each company or venture delivers into the pipeline system (see Table 6-1 below).[59]

Figure 6-1: The Crude Oils and Their Key Characteristics

Oil characteristics	API gravity	Sulphur content (%)	Volumes (bbl/d)	Main loading port
Kazakhstan				
Aqtobe	41.6°	0.73	120,000	Novorossiisk
CPC blend*	43.3°	0.59	650,000	Yuzhnaya Ozereevka
Karachaganak (condensate)	44.7°	0.81	200,000	Novorossiisk, Odessa
Kashagan	42–48°	0.80	75,000	Ceyhan
Kumkol	41.2°	0.11	200,000	Yuzhnaya Ozereevka, Batumi
Russia				
Siberian Light	35.1°	0.57	100,000	Tuapse
Sokol (Sakhalin I)	37.9°	0.23	40,000	DeKastri
Vityaz (Sakhalin II)	34.6°	0.22	33,000	Molikpaq-Prigorodnoye
Urals	31–32°	0.8–1.8	8,500,000	Primorsk, Novorossiisk, Odessa

Johnson reported the figure of 44.2° for the CPC blend in December 2011. Johnson, Christopher: Kazakhstan: Central Asia's oil and gas powerhouse, in: Reuters, 20 December 2011, http://www.reuters.com/article/2011/12/20/kazakhstan-oil-idAFL6E7NK2Y820111220.

Source: Own elaboration on data from Energy Intelligence Group: The Crude Oils and their Key Characteristics, 2007, http://www.energyintel.com/pages/Eig_Article.aspx?DocId=200017 (accessed: 30 May 2013).

---

57 British Petroleum had minority stakes in both LukARCO (46%) and Kazakhstan Pipeline Ventures (49%).
58 For the volumes of crude shipped at the CPC marine terminal, see CPC's official company: http://www.cpc.ru/EN/shippers/Pages/volumes.aspx (accessed 30 May 2013).
59 Information on the CPC quality bank available in Russian at http://www.cpc.ru/ru/shippers/pages/qb.aspx (accessed: 30 May 2013). It is worth noting that 'Russia has refused to create a quality bank until late 2004, through the facilitation of the 2003 Energy Strategy, in order to gain from blending its own (sour) oil with the Caucasian and Central Asian variety. As an example, the CPC collects and carries oils from the northeastern part of the Caspian (Tengiz) to Novorossiisk. The mixture measured in Astrakhan has an API gravity of 43.3° while the oil extracted in Tengiz has a medium gravity of 48.2°. This is not a small difference, because it means that the CPC blend yields approximately 0.3 barrels of oil per ton less than the Tengiz crude by itself would' (Sorbello, Paolo: The role of energy in Russian foreign policy towards Kazakhstan, Saarbrücken: Lambert Academic Publishing, 2011, p. 40).

6. Pipelines and Hegemonies in the Caspian: a Gramscian Appraisal 191

The management of the CPC has been undertaken by Russian general directors for almost its entire history, with the notable exception of the 2002–2006 period when the mandate was assigned to Chevron's Ian MacDonald. The Russian directors are generally closely connected to Transneft or Zarubezhneft, a Russian oil service and construction company. In 2007, the Russian Federation transferred its share to the state-owned company Transneft and purchased Oman's 7% through the ad hoc Caspian Pipeline Consortium Company, allegedly at a price lower than that offered by Hungary's MOL and Kazakhstan.[60] Moreover, the Russian participation in the consortium is felt through state-owned Rosneft',[61] which owns 51% of the Rosneft'-Shell joint venture, and the private company LUKOIL, which has owned 100% of LukARCO since 2009 when the partnership with BP ended. Thus, almost half the shares are controlled by Russian entities. In short, although the CPC pipeline from Tengiz to Novorossiisk was the first major private energy infrastructure built in the former Soviet Union, Russia has successfully managed to keep it under its control.

Talks of expansion have been constant in the past decade, the goal being to double the capacity to approximately 67mt per year. In this case, the successful ability of the Consortium to fill the pipeline at near-maximum capacity has precipitated the negotiations, while Russia's demands for higher transit fees have slowed down the decision. The project, agreed upon unanimously at the December 2010 stockholders' meeting, was first scheduled for 2012, but construction started only in July 2011,[62] therefore pushing back completion to 2015.[63] As William Simpson of CPC noted recently,

> the investment decision for the 'Future Growth Project', the expansion of CPC, has been made and ship-or-pay contracts have already been defined by the partners. With the coming in line of Kashagan in September 2013, hopes are high for the success of the project.[64]

Transit fees, expressed in dollars per tonne, have gone up since 2004, when they were first raised from US$26.32 per tonne to US$27.19,[65] with a further increase to US$29.50[66]

---

60   Babali, Tuncay: Prospects of export routes for Kashagan oil, in: Energy Policy, 2009 (vol. 37), no. 4, pp. 1298–1308, here p. 1301.
61   Rosneft' delivers oil and gas condensate through the pipeline since the end of 2004. It has exported an average of 6.3mt per year in the period 2008–2010 (http://www.rosneft.com/Downstream/crude_oil_sales/gas_condensate_exports/Export_terminals/Caspian_Pipeline_Consortium/, accessed 30 May 2013).
62   IHS Global Insight, CERA: Country Intelligence Report Kazakhstan, 1 October 2012, p. 24.
63   Şaban, Ilham: Kazakhstan's oil transportation via CPC to be increased by 7.5%, in: CaspianBarrel.org, 29 May 2013, www.caspianbarrel.org/index.php/en/2012-08-14-15-04-20/960-kazakhstan-s-oil-transportation-via-cpc-to-be-increased-by-7-5. Details can be found also on Transneft's official website, http://eng.transneft.ru/projects/119/10203/ (accessed 30 May 2013).
64   Interview by the author with William Simpson, Deputy General Director, Engineering and Projects, Caspian Pipeline Consortium, conducted in Almaty on 3 October 2013 during the 21st Kazakhstan International 'Oil & Gas' Exhibition and Conference.
65   Şaban, Ilham: Russia to increase profitability of Baku–Tbilisi–Ceyhan, in: Azerbaijan Today, 31 October 2005, http://bakudot.blogspot.com/2005_10_01_archive.html.
66   Kandiyoti, Rafael: Pipelines: oil flows and crude politics, London: IB Tauris, 2008, p. 187.

and a spike to US$38, which represented the last significant increase since 2007.⁶⁷ Being the principal beneficiary of higher fees, Transneft has had a counterproductive role in facilitating the negotiations for the expansion of the CPC primarily because it was not satisfied with the pace of the return on its investment in the pipeline. The prospect of a fee hike that would decrease the competitiveness of Kazakh oil (mostly extracted by the private companies of the Consortium) strangled the enthusiasm of the TNCs; however, the delay in the expansion prevented Russia from gaining larger inflows of money simply from the increased volume of shipped oil. Russia's veiled intention of using this tug-of-war over the expansion is to show its decisive role in the venture, even if the private business partners are the ones responsible for producing and shipping most of the CPC oil.⁶⁸

In the historical bloc that began forming in 1991, Kazakhstan realised the hegemonic project crafted by private multinational companies, significantly improving the business environment in terms conceived of by global capitalist institutions and rating agencies. The construction of the CPC under private ownership and the favour towards the rapid privatisation of the oil industry are two aspects of the same hegemonic configuration that characterised the 1990s.

6.5.2. The KCP

With the turn of the century, the domestic political situation, the agreement to form the Shanghai Cooperation Organisation, the 9/11 attacks in the United States, and the advent of Vladimir Putin as president of the Russian Federation had a sizeable impact on the process of the waning of 'private hegemony' on Kazakh foreign energy policy. Suddenly, the political discourse shifted towards the re-nationalisation of energy resources (mainly oil fields) and the investment climate worsened. New laws on the extractive sector were passed (Tax Code in 2001 and PSA legislation in 2005), leaving the existing three PSAs under threat of renegotiation.⁶⁹ In addition, the government consolidated the entire oil and gas sector under KMG in 2002. Finally, Chinese state-owned companies started successfully bidding to participate in new or existing oil and gas fields in Kazakhstan. Their welcome by Astana and the gradual ousting of a few Western companies was a sign of the changed material conditions under which oil politics influenced foreign policy. These new conditions, the new historical bloc,

---

67 Rodova, Nadia: CPC shareholders reach agreement to expand crude pipeline, in: Platts Oilgram News, 18 December 2008 (vol. 86), no. 250, p. 4.
68 Dellecker, Adrian: Caspian Pipeline Consortium: bellwether of Russia's investment climate?, Paris: Institut français des relations internationales, 2008, http://www.ifri.org/downloads/ifrirnvdelle ckercpcengjuin2008.pdf.
69 Maniruzzaman, A.F.M.: National laws providing for stability of international investment contracts: a comparative perspective, in: Journal of World Investment and Trade, 2007 (vol. 8), no. 2, pp. 1–9.

6. Pipelines and Hegemonies in the Caspian: a Gramscian Appraisal

created a hegemonic configuration by which resource nationalism and state capitalism overtook free-market and globalist orientations in the foreign policy decision-making circles of Kazakhstan.

The pipeline from Atasu to Alashankou is the main part of the KCP and has the ambitious goal of linking the Kazakh oil fields in the Caspian region with the Xinjiang region in China. Only two months behind schedule, Kazakh oil reached the western Chinese border via the Atasu–Alashankou pipeline at the end of July 2006. The 962.2 kilometre (km) long pipeline has a capacity of approximately 10mt per year. At full capacity, the pipeline only delivers approximately 2–2.4% of China's increasing oil demand. Nevertheless, it was considered an important project for diversification of imports and exports by both countries. The initial intergovernmental agreement was reached in September 1997, the partnership between KazTransOil and China National Oil and Gas Exploration and Development Corporation was established in June 2004 and in December 2005 the pipeline was officially opened by Nursultan Nazarbaev.[70]

The second stage of the KCP, the 794km Kenkiyak–Kumkol pipeline, was completed in 2009. Its capacity mirrors the Atasu–Alashankou section and talks of an expansion of both pipelines occurred in 2010 and 2011 (although the current throughput of the Kenkiyak–Kumkol pipeline is only 5.4mt per year).[71] This pipeline is linked to the existing infrastructure that sent oil from the Aktobe region to Atyrau via the 448.5km Kenkyiak–Atyrau pipeline, which has a capacity of 6mt per year and was completed in 2004. Kenkyiak can be considered the true hub of Kazakh oil transport because it is also connected to the production in the Aktyubinsk Region via the Alibekmola–Kenkyiak pipeline (3.6mt).[72] This network was built with the purpose of feeding west- and northward exports via either the CPC or the Atyrau–Samara pipeline. Experts consider that the flow of these pipelines could be reversed to serve the KCP, although several technical shortcomings could hinder flow reversion.[73]

Along with agreements on pipelines, in fact, the Chinese National Petroleum Company (CNPC) secured 60% of the shares of AqtobeMunaiGaz (AMG), which retained licenses for the exploitation of the Kenkiyak oil field as well as the Zhanazhol gas field.

---

70   Yermukanov, Marat: Atasu–Alashanko pipeline cements 'strategic alliance' between Beijing and Astana, in: Eurasia Daily Monitor, 3 January 2006 (vol. 3) No: 1, http://www.jamestown.org/single/?no_cache=1&tx_ttnews%5Btt_news%5D=31239.
71   Kazakhstan–China Pipeline Group: International financial reporting standards consolidated financial statements and independent auditor's report, 1 February 2013, http://kcp.kz/en/infor mation/.
72   Cf. the Stroitransgaz company website, http://www.stroytransgaz.com/projects/kazakhstan/kenkiak-atyrau, and the Kaztransoil company website, http://www.kmg.kz/en/manufacturing/oil/keniyak_atyrau/ (both accessed 30 May 2013).
73   Eurasia Group: China's overseas investments in oil and gas production. Report prepared for the US–China Economic and Security Review Commission, New York: Eurasia Group, 2006, p. 16, http://www.uscc.gov/Research/china%E2%80%99s-overseas-investments-oil-and-gas-produc tion.

Apart from the capital buyout, China met its promise to invest in the oil field development and to employ Kazakh workers. The successful operation pleased the Kazakh authorities, who did not oppose the further purchase of a 25% in AMG, agreed upon in 2003 for additional US$150 million. Without considering Kashagan, AMG is responsible for one-seventh of the total Kazakh oil production. In 2006, it produced 10mt of oil and 1.5 billion cubic metres of gas.[74]

The same idea was behind the Chinese acquisition of the Uzen field, which proved unprofitable due to the lack of substantial reserves and connection with the pipeline system. Only two years after CNPC secured the tender in 1997, Beijing pulled the plug and CNPC had to resell its stakes to KMG. In addition to unsuccessful ventures, China also experienced indirect pressure from the Kazakh government on a number of issues. The first concerned the deal for the Buzachi field, owned by Saudi and American companies before the takeover by CNPC in 2003, which was stopped and reversed under pressure from the government. This led to the entry of a LUKOIL–Mittal joint venture alongside the Chinese giant. This row occurred even though the field yields only small amounts of oil (between 0.5mt in 2007 and 2.2mt in 2012) and its production is exported west through the Atyrau–Samara pipeline.[75]

The second issue between Kazakhstan and China is the one that resonated the most during the past decade. Canadian Hurricane Petroleum had found significant reserves in the Kumkol region. The company, renamed Petrokazakhstan, was under pressure from the Kazakh government beginning in 2003, when fines and legal actions were directed against the management of the company.[76] The buyout of Petrokazakhstan in 2005 was possible only after the Chinese side bowed to the new oil law, which granted the Kazakh government the right to purchase up to 50% of any domestic venture. The price that CNOC had to pay (US$4.2 billion and half of Petrokazakhstan's shares in the Chimkent refinery) was regarded as too high by outside observers and represented the biggest sum ever paid by a Chinese company for a foreign acquisition.[77]

All in all, through the purchase of upstream operations and the construction of the KPC pipeline, China has managed to secure a working collaboration with the Kazakh government and its NOCs, diverting the heavy focus of Kazakh exports to and through Russia. The Chinese interest in Kazakh oil is justified by the increasing need of energy supplies for its supersized economy. This goes hand in hand with the new hegemonic configuration, through which Kazakhstan tried to diversify its foreign policy options

---

74 Peyrouse, Sébastien: Economic aspects of the Chinese–Central Asia rapprochement, Stockholm/Washington, DC: Central Asia-Caucasus Institute & Silk Road Studies Program, 2007, p. 51, http://www.silkroadstudies.org/new/docs/Silkroadpapers/2007/0709China-Central_Asia.pdf.
75 Ibid., p. 53.
76 Ostrowski, Wojciech: Politics and oil in Kazakhstan, London: Routledge, 2009, p. 147.
77 Peyrouse, Sébastien: Economic aspects of the Chinese–Central Asia rapprochement, Stockholm/Washington, DC: Central Asia-Caucasus Institute & Silk Road Studies Program, 2007, pp. 53–54, http://www.silkroadstudies.org/new/docs/Silkroadpapers/2007/0709China-Central_Asia.pdf.

by granting selective access to its oil wealth and re-gaining sovereignty over a number of oil fields in its territory. Kazakhstan was practising what in the 1990s was heralded as a 'multi-vector foreign policy'.[78] If during the 1990s, Kazakhstan was only juggling between Russia, the 'God-chosen partner and neighbour',[79] and Western TNCs, the situation changed in the 2000s. Already in 1997, President Nazarbaev said that China was poised to be 'the 21st century frontier'.[80]

6.6. Bridging Theory and Practice

Applying the Gramscian concepts of historical bloc and hegemony, one could try to track the dynamics involving the construction of both pipelines. The complete isolation of the two 'moments' is impossible given that Kazakhstan's relations with its neighbours do not happen in a historical void. One of the major tenets of Kazakh foreign policy since independence is its multi-vector approach. Although preserving the willingness to maintain friendly relations in the Eurasian region, President Nazarbaev strived to build new relationships with the West and China. The policy's effectiveness, after more than 20 years of independence, has not followed a clear path.[81] In the 1990s, the strongest external forces interested in Kazakh oil were Western TNCs, which had strong expertise and were eager to embark in new ventures to exploit the untapped riches of the post-Soviet oilfields. The material conditions were favourable to their ventures, as weak legal frameworks on ownership allowed them to acquire many promising oil and gas deposits. In addition, the export of natural resources allowed for non-transparent practices and lucrative contracts.[82]

Once the production sites were seized, Western oil companies lobbied for the construction of an export pipeline that was independent of Transneft's monopoly (the CPC). Kazakh oil exports were still reliant on the Soviet-era pipeline network and shipments via other means (tanker, rail) that would satisfy a smaller percentage of the volumes than the foreign ventures envisioned extracting. Russia, with its uncertain foreign policy stance for the most part of the 1990s, alternated a favourable position with

---

78 Kasenov, Oumirseric: The institutions and conduct of the foreign policy of postcommunist Kazakhstan, in: Dawisha, Adeed/ Dawisha, Karen (eds): The making of foreign policy in Russia and the new states of Eurasia, vol. 4, Armonk, NY: M.E. Sharpe, 1995, pp. 263–285.
79 Mansurov, Tair: Kazakhstan i Rossiya: suverenizatsiya, integratsiya opyt strategicheskogo partnerstva, Moscow: Russkii Raritet, 1997.
80 Nazarbayev's intervention, just one week before transferring the capital to Astana (then Akmola), at the 'Almaty Summit of the CCCBMA', in: Interfax Kazakhstan, 4 December 1997, (cited in Black, Joseph L.: Russia and Eurasia documents annual: 1997 (vol. 2), Central Eurasian States, Gulf Breeze, FL: Academic International Press, 1998, p. 240).
81 İpek, Pinar: The role of oil and gas in Kazakhstan's foreign policy: looking east or west?, in: Europe-Asia Studies, 2007 (vol. 59), no. 7, pp. 1179–1199.
82 Cf. Kleveman, Lutz: The new great game: blood and oil in Central Asia, New York: Grove Press, 2004; Ostrowski, Wojciech: Politics and oil in Kazakhstan, London: Routledge, 2009.

retaliatory actions, such as its uncompromising position on the status of the Caspian and Transneft's fabricated accusation over the environmental impact of the pipeline.[83]

With the formation of KMG, the amendment of the legislation for the extractive sector, and the new business configuration in the Kumkol region, the way was paved for the construction of the KCP, concretely signalling the new material conditions that formed the historical bloc of the 2000s. The new historical bloc is one in which the hegemonic idea is no longer free-market capitalism. Instead, the ideology that has shaped the last decade is one of 'state capitalism'. Domestically, it is a strategy that presupposes the control of key industries to ensure that economic growth is accompanied by political stability.[84] The key industry in Kazakhstan is the oil sector and this led to the analytical identity of the concept of 'state capitalism' with that of 'resource nationalism'.[85] For the purpose of this work, the two concepts remain separate. According to Bremmer, 'state capitalism' is typically a non-western approach,[86] which, in the case considered, signals the radical shift in the hegemonic force leading the new historical bloc. The case of Kazakhstan seems consistent with the Gramscian standpoint that hegemony is consolidated through consent, rather than imposition or domination.

6.7. Conclusion

In the formation of the historical bloc, Gramsci stressed the importance of ideas. It is thus relevant to note that Kazakhstan historically was distant from the original

---

83 Sorbello, Paolo: The role of energy in Russian foreign policy towards Kazakhstan, Saarbrücken: Lambert Academic Publishing, 2011, p. 43.
84 Hanson, Phil: The turn to statism in Russian economic policy, in: International Spectator, 2007 (vol. 42), no. 1, pp. 29–42. Ostrowski, Wojciech: State capitalism: an emerging regime, Dundee: POLINARES, 2012 (Working paper no. 51), http://www.polinares.eu/docs/d4-1/polinares_wp4_chapter1.pdf.
85 It is not the purpose of this work to explore or measure the extent to which actions of resource nationalism have been carried out and whether these are in line with aspects of 'state capitalism'. The question of 'resource nationalism' in Kazakhstan has been researched by several authors: Kalyuzhnova, Yelena/ Nygaard, Christian: Resource nationalism and credit growth in FSU countries, in: Energy Policy, 2009 (vol. 37), no. 11, pp. 4700–4711; Kalyuzhnova, Yelena: Economics of the Caspian Oil and Gas Wealth: Companies, Governments, Policies, Basingstoke: Palgrave Macmillan, 2008; Domjan, Paul/ Stone, Matt: A comparative study of resource nationalism in Russia and Kazakhstan 2004–2008, in: Europe-Asia Studies, 2010 (vol. 62), no. 1, pp. 35–62; Tussupov, Nurlan: The policy of resource nationalism: the case of Kazakhstan, in: OGEL, 2010 (vol. 8), no. 2, http://www.ogel.org/article.asp?key=3022; Kennedy, Ryan/ Nurmakov, Adilzhan: Resource nationalism trends in Kazakhstan, 2004–2009, Lysaker: Fridtjof Nansen Institute, 2010, p. 10 (RUSSCASP Working Paper), http://www.fni.no/russcasp/wp-nurmakov-kennedy-kazakhstan.pdf; Sarsenbayev, Kuanysh: Kazakhstan petroleum industry 2008–2010: trends of resource nationalism policy?, in: Journal of World Energy Law and Business, 2011 (vol. 4), no. 4, pp. 369–379. For a critique, see: Jones Luong, Pauline: Beyond 'resource nationalism': implications of state ownership in Kazakhstan's petroleum sector, in: PONARS Eurasia Policy Memos, no. 98, April 2010, http://www.ponarseurasia.org/sites/default/files/policy-memos-pdf/pepm_098.pdf.
86 Bremmer, Ian: The end of the free market: who wins the war between states and corporations?, in: European View, 2010 (vol. 9), no. 2, pp. 249–252.

Marxist-Leninist influences. Its map was drawn by Stalin and two powerful leaders have ruled it for the past 53 years, with two minor interruptions. Closer to a Brezhnevian populist conservatism than to Stalin's totalitarian regime, the regimes of Dinmukhammed Kunaev and Nursultan Nazarbaev have had a significant influence on society. Each had his own type of personality cult, and economic growth and tight control over the opposition has hindered any real debate over a political (or ideological) alternative. In the post-Soviet period, Nazarbaev was quick to embrace free-market and capitalist ideas as he realised that the profit from oil rents could strengthen his position at the vertex of the Kazakh political system.

A similar approach was taken by Yeltsin's Russia, which facilitated the inflow of foreign capital and prompted the interest of Western oil companies. This convergence of interests found common ideological grounds and flourished, albeit with occasional rows over who was to receive the larger share of the pie. However, the historical bloc that formed with the marriage of Western capital to underground resources led to a peaceful and consensual support for Western practices on the part of the Kazakh leadership. In 1997, before being demoted and forced into exile, Prime Minister Akezhan Kazhegeldin wrote the book 'Kazakhstan: Entering the Future'[87] on the benefits of free-market and liberalism that Kazakhstan should slowly adopt. In the transition period, he argues, Kazakhstan should implement a 'mixed economy' that takes into account the country's specific characteristics. Kazhegeldin was nominated reformer of the year by the Adam Smith Institute in London in 1996, before being ousted on charges of money laundering in 1998.[88]

Other instances of internal clashes, transnational influence, technical difficulties and diplomatic negotiations have contributed to the shift in the material conditions, which has arguably led to the formation of a new historical bloc after the turn of the century. The Kazakh government consolidated its energy enterprises into a giant national company, established an oil fund, and substantially modified legislation dealing with hydrocarbon resources. From this stronger position, it reshaped its relationship with China. It is crucial to note the new regional situation in term of security cooperation (i.e., Shanghai Cooperation Organisation) and trade, with China eroding the shares of the Central Asian market previously pertaining to Russia. This has led to new relations in terms of upstream operations (PetroKazakhstan) and to the success of the intergovernmental negotiations over the construction of the eastward KCP pipeline. The argument is that the TNCs and Russia have lost their material capability to trigger Kazakh consent, thus providing legitimacy to their projects. The surge of Kazakhstan as a solid actor, capable of shaping the social configurations (from contracts to technical expertise) that concern its energy policy, had a critical impact on its foreign policy output.

---

87 Kazhegeldin, Akezhan: Kazakhstan: entering the future, London: British Academic Press, 1997.
88 Ostrowski, Wojciech: Politics and oil in Kazakhstan, London: Routledge, 2009, p. 128.

Part III. The Southern Energy Corridor

Rufat Rustamov

7. The Trans-Adriatic Pipeline and Nabucco West Pipeline Projects: Advantages and Disadvantages for Azerbaijan

7.1. Introduction

The rising demand for and scarcity of natural gas reserves in Europe, as well as the European Union's (EU) dependence on a single dominant supplier, have led the European Commission (EC) to enact the Southern Energy Corridor. In 2008, the EC used the term 'Southern Gas Corridor' for the first time (in the EU Energy Security and Solidarity Action Plan) to describe infrastructure projects that would bring gas from sources in the Caspian Sea and the Middle East to the EU with the aim of improving the gas supply security. In the document, the EC argued that the development of the corridor was one of the EU's highest energy security priorities.[1]

On 13 January 2011, the EC President, José Manuel Barroso, and the President of Azerbaijan, Ilham Aliyev, signed a Joint Declaration on the Southern Corridor committing the two parties to the development of the corridor and engaging the Republic of Azerbaijan as a substantial contributor to (and enabler of) the initiative.[2] Azerbaijan is currently the only country that has been identified as a potential natural gas supplier for the Southern Gas Corridor. Azerbaijan is not a major player in the world's gas scene, but Azerbaijani production is expected to grow considerably in the next few years as a result of Azerbaijan's expected commencement of the operation of the Shah Deniz field's second phase, possibly by 2018. In December 2011, the Azerbaijani and Turkish governments signed a memorandum of understanding on the establishment of a consortium comprising the State Oil Company of the Azerbaijan Republic (SOCAR), the Turkish state-owned TPAO energy firm, and TPAO's pipeline subsidiary, BOTAS, to construct the Trans-Anatolia gas pipeline (TANAP).

This project was the starting point for the further development of the Southern Gas Corridor and boosted the development of additional natural gas pipeline projects, such as the Trans-Adriatic Pipeline (TAP) and Nabucco West. Subsequently, the Shah Deniz Consortium, in which SOCAR is a shareholder, considered having both the TAP and Nabucco West pipeline projects connect all planned natural gas pipelines and provide European consumers with natural gas.

---

1   European Commission: Second Strategic Energy Review: an EU energy security and solidarity action plan, COM/2008/781 final, 13 November 2008, p. 4, http://eur-lex.europa.eu/LexUriServ/LexUriServ.do?uri=COM:2008:0781:FIN:EN:PDF.
2   Joint Declaration on the Southern Gas Corridor, Baku, 13 January 2011, http://ec.europa.eu/energy/infrastructure/strategy/doc/2011_01_13_joint_declaration_southern_corridor.pdf.

This chapter focuses on identifying the advantages and disadvantages of the two natural gas export pipeline projects, TAP and Nabucco West, for Azerbaijan. Special priority is given to identifying the foreseeable perspectives for Azerbaijan in the event of the construction of either of these pipelines. What challenges and difficulties could face this young independent country with vast energy resources?

7.2. Azerbaijan and Natural Gas Pipeline Projects

After the demise of the Soviet Union, Azerbaijan assumed a new role in the Caucasus, which serves as a corridor between Asia and Europe, linking the Caspian Sea region and Central Asia with Europe. Subsequently, Russia, the United States (US) and other Western countries began expressing their interest in the region. The US's long-term strategy has been to ensure the independence of Azerbaijan and the other Caucasus countries while sustaining democratisation and promoting regional integration. The more democratised the three southern Caucasian countries are, the fewer chances there will be for the newly emerged states to fall under Russian influence.[3]

Upon gaining independence, Azerbaijan acquired sovereignty over its natural resources and began to attract the attention of energy-importing countries. At this time, the Azerbaijani government was certain that the possession of hydrocarbon resources was an advantage in rapidly developing relations with Western countries. However, there were particular difficulties in establishing a route for transporting hydrocarbons from the landlocked Caspian Sea region to international markets. The fact that the Caspian Sea has no access to the sea has always been perceived as a problem with regards to distributing oil and natural gas to countries seeking those hydrocarbons. In a 1997 speech, Zbigniew Brzezinski outlined how important it would be for the American government to support the construction of multiple pipelines in the region, saying that 'such economic interrelationships would create a greater degree of shared interest in peaceful accommodation'.[4]

Furthermore, Azerbaijan continued this policy strategy by partnering with Turkish and American companies to construct the Baku–Tbilisi–Ceyhan (BTC) oil pipeline and later the Baku–Tbilisi–Erzurum (BTE) gas pipeline (also known as the South Caspian Pipeline, or SCP), which allowed Azerbaijan to export its growing oil and gas output to world markets without passing through Russia or Iran. Russian control of Caspian Sea energy resources began to weaken with the opening of the Southern Energy Corridor (which, to a great extent, was supported by the US President Clinton and his

---

[3] Inessa Baban/ Zaur Shiriyev: The U.S. South Caucasus strategy and Azerbaijan, in: Turkish Policy Quarterly, 2010 (vol. 9), no. 2, pp. 93–103, here p. 93.
[4] Brzezinski, Zbigniew: The Caucasus and new geo-political realities: how the West can support the region, in: Azerbaijan International, 1997 (vol. 5), no. 2, pp. 42–45, http://www.azer.com/aiweb/categories/magazine/52_folder/52_articles/52_caucasus.html.

7. The Trans-Adriatic Pipeline and Nabucco West Pipeline Projects 203

administration), allowing Azerbaijani oil and natural gas exports to reach the West while bypassing Russian territory.[5] Azerbaijan, Georgia and Turkey were connected by the BTC oil pipeline, whose construction (from an American perspective) established a solid axis linking these three countries.[6] Thus, determining the route of a pipeline becomes not only an economic but also a geopolitical issue.

These pipeline projects helped Azerbaijan to foster integration with Europe by attracting the attention of the EU and the US to the region and helped to tighten Azerbaijan's relations with Turkey and Georgia. The mass media frequently stressed the role of Azerbaijan for the energy security of Europe. Azerbaijan, with the aid of its energy resources, was initiating a balancing policy in the region, having stable relations with Russia and Iran and at the same time cooperating with the EU and US.

The EU itself, by launching the Southern Gas Corridor, supported the further political and economic development of the Caucasus region with the aim of lessening the EU's dependency on a single supplier of natural gas. To maintain a stable and lasting supply of natural gas, the EU will have to cooperate with new supplier countries and guarantee the security and protection of pipelines in the Caspian region as much as possible. Azerbaijan's energy resources helped to strengthen relations with the EU and the US and to enhance the support of Western countries for the realisation of the country's huge energy projects both domestically and internationally.

According to BP's statistical review of world energy in 2012, along with oil reserves, Azerbaijan possesses nearly 1.3 trillion cubic metres in natural gas reserves. Shah Deniz, the main natural gas field in Azerbaijan, is one of the world's largest gas-condensate fields, with nearly 1 trillion cubic metres of gas. It is located on the deep-water shelf of the Caspian Sea, 70 kilometres (km) southeast of Baku. BP started operating the Shah Deniz gas fields on behalf of its partners in the Shah Deniz Consortium.[7] In March 2007, for the first time, Azerbaijan started to export 11 billion cubic metres (bcm) of natural gas from Shah Deniz I to Georgia; in July 2007, Azerbaijan also began to export natural gas to Turkey. However, the EU is attempting to gain access to gas from Shah Deniz's second stage. In the case of Shah Deniz II, field exploration is expected to extract approximately 16bcm of natural gas per year. In June 2012, the prospects of the exploration of vast natural gas reserves brought the Azerbaijani and Turkish presidents and the heads of the respective oil firms to sign accords to build TANAP, which will cross

---

5   Baban, Inessa/ Shiriyev, Zaur: The U.S. South Caucasus strategy and Azerbaijan, in: Turkish Policy Quarterly, 2010 (vol. 9), no. 2, pp. 93–103, here p. 97.
6   Nuriyev, Elkhan: The South Caucasus at the crossroads. Conflicts, Caspian oil and great power politics, Berlin: LIT, 2007, p. 288.
7   The shareholders of this consortium are BP and Statoil (with a share of 25.5% each), SOCAR, LUKOIL, Total and Naftiran Intertrade Co. (with 10% each), and Türkiye Petrolleri Anonim Ortakl (TPAO) with a 9% stake (see http://www.bp.com/shahdeniz/stage1).

Turkey from east to west and will transport Caspian gas in the direction of Europe.[8] Regarding supply capacity, Turkey and Azerbaijan reached an agreement for 16bcm of Shah Deniz II gas: 6bcm will be purchased by Turkey and 10bcm will be transported to European markets through Turkey.[9]

The Shah Deniz gas field is very important, not only because of its size but also because the opening of the Southern Corridor unlocks the entire Caspian Sea reserves. As EU Energy Commissioner Günther Oettinger stated in his speech in Frankfurt,

> the 10bcm annual gas that the EU will import from Shah Deniz II is small compared to the overall needs of the EU. This quantity could not justify all the efforts put forward by the EU, but with the Shah Deniz gas, the Southern Corridor will open access to gas supplies from the Caspian Region.[10]

The realisation of these natural gas pipeline projects stretching to Europe helped Azerbaijan to promote a balanced policy toward Russia and Western countries in the Caucasus region, and this co-operation was intended to attract their attention to related local problems.

7.3. Trans-Adriatic and Nabucco West Pipeline Projects

According to the PRIMES models, there is little expected growth in European gas demand up to 2030. Annual gas demand is expected to increase from 527–531bcm in 2010 to 479–538bcm in 2020 and to then fall to 457–510bcm in 2030.[11] Subsequently, the EU has had to address two outstanding challenges in recent decades: negotiating with supplier countries to obtain sufficient amounts of natural gas for Europe at reasonable prices and promoting the proliferation of new supply routes. One of the most important processes in the path to the realisation of the Southern Gas Corridor was the consideration by the Shah Deniz Consortium of the two rival gas pipeline projects. The Consortium has begun evaluating binding transportation offers from Nabucco West and TAP consortia, offering to carry Azerbaijani gas into Europe by the

---

8   Rzayeva, Gulmira/ Tsakiris, Theodoros G.R.: Azerbaijani gas strategy and the EU's Southern Corridor, Baku: Strategic Imperative, 2012, p. 10, http://www.eliamep.gr/wp-content/uploads/2012/08/tsakiris1.pdf.
9   'Turkey's energy strategy', http://www.mfa.gov.tr/turkeys-energy-strategy.en.mfa (accessed 26 May 2013).
10 Günther Oettinger, Keynote Speech delivered at the First Annual Frankfurt Gas Forum, Frankfurt, Germany, 29 November 2012, quoted in: Livanios, Anthony: The conundrum of the Southern Gas Corridor: what are the risks for Europe and Azerbaijan? The viewpoint of an insider. Paris: Institut Français des Relations Internationales, 2013, p. 3, http://www.naturalgaseurope.com/pdfs/IFRI_actuelleslivanios17413.pdf.
11 Matt MacDonald: Supplying the EU natural gas market: final report, Croydon: Matt MacDonald, 2010, http://ec.europa.eu/energy/international/studies/doc/2010_11_supplying_eu_gas_market.pdf.

7. The Trans-Adriatic Pipeline and Nabucco West Pipeline Projects

end of March, and the Shah Deniz Consortium finally considered the proposals at the end of June 2013.[12]

The Nabucco West pipeline is the scaled-back proposal following Nabucco's original route, which was proposed a few years ago, running from Turkey across Bulgaria, Romania and Hungary to deliver gas to the EU. The original Nabucco pipeline was supposed to be more than 3,000km in length and to transport 31bcm of gas from Central Asia, the South Caucasus and the Middle East. However, the geopolitical situation and the complete absence of export routes to Europe in these regions have compromised the realisation of the original Nabucco project.

The Nabucco West project offered direct and immediate diversification to countries in Central and Southeastern Europe that need another source of natural gas besides Russia, especially Bulgaria (which imports 89% of its gas from Russia), Hungary (57%), Romania (23%) and Austria (67%). After the construction of short- and medium-distance interconnectors, Slovakia and the Czech Republic could join that list of countries. Nabucco West clearly advances US foreign policy interests, directly providing energy to the countries in Central and Southeastern Europe with the greatest need to diversify away from Russian supply. As the pipelines are currently planned, Nabucco West is clearly superior to TAP for US foreign policy interests in the region.[13] Most critically for the situation in this region, Nabucco West could introduce international competition, which would improve the consumer countries' negotiating position with Russia, reduce the potency of supply disruption threats, and bolster the internal stability of the North Atlantic Treaty Organization's (NATO) friends and allies.

Russian influence and its role in this region are rigid, so any interests of the Shah Deniz Consortium in the Central and Southeastern Europe gas market face confrontation with Russia. The Russian state-owned gas monopoly Gazprom, backed by Russian foreign policy, is working hard on the construction of the South Stream gas pipeline, which is a direct rival to Nabucco West. The South Stream gas pipeline is planned to connect Russia and the Balkans through the Black Sea, pumping Russian gas to Southern and Central Europe, with Bulgaria as the main gas distribution centre, and further transit will transverse Serbia, Hungary, and Slovenia, ending in Austria.[14] The consideration of the Nabucco West pipeline project by the Shah Deniz Consortium has kept Azerbaijani–Russian relations tense.

---

12 'Shah Deniz Consortium begins evaluating binding transportation offers from Nabucco West and TAP', in: Trend.az, 2 May 2013, http://en.trend.az/capital/energy/2146122.html.
13 Committee on Foreign Relations, United States Senate: Energy and security from the Caspian to Europe, Minority report, 112th Congress, second session, 12 December 2012, pp. 26–29, http://www.foreign.senate.gov/publications/download/energy-and-security-from-the-caspian-to-europe.
14 Cf. South Stream company information, http://www.south-stream.info/en/pipeline/ (accessed 28 May 2013).

While considering the implementation of the Nabucco West project, some international companies within the Shah Deniz Consortium did not want to worsen their relations with Russia. However, from the perspective of gas supply, the markets along the route of Nabucco West are small, and their gas import volumes are limited. It is very difficult to foresee the entire Balkan markets being able to absorb more than 10bcm per year in new gas supplies beyond 2020. Moreover, the construction of additional interconnectors to the Balkans and the allocation of new sources of shale gas reserves could eventually diminish the final netback price of the natural gas in this region. New discoveries of shale gas on the Romanian coast of the Black Sea could introduce some concerns about the market's ability to support this project. According to US Energy Information Administration estimates, Romania, Bulgaria and Hungary might have 538bcm of shale gas between them, which is slightly more than Europe's annual consumption and is sufficient to meet Romania's demand for almost 40 years.[15] As declared repeatedly by Azerbaijani authorities, energy projects implemented with its participation were considered on the basis of national interests and economic efficiency.[16] This includes consideration from various political and economic points of view. Thus, the Nabucco West project was not attractive enough to Azerbaijan and to the Shah Deniz Consortium, in which SOCAR is a shareholder.

The TAP project is an important part of the Southern Gas Corridor route, and this project was more attractive to the Shah Deniz Consortium than Nabucco West. TAP is actually shorter and less costly than Nabucco West. The project was designed to be extended from Turkey to Greece and Albania, under the Adriatic Sea, to Italy, and then finally to a distribution hub in Austria; thus, it requires less complex engineering than Nabucco West and has an initial capacity of 10bcm of natural gas. The short distance for such a scaled project means less effort in final financing, and such a project would be less affected by changes in gas prices. From the economic perspective of the TAP project, Azerbaijan will receive more revenue than from the Nabucco West project. The gas price on the Italian hub is 20% higher than in Western Europe and approximately 5% higher (depending on seasonal fluctuations) than in Baumgartner in Austria (the end point of Nabucco West). TAP will be well connected to the Italian Trans-Alpine gas pipeline and will be capable of reaching other Central European markets (as well as other major consumers in Switzerland and France) through Germany. These markets have the ability to absorb several tens of billions of cubic metres in the long term.[17]

---

15 Patran, Ioana: Chevron plans to start Romania shale gas exploration, in: Reuters, 9 May 2013, http://www.reuters.com/article/2013/05/09/romania-shale-chevron-idUSL6N0DQ1M420130509.
16 Cf. SOCAR company information, http://www.socarplus.az/en/article/231/nabucco-West-or-tap (accessed 4 November 2013).
17 'Farhad Mammadov about advantages and disadvantages of TAP and Nabucco West', in: contact.az, 23 November 2012, http://www.contact.az/docs/2012/Economics&Finance/112300019 109en.htm#.UZhlQLVTB9J.

7. The Trans-Adriatic Pipeline and Nabucco West Pipeline Projects

Additionally, the TAP project is expected to promote the direct access of SOCAR to the Balkan market where gas grids have not been developed. SOCAR was also interested in gaining direct access to gas-purchasing companies in three Balkan countries (Albania, Montenegro and Kosovo). Albania is one of the few countries in southeast Europe (along with Kosovo, Montenegro, and Macedonia to a certain extent) that do not have a working gas distribution network.[18] In February 2011, the TAP joint venture[19] signed two Memoranda of Understanding and Cooperation with the Croatian and Bosnian system operators, Plinacro and BH-Gas, which both are promoting the Ionian–Adriatic Pipeline (IAP). Starting at a tie-in point to TAP in Albania, IAP aims to deliver gas to Albania, Montenegro, Bosnia and Herzegovina, Slovenia and Croatia. In effect, TAP will be able to guarantee that Azerbaijani gas can reach the Western Balkans. The TAP project will also increase the cooperation of Azerbaijan with the countries where this pipeline will lay and simultaneously increase the benefits for these countries as well. Turkey, a big consumer of natural gas, will also benefit. Greece will earn 320 million Euros, and approximately 2,700 new jobs will be created by the project during the construction period between 2015 and 2018. Over the next 50 years, the TAP project will earn approximately 35 billion Euros, posting 5 billion Euros of net profit.[20] Therefore, the last obstacles were overcome by the approval of an intergovernmental agreement among Albania, Italy and Greece on 9 April 2013 on the construction and operation of the TAP project, and the Greek Parliament subsequently ratified the agreement.[21]

If considered from the economic and political perspectives of the Shah Deniz Consortium, the TAP project is more attractive than Nabucco West as some Shah Deniz Consortium members have shares in the TAP project. Additionally, SOCAR, apart from gas supplies, is also interested in investing in the infrastructure of the downstream market; it acquired 65% in the Greek gas network operator DESFA. Athens also expected that the DESFA sale would raise the chances of TAP becoming a conduit for Azerbaijani gas to Western Europe.[22] With the acquisition of DESFA, SOCAR will control the flow of natural gas starting from Azerbaijan and ending in Italy. Despite Russian interest in acquiring facilities in Greece, its decision to abandon the South Stream's southern branch (from Greece via the Adriatic seabed to Italy) was seen as a hint that Russia plans

---

18 SeeNews, Research on demand: Albania Gas system, Sofia: SeeNews, 2010.
19 The Trans Adriatic Pipeline (TAP) AG is a joint venture company consisting of BP, SOCAR, and Statoil (with a stake of 20% each), Fluxys (16%), Total (10%), E.ON (9%), and Axpo (5%) (see http://www.trans-adriatic-pipeline.com/about-us/tap-ag-company).
20 Kömürcüler, Güneş: Italy, Greece welcome selection of TAP gas route, in: Hürriyet Daily News, 28 June 2013, http://www.hurriyetdailynews.com/italy-greece-welcome-selection-of-tap-gas-%20route.aspx?pageID=238&nID=49620&NewsCatID=348.
21 'Greek Parliament ratifies intergovernmental agreement on the Trans Adriatic Pipeline', TAP press release, 10 April 2013, http://www.trans-adriatic-pipeline.com/news/news/detail-view/article/391/.
22 'Azerbaijan's SOCAR buys Greek gas operator DESFA', EurActiv, 19 June 2013, http://www.euractiv.com/energy/azerbaijan-socar-buys-greek-gas-news-528689.

to concentrate on the gas market of southeastern Europe instead of Italy. Gazprom's withdrawal from the South Stream's southern line cleared the field for TAP in exporting gas to the Italian market.[23]

The possibility of the Italian market becoming oversupplied, which could happen because of the financial crisis and a projected decrease in demand, could be considered a disadvantage of the TAP project. Furthermore, there could be an additional volume of 20bcm per year in the south of the country if all of the liquefied natural gas (LNG) and pipeline import projects (e.g., the GALSI project) are completed. Another difficulty that might arise is that TAP will end on the Italian border; Snam Rete Gas, the Italian transmission system operator, will then have to ship gas to northern Italy.[24]

When the Shah Deniz Consortium opted for TAP over the Nabucco West pipeline project, the EU and the US remained neutral. From a geopolitical and strategic point of view, this political decision hypothetically concerned their interests in the region. Actually, TAP does not reflect the strategic goals and priorities that had initially inspired the EC to design the Southern Gas Corridor. Furthermore, the strategic benefit that the EU and the US derive from facilitating TAP will be exceedingly narrow, resulting principally in a gas glut in Italy. In fact, some supporters of TAP argue that the principal benefits would be the promotion of price competition in Italy and the enabling of additional (likely Russian) gas flow to Western European markets.[25] Hypothetically, the TAP project will halve the role of Azerbaijan in the EU's energy security concept.

However, the TAP project will be a step forward in the opening of the Southern Energy Corridor, and the completion of its associated export transit infrastructure will increase European gas supply diversification. Azerbaijan will gain from these projects, enhancing Azerbaijan's long-run potential as a crucial gas supplier and gas transit country for Central Asian gas resources. SOCAR will become one of the key players in shaping the Southern Corridor project, gaining economic and political leverage that could be used to involve the EU in addressing stagnated, unresolved conflicts and threats in the Caucasus.

---

23 Socor, Vladimir: TAP Project surging ahead of rival Nabucco-West (Part One), in: Eurasia Daily Monitor, 26 March 2013, (vol. 10), no. 56, http://www.jamestown.org/programs/edm/single/?tx_ttnews[tt_news]=40644&tx_ttnews[backPid]=685&no_cache.
24 Rzayeva, Gulmira / Tsakiris, Theodoros G.R.: Azerbaijani gas strategy and the EU's Southern Corridor, Baku: Strategic Imperative, 2012, p. 16, http://www.eliamep.gr/wp-content/uploads/2012/08/tsakiris1.pdf.
25 Committee on Foreign Relations, United States Senate: Energy and security from the Caspian to Europe, Minority report, 112th Congress, second session, 12 December 2012, http://www.foreign.senate.gov/publications/download/energy-and-security-from-the-caspian-to-europe.

7.4. The Regional Security of Azerbaijan's Natural Gas Pipelines

Building and operating energy transport pipelines and infrastructures create political and physical security concerns. This process requires the creation of a safe and stable environment and steps for securing energy pipelines. Initially, significant steps are required to establish proper security measures based on an intergovernmental agreement between the countries through which the pipeline passes. Further steps should be taken to attract the interest of international institutions related to the security of the region and, more explicitly, of the pipelines.

Because of the importance of energy infrastructure for the economic prosperity of Azerbaijan and for the energy security of the EU, these assets could become a target for conflicts or struggles for control at any time. The BTE is currently the only major natural gas pipeline in the region running parallel to the BTC oil pipeline bypassing Russian territory. These two major pipelines run through difficult and dangerous territory, close to the region's frozen conflicts and hotspots, such as Nagorno-Karabakh, South Ossetia, the North Caucasus, Abkhazia, and the Armenian enclaves in southern Georgia.[26]

If we consider Russia's 88,000 troops stationed in the South and North Caucasus and Russia's naval forces (the Caspian Sea Flotilla is based in Astrakhan, while some naval forces of the Black Sea Fleet are docked at the port of Ochamchira in Abkhazia, Georgia), then there is reason for concern.[27] Furthermore, Russia maintains a contingent in Armenia on military bases in Gumry. The only ally of Russia in this region is Armenia, which agreed in August 2010 to extend the lease for the military bases to the year 2044. According to this agreement, Russian forces will help safeguard Armenia's national security and will supply modernised weaponry to Armenia's armed forces. Georgian Foreign Minister Grigol Vashadze, however, criticised the accord as strengthening Russia's military influence in the region, thus compromising Armenia's independence and raising tensions inimical to the settlement of the Nagorno-Karabakh conflict.[28]

In general, the stability of the Caucasus is being weakened, and Azerbaijan is becoming entangled in a hostile triangle of countries, such as Iran, Armenia, and Russia, that could affect the fragile security of the entire region with its aggressive actions. One of the main impediments to the peaceful development of the South Caucasus is the Nagorno-Karabakh conflict between Azerbaijan and Armenia, which remains

---

26 O'Kelly, Cillian: The Russian state and Gazprom: a study in the politics of Russia's natural gas, MA thesis, University College Cork, Ireland, October 2010, http://www.fintanhastings.eu/internationalrelations/wp-content/uploads/2010/12/The-Russian-State-and-Gazprom.pdf.
27 Nichol, Jim: Armenia, Azerbaijan, and Georgia: political developments and implications for U.S. interests, Washington, DC: Congressional Research Service, 2013, p. 11, http://fpc.state.gov/documents/organization/203729.pdf.
28 Kucera, Joshua: Armenia boosting relations with both NATO and Russia, in: Eurasianet.org, 19 November 2012, http://www.eurasianet.org/node/66196.

unresolved. The United Nations General Assembly adopted Resolution 62/243, 'The Situation in the Occupied Territories of Azerbaijan', which addresses the situation in Nagorno-Karabakh, on 14 March 2008 during the 62nd session. The resolution reaffirmed the 'continued respect and support for the sovereignty and territorial integrity' of Azerbaijan 'within its internationally recognized borders' and demanded the 'immediate, complete and unconditional withdrawal of all Armenian forces from all the occupied territories' of Azerbaijan.[29] As of now, none of the four previous resolutions of the United Nations related to Nagorno-Karabakh, including 62/243, have been implemented.

Recently, an additional new concern arose from a speech by the head of the operational department of the Armenian army, General Davtyan, about Armenia's capability of attacking Azerbaijan's oil and natural gas facilities in the case of war. In this speech, he referred to new missiles unveiled by Armenia last year: upgraded Scuds and Tochka U, which have ranges of 300km (which covers most of Azerbaijan, in theory).[30] However, the BTC, BTE and future pipelines will run just 30km to the north of the disputed Nagorno-Karabakh region. If left unresolved, any future conflict in the region could have an effect on the energy production of the Caspian Basin and could ultimately affect energy supplies to the international market. Peaceful settlement of three conflicts—Nagorno-Karabakh, Abkhazia and South Ossetia—will also boost the stability of the region, strengthen regional security and cooperation and, in the long run, improve energy security.[31]

International organisations such as NATO and the EU could double their commitments to stability and democracy in countries in the region, as well as their involvement in the search for acceptable solutions to long-standing conflicts. NATO's representative for the Caucasus, William Lahue, recently said that a security infrastructure in the Caspian region is in the interests of NATO members and partners. NATO is interested in helping its allies and partners protect their energy infrastructure for sustainable energy supplies. Furthermore, he highlighted the co-operation of individual NATO members and Azerbaijan in this direction and on a bilateral basis. In particular, the US is co-operating with Azerbaijan to protect it from the threats of offshore platforms.[32] The security environment of the Caucasus and the threats to the pipeline were highlighted in a communication from the EC in December 2006, which called on

---

29 United Nations, General Assembly: Resolution 62/243—The situation in the occupied territories of Azerbaijan, 25 April 2008, http://www.un.int/azerbaijan/pdf/N0747835_occupied_terr.pdf.
30 Kucera, Joshua: Armenia boosting relations with both NATO and Russia, in: Eurasianet.org, 19 November 2012, http://www.eurasianet.org/node/66196.
31 German, Tracey C.: Corridor of power: the Caucasus and energy security, in: Caucasian Review of International Affairs: CRIA, 2008 (vol. 2), no. 2, pp. 64–72, http://www.cria-online.org/3_1.html.
32 'NATO promises to protect the energy infrastructure of Azerbaijan', in: contact.az, 22 November 2012, http://www.contact.az/docs/2012/Politics/112200018980en.htm#.UZjE5bVTB9I.

the EU to be more active in addressing frozen conflicts in the South Caucasus as they threatened to produce unreliable energy supplies.[33] However, the presence of these organisations in the Caucasus only adds a formal character to the establishment of environments that are already possibly secure; their presence does not increase the security of the region.

Azerbaijan should attempt to gain political benefits by enhancing its strategy regarding the utilisation of its growing energy power and its attempts to resolve the Nagorno-Karabakh conflict. Baku will benefit politically from these energy projects by gaining influence in various international organisations while also accumulating diplomatic capital in the US and several EU states, whose support will be crucial for resolving the decades-long conflict between Armenia and Azerbaijan.[34]

The completion of the Shah Deniz project and collaboration in the frame of pipeline projects in the South European Energy Corridor strengthened ties between Turkey, Georgia and Azerbaijan. These countries co-operate in different spheres, including in regional security. Based on these relations, Azerbaijan's parliament ratified the Agreement on Strategic Partnership and Mutual Support in December 2011, which pledges that Turkey and Azerbaijan will support one another 'using all possibilities' in the case of a military attack or aggression against either of the countries.[35] Geopolitically, the development of Shah Deniz and pipeline politics will also help Azerbaijan establish closer security cooperation with the US, the EU, and NATO, which could in turn be instrumental for balancing the regional influence of Russia and Iran.

The military conflict between Russia and Georgia in August 2008, in which Russian troops reportedly advanced to within 25 km of the pipeline, refocused attention to energy security in the Caucasus.[36] The possibility of energy infrastructure becoming a target in a military conflict has become plausible rather than just hypothetical. The war in Georgia demonstrated that regional energy infrastructure could be targeted during conflicts. Russian manoeuvres during the Georgian war were signs of a militaristic Russian foreign policy in the Caucasus, which seemed to be an adverse reaction to NATO enlargement toward the Caucasus. Russia mostly tried to protect its sphere of interests in the region, but we have to be sure that the energy infrastructure and pipeline projects are beyond Russian influence due to the involvement of the EU and

---

33 European Commission: On strengthening the European neighbourhood policy, COM/2006/726 final, Brussels, 4 December 2006, http://ec.europa.eu/world/enp/pdf/com06_726_en.pdf.
34 Rzayeva, Gulmira/ Tsakiris, Theodoros G.R.: Azerbaijani gas strategy and the EU's Southern Corridor, Baku: Strategic Imperative, 2012, p. 12, http://www.eliamep.gr/wp-content/uploads/2012/08/tsakiris1.pdf.
35 Abbasov, Shahin: Azerbaijan–Turkey military pact signals impatience with Minsk talks—Analysts, in: Eurasianet.org, 18 January 2011, http://www.eurasianet.org/node/62732.
36 O'Kelly, Cillian: The Russian state and Gazprom: a study in the politics of Russia's natural gas, MA thesis, University College Cork, Ireland, October 2010, pp. 1–73, http://www.fintanhastings.eu/internationalrelations/wp-content/uploads/2010/12/The-Russian-State-and-Gazprom.pdf.

NATO making an open Russian intervention against these projects unrealistic or even impossible.

Coincidentally, prior to the outbreak of hostilities in Georgia, the BTC pipeline came under terrorist attack on 5 August 2008. An explosion on Turkish soil disrupted the supply of Azerbaijani oil to Ceyhan harbour. The outlawed Kurdistan Workers' Party (PKK) claimed responsibility for the explosion at one of the valves of the BTC pipeline, disrupting the transportation of oil for 14 days, which caused an increase in oil prices by more than US$2.[37] There are various reasons why terrorist groups around the world often attack energy pipelines and pipeline personnel. First, the direct and indirect effects of pipelines on society, the economy and diplomacy make them highly valuable targets. Second, given that petroleum and natural gas can easily ignite, terrorists can attack them with explosives. Because securing the infrastructure of pipelines is extremely difficult, the physical vulnerability of pipelines and related facilities makes them easy targets.[38]

Azerbaijan, Georgia and Turkey have to consider protecting the critical regional energy infrastructure from both conventional and unconventional attacks (including terrorist or criminal threats). Faced with a host of potential threats, the BTC Consortium has implemented stringent security measures. The pipeline route is currently monitored and patrolled extensively: sensors along the entire length allow for any disruptions to be immediately identified on a constantly monitored digital map of the pipeline.

The security consequences in the region require that Brussels and Washington assist Baku, Tbilisi and Ankara in analysing and implementing the most practical and affordable ways of protecting key energy nodes. This is one area in which NATO (and perhaps the EU) could be more involved in the security of the Caucasus as energy supplied by the Caucasus is critical for European energy security.

7.5. Conclusion

Natural gas pipeline projects offer Azerbaijan broad room for manoeuvring in foreign policy and the ability to simultaneously decrease its dependence on Russia. Due to international pipeline projects, Azerbaijan has gained economic, political and partially geopolitical benefits. Historically, the Caucasus has long been central to the political interests of powerful countries, and therefore, the stability and security of the region

---

37   Ozcan, Nihat Ali: Energy security and the PKK threat to the Baku–Tbilisi–Ceyhan pipeline, in: Terrorism Monitor, 2008 (vol. 6), no. 18, http://www.jamestown.org/programs/tm/single/?tx_ttnews[tt_news]=5170&tx_ttnews[backPid]=167&no_cache=1; İsmail Altunsoy: PKK claims responsibility for BTC pipeline explosion, in: todayszaman.com, 8 August 2008, http://www.todayszaman.com/newsDetail_getNewsById.action?load=detay&link=149686.
38   Ozcan, Nihat Ali: Energy security and the PKK threat to the Baku–Tbilisi–Ceyhan pipeline, in: Terrorism Monitor, 2008 (vol. 6), no. 18, http://www.jamestown.org/programs/tm/single/?tx_ttnews[tt_news]=5170&tx_ttnews[backPid]=167&no_cache=1.

7. The Trans-Adriatic Pipeline and Nabucco West Pipeline Projects

are variable. Because of the complexity of the region, unresolved conflicts and energy politics still remain trump cards in the hands of foreign powers interested in the region.

Azerbaijan could supply the EU for a long time with its natural gas because of existing demand and Azerbaijan's leading position in the exploration of gas reserves. All of these energy pipelines are planned as long-term projects if one considers that the export capacity, in the estimation of the Azerbaijani government, is expected to double by 2025.[39] According to potential gas extraction volumes, Azerbaijan will be one of the main suppliers of natural gas to Europe, as well as other countries, for at least 70 years. TANAP is an extremely ambitious project supported by Turkey and Azerbaijan that has led to sustaining the South European Corridor. Because of this project, TAP has become competitive and will be connected with TANAP and thus European markets.

Energy pipelines have strengthened the economic security of both Azerbaijan and Georgia, helping the two countries move away from the Russian sphere of influence and more firmly orient themselves with the West, as well as attract the attention of international institutions to the region. This process will be long-term because Azerbaijan has become, in the eyes of European gas consumers, enshrined as a gateway to Caspian and Middle East gas reserves. Therefore, any action intended to disrupt and suspend the supply of natural energy resources will face high resistance not only from regional countries, such as Azerbaijan, Georgia and Turkey, but also from international organisations. Because of the interests and impact of Western countries on the region, conquering or going to war with Azerbaijan (or even carrying out any action near gas pipelines) would face rebuff from the EU and NATO countries. In the near future, there will be neither peace nor any major wars or clashes in this region because war would be a grave threat to the security of the main energy supply for Europe.

---

39 Blank, Stephen: Azerbaijan breaks through into Eastern Europe, in: CACI Analyst, 2 June 2013, http://old.cacianalyst.org/?q=node/5921.

Irina Kustova

8. EU Energy Policy Towards the Caspian Region: Assessing the Southern Gas Corridor

8.1. Introduction

Since the end of the Cold War, the transportation of energy resources from the Caspian region to European and world energy markets has become an important issue for Western energy importers. To connect the landlocked countries of the Caspian region and Central Asia with European markets and bypass Russian territory, new infrastructure had to be built. This strategy was successful in the case of oil: the Baku–Tbilisi–Ceyhan (BTC) pipeline was built by a BP-led consortium with strong support from the US. The construction of new gas pipelines has proved more challenging, however, and only the South Caucasus Pipeline (also called the Baku–Tbilisi–Erzurum, or BTE) has been built, in parallel with the BTC by the BP- and Statoil-led consortium in 2004–2006.

In recent years, access to the gas resources of the Caspian region (primarily those of Azerbaijan and Turkmenistan) has become a priority issue on the European Union's (EU's) agenda for several reasons. First, the EU is highly dependent on external supplies of energy, especially natural gas.[1] Second, the interruption of the gas supply due to disputes between Russia and Ukraine has catalysed the EU's internal debates about the need to diversify its gas supplies and reduce its dependence on Russia. Since 2006, a number of initiatives have been developed within the EU to solve the problem of the EU's dependence on Russia for gas. The 'Southern Gas Corridor', an initiative aimed at importing gas from the Caspian region and the Middle East, has been labelled as 'one of the EU's highest energy security priorities'.[2] The Southern Gas Corridor is not only the backbone of an emerging EU external energy policy but also a project that might reconfigure the balance of the European gas market, directly connecting the gas fields of the Caspian Region and Central Asia to European consumers.

---

1   According to Eurostat, '[t]he security of the EU's primary energy supplies may be threatened if a high proportion of imports are concentrated among relatively few partners. Close to three quarters (74.4%) of the EU-27's imports of natural gas in 2010 came from Russia, Norway or Algeria—as such there was a diversification of imports as in 2009 the same three countries accounted for 79.2% of natural gas imports' (Eurostat: Energy production and imports, August 2012, http://epp.eurostat.ec.europa.eu/statistics_explained/index.php/Energy_production_and_imports).
2   European Commission: Second Strategic Energy Review: an EU energy security and solidarity action plan, COM/2008/781 final, 13 November 2008, p. 4, http://eur-lex.europa.eu/LexUriServ/LexUriServ.do?uri=COM:2008:0781:FIN:EN:PDF.

Since 2008, a number of infrastructure projects have been proposed for the Corridor, all of them with different degrees of commercial and political feasibility. These projects were designed to bring natural gas primarily from Azerbaijan because its Shah Deniz II gas field has remained the main and likely the only feasible source of supplies from the region until now. In the mid-term, gas supplies from Turkmenistan might also be considered as an option. However, legal disputes regarding the legal status of the Caspian Sea in addition to the lack of commercial interest demonstrated by suppliers, first of all, Turkmenistan, leave the Trans-Caspian gas pipeline, designed to connect Turkmenistan with Azerbaijan beneath the Caspian Sea, in limbo.

Until 2012, the Nabucco pipeline, a project under the Southern Gas Corridor umbrella, was the first priority project to connect Azerbaijan with Austria via Turkey and enjoyed strong support from the European Commission (EC). Due to the lack of financing commitments on the part of its participants, however, the project was reduced to bureaucracy and PR campaigns and, in 2012, was pared down to the Nabucco West, its shorter version, because Azerbaijan and Turkey reached an agreement to build the Trans-Anatolia Pipeline (TANAP). Among the projects that were highly publicised but never implemented are the Interconnector Turkey–Greece–Italy (ITGI) and the South East Europe Pipeline (SEEP), backed by BP, which dropped its proposal in 2012 to support Nabucco West.[3] In 2013, the latter, despite strong support from the EC,[4] lost out to the Trans-Adriatic Pipeline (TAP), which was selected by the Shah Deniz Consortium to connect Turkey with Southern Italy via Greece and Albania.

The results of the Southern Gas Corridor have been rather modest until now—the Nabucco, like many other projects, has failed to come to fruition, the Trans-Caspian gas pipeline remains a long-shot, and the TAP has upset Nabucco West, inhibiting EU-backed initiatives to diversify gas supplies. This chapter seeks to answer the following question: why has EU energy policy aimed at launching direct gas supplies from the Caspian region, the Southern Gas Corridor in particular, been ineffective until now?

The limited success of these pipeline projects has been most often discussed from the geopolitical perspective, and analysis has usually concentrated on mapping out proposed pipeline projects and determining the stakes of all the involved parties. From this perspective, access to Caspian resources is often viewed as a new 'Great Game', where Russia attempts to 'lock in' the resources of the former Soviet Union exporters while the other players (first of all, the USA and the EU) try, in turn, to 'lock Russia out'.[5]

---

3 'SEEP out of the running for Shah Deniz', in: Natural Gas Europe, 28 June 2012, http://www.naturalgaseurope.com/seep-tap-final-shah-deniz-contest.
4 'European Union sets forward its natural gas priorities', in: Natural Gas Europe, 1 May 2013, http://www.naturalgaseurope.com/european-union-sets-forward-its-natural-gas-priorities.
5 Dellecker, Adrian/ Gomart, Thomas: Conclusion, in: Dellecker, Adrian/ Gomart, Thomas (eds): Russian energy security and foreign policy, London: Routledge, 2011, pp. 203–208, here p. 203; Rutland, Peter: US energy policy and the former Soviet Union, in: Perovic, Jeronim/ Orttung, Robert W./ Wegner, Andreas (eds): Russian energy power and foreign relations: implications for

8. EU Energy Policy Towards the Caspian Region

Therefore, the failures or limited success of the EU policies in this region are explained by Russia's ability to block alternative export routes for Caspian energy resources.[6] The ambivalence of the Southern Gas Corridor is attributed to Russia's position, and policy recommendations concentrate on ways to overcome Russia's influence in the region.

While the geopolitical perspective is important in accounting for the configuration of gas export routes in the region and explaining the EU's limited success, the assessment of developments internal to the EU can shed light on additional aspects of the problem that are not captured by the geopolitical approach. Instead of analysing the stakes of the parties involved in geopolitical games, this chapter shifts the focus to the internal inconsistencies of EU policies both in terms of integration processes in the energy domain and the commercial interests of energy companies.

Energy policy has remained one of the politically sensitive areas of European integration. Until the 2000s, most member states opposed any significant competence transfer to the supranational level and continued to favour long-lasting bilateral relations with external suppliers. During the 1990s, because there was no explicit legal basis for energy policy in the EU treaties, energy issues were allocated to three interconnected policy areas that fell under the legislative competence of the EC: the environment (sustainable development), the single market (competition), and external relations (security of energy supply).

The EU is still not a unitary actor in energy policy, especially in its external dimension. On the one hand, a number of epochal decisions have been made regarding EU energy policy, such as the introduction of the chapter on energy in the Treaty of Lisbon and the 2009 Third Energy Package for the Internal Energy Market. The EU has launched a number of external initiatives towards its neighbourhood, including the EU–Russia Energy Dialogue, the Black Sea Synergy, the Eastern Partnership, and the Energy Community of South East Europe. Gradual consolidation of the Internal Energy Market started with the 1996 and 1998 Directives for Electricity and Gas, while the Second Energy Package was adopted in 2003 and the Third Internal Market Energy Package came into force in 2009, thus establishing the framework for the completion of the Internal Energy Market.[7] On the other hand, many initiatives that might have

---

conflict and cooperation, London: Routledge, 2009, pp. 181–200; Jaffe, Amy M./ Soligo, Ronald: Energy security: the Russian connection, in: Moran, Daniel/ Russel, James A. (eds): Energy security and global politics: the militarization of resource management, London: Routledge, 2009, pp. 112–134, here p. 128.

6   Smith, Keith C.: Security implications of Russian energy policies, Brussels: Centre for European Policy Studies, 2006 (CEPS Policy Brief, no. 90), http://www.ceps.be/book/security-implications-russian-energy-policies; Bilgin, Mert: Geopolitics of European natural gas demand: supplies from Russia, Caspian and the Middle East, in: Energy Policy, 2009 (vol. 37), no. 11, pp. 4482–4492.

7   Eikeland, Per Ove: The Third Internal Energy Market Package: new power relations among member states, EU institutions and non-state actors?, in: Journal of Common Market Studies, 2011 (vol. 49), no. 2, pp. 243–263.

led to progress in developing a common EU external energy policy have stalled during the decision-making process. Despite the importance ascribed to energy security and its visibility in the EU political agenda, the directions for the development of a common energy policy—and especially its external dimensions—remain unclear.[8] As Matláry notes, 'the conventional academic view has been that energy policy is one of the weakest policy areas of the EC/EU'.[9] Currently, the fragmented EU energy policy is characterised by a gap between a more or less unified Internal Energy Market and the absence of a common EU position in the international arena.

This chapter frames the debate about whether the EU should 'speak with a single voice' in external energy policy within the intergovernmental–supranational dichotomy: EU member states have different national energy strategies and tend to favour bilateral relations with their energy suppliers, while the EC seeks to acquire a wider range of competences in this area.[10] Many scholars underline the dominance of national policies in this area, with member states guarding their competences. The dynamic of EU energy policy development can be described as an 'expansion of the implied powers of the Community from internal to external matters',[11] while the member states are reluctant to accept this expansion, preferring intergovernmental co-operation. The lack of a comprehensive energy policy at the EU level results in both 'foreign policy side effects and the economic challenges of the sector'.[12] Eikeland adopts the historical institutionalist framework to assess the developments of the Internal Energy Market and the ability of the EC to drive the adoption of the Third Internal Energy Market Package.[13] Natorski and Herranz Surralés apply the securitisation approach to study the EC's attempts to frame energy policy as an urgent security issue from 2006–2008 in the immediate aftermath of the gas dispute between Russia and Ukraine.[14] The

---

8 Tekin, Ali/ Williams, Paul A./ Sever, Seda Duygu: Evolution of EU energy policy, in: Tekin, Ali/ Williams, Paul A. (eds): Geo-politics of the Euro-Asia energy nexus: the European Union, Russia and Turkey, New York: Palgrave Macmillan, 2011, pp. 13–36, here p. 13.
9 Matláry, Janne Haaland: Energy policy in the European Union, New York: St. Martin's Press, 1997, here p. 13.
10 Mayer, Sebastian: Path dependence and Commission activism in the evolution of the European Union's external energy policy, in: Journal of International Relations and Development, 2008 (vol. 11), no. 3, pp. 251–278; Neuman, Marek: EU–Russian energy relations after the 2004/2007 EU enlargement: an EU perspective, in: Journal of Contemporary European Studies, 2010 (vol. 18), no. 3, pp. 341–360.
11 Haghighi, Sanam S.: Energy security and the division of competences between the European Community and its member states, in: European Law Journal, 2008 (vol. 14), no. 4, pp. 461–482, here p. 465.
12 Hadfield, Amelia: EU foreign energy policy: in the pipeline?, in: CFSP Forum, 2006 (vol. 4), no. 1, pp. 1–5, here p. 2.
13 Eikeland, Per Ove: EU internal market energy policy: achievements and hurdles, in: Duffield, John S./ Birchfield, Vicky L. (eds): Toward a common European Union energy policy: problems, progress, and prospects, New York: Palgrave Macmillan, 2011, pp. 13–40, here p. 16.
14 Natorski, Michal/ Herranz Surralés, Anna: Securitizing moves to nowhere? The framing of the European Union energy policy, in: Journal of Contemporary European Research, 2008 (vol. 4),

concept of supranational spill-over has also been applied to EU rule promotion in its neighbourhood, such as the Energy Community of South East Europe;[15] however, the external dimensions of EU energy policy remain an intergovernmental domain with limited competences for the Commission.

This chapter argues that the EU's internal inconsistencies in decision-making and its disregard for commercial aspects of the pipeline projects have seriously impeded the successful implementation of the Southern Gas Corridor. It additionally undermines the EU's credibility as an important regional actor. This research is based on EU documents and publicly available publications concerning EU decision-making on energy policy and pipeline projects.

The chapter is organised into two sections. The first discusses the role of the EC as a promoter of the Southern Gas Corridor and its role in the EU's internal developments. The second analyses the role of energy companies' commercial interests in making pipeline projects viable and how these interests can be influenced by external decisions, paying special attention to the changes in gas pricing mechanisms in Central Asia.

8.2. The European Commission as a Promoter of the Southern Gas Corridor

This section argues that ineffectiveness of EU energy policy towards the Caspian region can be attributed to a certain extent to the EU's internal inconsistencies and complex integration processes. Intra-institutional competition within the EU can impede the coherent incorporation of different aspects of pipeline construction in the EU's external energy policy. Two issues can be underlined in this regard. First, the EU's incoherent process of gradual integration in the energy sector potentially weakens its position in the international arena. Second, internal competition between the EU institutions can overemphasise some issues over others.

Despite the EC's attempts to call for a common security of supplies in its 2000 Green Paper and the 2003 European Security Strategy,[16] both documents only mention the problem of energy dependency; energy security was not considered as 'part

---

no. 2, pp. 71–89.
15 Renner, Stephan: The energy community of Southeast Europe: a neofunctionalist project of regional integration, in: European Integration Online Papers, 2009 (vol. 13), no. 1, http://eiop.or.at/eiop/index.php/eiop/article/view/2009_001a.
16 European Commission: Green Paper: towards a European strategy for the security of energy supply, COM/2000/769 final, 29 November 2000, p. 27, http://aei.pitt.edu/1184/1/enegy_supply_security_gp_COM_2000_769.pdf (for a publicised version, see http://ec.europa.eu/energy/green-paper-energy-supply/doc/green_paper_energy_supply_en.pdf); European Union: European security strategy: a secure Europe in a better world, Brussels, 12 December 2003, http://www.consilium.europa.eu/uedocs/cmsUpload/78367.pdf.

of strategic culture to be developed by the EU to strengthen the instruments and capabilities of the CFSP'.[17] These careful proposals for further co-operation from the early 2000s were followed in 2005–2007 by more active involvement by the Commission, which mainly aimed at incorporating supply security in the EU common energy framework.[18] The window of opportunity for proposing a more integrated—both internal and external—energy policy was granted to the EC after the Russian–Ukrainian gas crisis in January 2006.[19] The measures taken by the Russian authorities affected not only its nearby neighbours: the non-deliveries of gas also made the energy situation in several EU member states critical. The existing multilateral frameworks, such as the Energy Charter Treaty, failed to address such a crisis. Notwithstanding the causes of the dispute between Russia and Ukraine, it became a catalyst for debates within the EU about further integration in the area of energy. Even if the security of supplies and relations with suppliers remained a clearly exclusive prerogative of member states, the call for a more coherent external energy policy emerged on the EU agenda and received at least rhetorical support among member states.

During the discussions following the 2006 Russian–Ukrainian gas crisis, the positions of the EU member states were more or less unified regarding the understanding of the security of supplies, not only in economic but also in strategic terms.[20] Many of them, however, remained reluctant to implement any type of competence transfer from the national to the supranational level, favouring a certain degree of co-operation either at the EU or the regional level, such as the framework of an 'Energy OSCE', proposed by Germany.[21] Some member states, however, insisted on a more coherent

---

17 Hadfield, Amelia: EU foreign energy policy: in the pipeline?, in: CFSP Forum, 2006 (vol. 4), no. 1, pp. 1–5, here p. 2.
18 Van der Linde, Coby: External energy policy: old fears and new dilemmas in a larger union, in: Sapir, André (ed.): Fragmented power: Europe and the global economy, Brussels: Bruegel Books, 2007, pp. 266–307, here p. 266.
19 Bahgat, Gawdat: Europe's energy security: challenges and opportunities, in: International Affairs, 2006 (vol. 82), no. 5, pp. 965–967; Natorski, Michal/ Herranz Surralés, Anna: Securitizing moves to nowhere? The framing of the European Union energy policy, in: Journal of Contemporary European Research, 2008 (vol. 4), no. 2, pp. 71–89, here p. 72.
20 Council of the European Union: A new energy policy (contribution from Italy to the debate on a New Energy Policy for Europe), 5944/06, 2 February 2006, http://register.consilium.europa.eu/pdf/en/06/st05/st05944.en06.pdf; Council of the European Union: A new energy policy (contribution from Belgium to the debate on a New Energy Policy for Europe), 6014/06, 3 February 2006, http://register.consilium.europa.eu/pdf/en/06/st06/st06014.en06.pdf; Council of the European Union: A new energy policy (contribution from Spain to the debate on a New Energy Policy for Europe), 6084/06, 7 February 2006, http://register.consilium.europa.eu/pdf/en/06/st06/st06084.en06.pdf.
21 Council of the European Union: A new energy policy for Europe (presidential note), 5401/06, 16 January 2006, http://register.consilium.europa.eu/pdf/en/06/st05/st05401.en06.pdf; Council of the European Union: A new energy policy (contribution from Germany to the debate on a New Energy Policy for Europe), 6009/06, 3 February 2006, http://register.consilium.europa.eu/pdf/en/06/st06/st06009.en06.pdf; Council of the European Union: French memorandum for

8. EU Energy Policy Towards the Caspian Region

external energy policy, arguing the importance of the inclusion of a military component; Poland, for example, proposed the 'Energy NATO' with a collective defence clause in case of energy supply interruptions.[22]

The Green Paper of 2006,[23] besides discussing initiatives for completing the Internal Energy Market and providing environmental sustainability, paid special attention to the importance of solidarity among the member states to ensure energy supply security. Among six proposed areas for future strategy, three—solidarity, the diversification of the energy mix, and external policy—were directly linked to the need for a more cohesive external energy policy as part of the EU's Common Foreign and Security Policy (CFSP).

Despite the consensus over the conceptualisation of energy as a security issue, the EC and the member states have failed to find a common approach to address the problem. The Commission's proposals did not result in any substantial measures to improve and co-ordinate the external energy policy. The priority of common energy policy was 'equivocal', and member states tended to expand bilateral policies while rhetorically committing to 'the outward expansion of EU internal market norms'.[24] Eventually, however, several initiatives were finally agreed upon, including the establishment of more effective crisis response mechanisms and a network of energy correspondents in 2007.[25] With regards to the external dimension, the Second Strategic Energy Review launched the Southern Gas Corridor initiative to bring gas from the EU's Southern neighbours.[26] More an umbrella initiative for gas pipeline projects bypassing Russia, it was labelled 'the New Silk Road' at a summit in May 2009.[27] It complemented existing EU-backed initiatives in the Black Sea and Caspian regions, namely,

---

revitalising European energy policy, 5724/06, 26 January 2006, http://register.consilium.europa.eu/pdf/en/06/st05/st05724.en06.pdf.

22 Council of the European Union: Proposal for a European energy security treaty (presentation by the Polish delegation), 7160/06, 9 March 2006, p. 3, http://register.consilium.europa.eu/pdf/en/06/st07/st07160.en06.pdf.

23 European Commission: Green Paper: a European strategy for sustainable, competitive and secure Energy, COM/2006/105 final, 8 March 2006, p. 5, http://eur-lex.europa.eu/LexUriServ/LexUriServ.do?uri=COM:2006:0105:FIN:EN:PDF.

24 Youngs, Richard: Energy security: Europe's new foreign policy challenge, London: Routledge, 2009, here p. 22.

25 Council of the European Union: Council conclusions 'A New Energy Policy for Europe': contribution of the Energy Ministers to the 2006 Spring European Council, 6878/06, 2 March 2006, http://register.consilium.europa.eu/pdf/en/06/st06/st06878.en06.pdf; Council of the European Union: Presidency conclusions of the Brussels European Council, 23/ 24 March, 7775/1/06 REV 1, 18 May 2006, http://register.consilium.europa.eu/pdf/en/06/st07/st07775-re01.en06.pdf.

26 European Commission: Second strategic energy review: an EU energy security and solidarity action plan, COM/2008/781 final, 13 November 2008, http://eur-lex.europa.eu/LexUriServ/LexUriServ.do?uri=COM:2008:0781:FIN:EN:PDF.

27 Czech Presidency of the European Union: Declaration of the Southern Corridor summit, press release, 8 May 2009, http://www.eu2009.cz/en/news-and-documents/press-releases/declaration---prague-summit--southern-corridor--may-8--2009-21533/index.html.

the Interstate Oil and Gas Transport to Europe (INOGATE) programme, launched in 1997 to facilitate co-operation on energy matters among the Black Sea states, and the 2004 Baku Initiative, aimed at connecting the regional energy markets with Europe. Despite the EC's claims for solidarity and a common position on relations with suppliers, member states have inevitably privileged their national interests over the common EU stance. Thus, some member states opted to support the pipeline projects that were not defined as EU priorities. Italy, for example, has long been a proponent of the South Stream gas pipeline, a project to directly connect Russia and the EU beneath the Black Sea, bypassing Ukraine. Germany has completed the Nord Stream gas pipeline, a German–Russian project that was unwelcome to a number of Central European member states. Because the competence of the EC in external energy policy reflects to a large extent that of the lowest common denominator, the Southern Gas Corridor remains no more than an umbrella of numerous pipeline projects, often competing with each other. Because there is no unanimous support for the initiative from member states or at least no desire to commit to it through action, the Corridor lacks real progress towards implementation.

At the same time, there has been some success in the integration of the EU's external energy policy. First, the Lisbon Treaty has established energy as one of the new shared competences between the EU and its member states and included a reference to energy solidarity among member states. Even if member states have retained the right to define their national energy mixes, the 'codification' of the existing competences at the supranational level can be interpreted as a next step to the member states' commitment to work towards a more coherent energy policy.[28] Still, most EU initiatives have concentrated on energy efficiency and environmental protection—issues that are obviously important but not closely connected to security of supply.[29]

Second, the gradual but mostly symbolic competence transfer to the EC is exemplified by the fact that on 12 September 2011, the Commission received a mandate from the European Council for negotiations over the Trans-Caspian gas pipeline with Azerbaijan and Turkmenistan. This marked the first time that the EC was deputised to speak for the entire EU in gas pipeline negotiations.[30] Moreover, in January 2011, the joint declaration on the Southern Gas Corridor was signed by the EC President and the

---

28   Tekin, Ali/ Williams, Paul A./ Sever, Seda Duygu: Evolution of EU energy policy, in: Tekin, Ali/ Williams, Paul A. (eds): Geo-politics of the Euro-Asia energy nexus: the European Union, Russia and Turkey, New York: Palgrave Macmillan, 2011, pp. 13–36, here pp. 13–15.
29   Duffield, John S./ Birchfield, Vicky L.: Introduction: the recent upheaval in EU energy policy, in: Duffield, John S./ Birchfield, Vicky L. (eds): Toward a common European Union energy policy: problems, progress, and prospects, New York: Palgrave Macmillan, 2011, pp. 1–12, here pp. 5–6.
30   European Commission: EU starts negotiations on Caspian pipeline to bring gas to Europe, press release, IP/11/1023, 12 September 2011, http://europa.eu/rapid/press-release_IP-11-1023_en.htm?locale=en.

8. EU Energy Policy Towards the Caspian Region 223

President of Azerbaijan, symbolising further co-operation in gas route development.[31] Accordingly, Azerbaijan agreed to bring 10 billion cubic metres of gas per year to Europe.[32]

Another highly symbolic transfer of competence to the supranational level is the final adoption of the legislation on the disclosure of bilateral agreements on energy with third parties by member states. In September 2011, the EC proposed an information exchange mechanism for intergovernmental energy agreements between Member States and third countries,[33] which it argued was essential for external energy policy. Finally, in October 2012, this information exchange mechanism was adopted.[34]

With these successes, the Southern Gas Corridor remains a more politically motivated initiative, an umbrella for various incoherent projects, limited to the narrow competence of the EC and its political proclamations. In this situation, member states remain the key actors who decide whether to support and finance one project or another, or to drop less feasible projects. Therefore, the next section will discuss the argument that economic rationales play an important role in implementing projects and that political proclamations alone cannot propel projects towards implementation.

8.3. Commercial Interests of Energy Companies Disregarded?

The construction of new pipelines is usually discussed from the geopolitical point of view: in the most general terms, the most crucial aspect of pipeline construction is how to reach an agreement between all the states whose territory the pipeline transverses and overcome the resistance of other states. While political rationales for pipeline projects are indispensable, the commercial aspects of the pipeline business are essential in the assessment of pipeline projects' prospects for success. As Guillet underlines, 'while politics and diplomacy can usually kill pipeline projects and, at best, help make a choice between two viable projects, they cannot make happen projects that do not have a strong underlying financial and business rationale'.[35]

---

31 Joint Declaration on the Southern Gas Corridor, Baku, 13 January 2011, http://ec.europa.eu/energy/infrastructure/strategy/doc/2011_01_13_joint_declaration_southern_corridor.pdf.
32 'Barroso tops Azeri gas deal with visa facilitation', in: EurActiv, 14 January, 2011, http://www.euractiv.com/energy/barroso-tops-azeri-gas-deal-visa-news-501255.
33 European Commission: On security of energy supply and international cooperation: the EU energy policy—engaging with partners beyond our borders, COM/2011/539 final, 7 September 2011, http://eur-lex.europa.eu/LexUriServ/LexUriServ.do?uri=COM:2011:0539:FIN:EN:PDF.
34 'Decision no. 994/2012/EU of the European Parliament and of the Council of 25 October 2012 establishing an information exchange mechanism with regard to intergovernmental agreements between Member States and third countries in the field of energy', in: Official Journal of the European Union, L/299, 27 October 2012, pp. 13–17, http://eur-lex.europa.eu/LexUriServ/LexUriServ.do?uri=OJ:L:2012:299:FULL:EN:PDF.
35 Guillet, Jérôme: How to get a pipeline built: myth and reality, in: Dellecker, Adrian/ Gomart, Thomas (eds): Russian energy security and foreign policy, London: Routledge, 2011, pp. 58–73, here p. 58.

In general, pipelines have high initial investment costs with a relatively long period—up to several decades—of return, meaning that energy companies need guarantees for a return on investment. These considerations make the investment decision likely to be the most important stage of pipeline construction. The ability of energy companies to form a consortium—be it as technical operators of the pipeline or the parties involved in the commercial arrangement—is a crucial part of building a pipeline.[36] Because the Gas Sales and Purchase Agreement is the cornerstone of pipeline construction, the most important, if not the most obvious, elements are that gas should be available and that there should be a buyer.[37]

For instance, the commercial considerations of the energy companies that participated in the Nabucco project were not properly reflected during the negotiations at the EU level. There was no cohesion, and a gap existed between the level of the policy formation behind the Southern Gas Corridor and the level of member states and their energy companies on the ground. Because financing for pipeline projects at the EU level is limited to financial aid within the Trans-European Energy Networks (TEN-E) programme, mostly for project assessment, member states' decisions are essential for projects, as all costs are borne by companies. The EU campaign for the Southern Gas Corridor, however, has not coincided with the considerations of member states and energy companies. The Southern Gas Corridor collects every single project under its umbrella but fails to reflect adjustments caused by changes in the business environment.

Ignoring the project's commercial feasibility is exactly what seems to have happened with the Nabucco, the backbone of the Southern Gas Corridor, despite the EC's political support. From 2002, when the initial agreement between OMV (Austria) and BOTAŞ (Turkey) was made,[38] there were serious concerns about the available volumes of gas. In the end, the Shah Deniz II gas field in Azerbaijan remains the only feasible option, and Turkmenistan and northern Iraq are not considered feasible options. With such uncertain prospective gas suppliers, many participants gradually withdrew from the project, which was ultimately reduced to its shorter version, the Nabucco West, running from the Turkish border to Austria. In the end, there was no cohesion among member states—some opted to join the competing South Stream project, led by Gazprom—and energy companies were unsure about the commercial viability of the Nabucco pipeline.[39]

---

36 Ibid., p. 68.
37 Ibid., p. 62.
38 See http://www.nabuccopipeline.com/portal/page/portal/en/company_main/about_us, accessed 29 May 2013.
39 In 2013, RWE sold its share to Austrian OMV. 'Germany's RWE to give up Nabucco project', RT, 3 December 2012, http://rt.com/business/rwe-nabucco-project-quit-131/.

8. EU Energy Policy Towards the Caspian Region

Steering efforts by the Commission and financial sponsorship have brought few results. Even widely publicised support by the EC has not helped. As Guillet notes,

> it is rather safe to assume that the more one talks about a given project, the less likely is to happen: in fact, it may instead be said that the more numerous such public announcements are, the more it implies that the project is unable to move towards the next steps [...] and is stuck in what may be dubbed the 'diplomacy-and-PR-heavy, partner-seeking phase'.[40]

Following its predecessor, which was more a 'PR concept' than a commercially sustainable project,[41] the Nabucco West pipeline finally lost out to the TAP in the Shah Deniz Consortium's final decision in 2013.[42]

In the Trans-Caspian pipeline situation, gas pricing mechanisms for Central Asian suppliers are also a crucial commercial aspect. As Konoplyanik argues, changes in gas pricing are essential for further estimations of whether gas might be supplied to Europe.[43] Before 2009, Central Asian gas prices in contracts with Russia were defined according to the 'cost-plus' principle at the borders of the exporters. The difference between the much higher netback replacement value price in the European markets and the low cost-plus price at the border of Central Asia exporters—the so-called Hotelling rent—'explains the economic reasoning of the intensive fight of the EU companies, supported by the Commission, for direct access to Turkmen gas in particular and the gas from the right bank of Caspian in general'.[44] Since netback pricing was adopted for Central Asian gas in 2009, the previous difference in the gas price for EU end-users and the cost-plus price at the border of Central Asian exporters 'that would have stimulated construction of alternative pipelines and that would have been available for EU companies to pay-back such huge investments'[45] is no longer available. These changes can also influence the intentions of commercial players (European energy companies) to participate in the projects, such as the Trans-Caspian pipeline, over which the EC is negotiating with Turkmenistan and Azerbaijan.

---

40 Guillet, Jérôme: How to get a pipeline built: myth and reality, in: Dellecker, Adrian/ Gomart, Thomas (eds): Russian energy security and foreign policy, London: Routledge, 2011, pp. 58–73, here p. 61.
41 Ibid., p. 59.
42 BP: Shah Deniz targets Italian and Southeastern European gas markets through Trans Adriatic Pipeline, press release, 28 June 2013, http://www.bp.com/genericarticle.do?categoryId=9006615&contentId=7086179.
43 Konoplyanik, Andrei A.: Na razvilke trekh dorog, in: Neft' Rossii, 2011 (vol. 2), pp. 50–55.
44 Konoplyanik, Andrey: Gas export pricing & alternative gas export routes (Central Asia case), in: OGEL (Oil, Gas & Energy Law), 2011 (vol. 3), here p. 10, cited from the personal website of Andrey Konoplyanik, http://konoplyanik.ru/ru/publications/articles.html.
45 Ibid., here p. 11.

8.4. Conclusion

The Nabucco pipeline project is a part of history, and the Nabucco West, its shorter version, was upset by the TAP in 2013; the South Stream pipeline, however, which the EC often stressed as not being a solution for the EU's ambition to diversify its gas supplies,[46] is under construction. The limitations of the Southern Gas Corridor have become even more apparent. The TAP, even if it is part of the Corridor, is primarily aimed at the Italian, not the Central European, market, limiting the EU's efforts for diversification. Some results have certainly been achieved, such as the agreement with Azerbaijan to supply natural gas, and the real prospects of opening the Caspian region and receiving physical supplies of natural gas after the construction of the TANAP pipeline.

According to this chapter, which has concentrated on the internal developments of the EU, several conclusions can be drawn concerning the prospects of the Southern Gas Corridor.

First, EU integration is a gradual and multidirectional process. The complex system of EU decision-making makes it possible for many decisions to be stalled by red tape, disregarded by numerous participants, or significantly altered during debate. In a number of cases—and energy policy, with its complex web of decision-making and competences, is a good example—the final decisions reflect only the lowest common denominator and are limited to more rhetorical or general measures. The complex system of EU decision-making has impeded effective responses to events such as gas supply crises. During 2006–2008, only limited actions were taken due to the EU's internal inconsistencies. On the one hand, there was little agreement among member states about common policies, especially in the external dimension. Being reluctant to hand over competences to the EU level, member states often pursue competing and even conflicting policies in the view of the EU. On the other hand, the EC has managed to support initiatives that have definitely pushed further integration in the EU energy sector, namely, the Third Energy Package, the disclosure of bilateral agreements of member states with third parties, and the negotiations on the Trans-Caspian gas pipeline.

Second, during the inter-institutional battles, supranational institutions may continuously favour projects that are commercially unattractive from their very beginning. Strangely enough, the EC continues to insist that the Nabucco West is a viable, priority project.[47] While this may be a mere rhetorical commitment, this support demonstrates short-sighted vision and seriously challenges reasonable policy making in EU energy policy.

---

46 'Oettinger: South Stream is not our top priority', in: New Europe Online, 25 May 2011, http://www.neurope.eu/article/oettinger-south-stream-not-our-top-priority.
47 'Failed Nabucco West plan still on EU priority list—sources', in: Reuters, 20 September 2013, http://www.reuters.com/article/2013/09/20/eu-energyprojects-idUSL5N0HG26S20130920.

Third, while cohesion within the EU should not be completely disregarded, it is important to underline that investment decisions are still made by energy companies and national governments. The EU's financial instruments are mostly limited to financial aid to trans-European networks, and while EU-backed support creates necessary publicity around a project, it is unlikely to make such project come into reality in cases where participating energy companies and national governments find a project commercially unsustainable. The prospects of pipeline projects can also be challenged by member states' interests in launching or joining competing projects included in the framework of the Southern Gas Corridor. The changes in pricing mechanisms in Central Asian export contracts is another factor that might inhibit the intentions of energy companies to participate in pipeline projects.

Concerning the future developments of the Southern Gas Corridor, it is necessary to underline that progress in EU integration is visible and gradual, and in the future, it is likely to deepen. The other question is whether further integration and reallocation of competences to the supranational level will make external energy policy effective. Without the predisposition of a common EU position, the EU is unlikely to be able to pursue any effective policies in the Caspian region.

Lusine Badalyan[1]

9. The EU's Democracy Promotion in its Eastern Neighbourhood: Consistency in the Shadow of Energy Trade and the EU's Ability

9.1. Introduction

The promotion of democracy, human rights and the rule of law are the self-declared core values that are supposed to shape the European Union's (EU's) external actions in its neighbourhood.[2] Yet despite its declared policy commitments, the EU's actual support for democratisation processes has frequently been contended in the literature.[3] The EU's successful impact on democratisation processes in its Eastern Partnership (EaP)[4] is not self-evident, forming another democracy-reform dilemma. This chapter aims to address the EU's democracy-promotion policies in its EaP and seeks to explore to what extent and under what conditions the EU is effective in its efforts to promote democracy.

Although endogenous factors such as domestic institutions and national elites are recognised as significant aspects for the process of promoting democracy, in recent years a growing body of literature has acknowledged the considerable effect of external factors on democratisation processes. In this line, the EU's *sui generis* nature as an international actor and democracy promoter has raised special attention and become the subject of several studies. Research concentrating particularly on the EU's democracy promotion actions in its neighbourhood, where a perspective of EU membership

---

1 This chapter is part of my doctoral project. The research is funded by a grant from the Volkswagen Foundation as part of a research project on domestic debates and foreign policy-making in the Caspian region.
2 European Commission: Wider Europe–neighbourhood: a new framework for relations with our eastern and southern neighbours, COM/2003/104 final, 11 March 2003, p. 3, http://eur-lex.europa.eu/LexUriServ/LexUriServ.do?uri=COM:2003:0104:FIN:EN:PDF.
3 Youngs, Richard: The European Union and the promotion of democracy, New York: Oxford University Press, 2001; Schimmelfennig, Frank/ Sedelmeier, Ulrich: Governance by conditionality: EU rule transfer to the candidate countries of central and eastern Europe, in: Journal of European Public Policy, 2004 (vol. 11), no. 4, pp. 661–679; Grabbe, Heather: The EU's transformative power: Europeanization through conditionality in central and eastern Europe, Basingstoke: Palgrave McMillan, 2006; Hyde-Price, Adrian: Normative power Europe: a realist critique, in: Journal of European Public Policy, 2006 (vol. 13), no. 2, pp. 217–234.
4 This programme was put forward in 2008 to enhance the EU's relationship with Armenia, Azerbaijan, Belarus, Georgia, Moldova, and Ukraine. Among other things, the Partnership intends to promote democracy and good governance, strengthen energy security, and support economic and social development (for more information, see http://eeas.europa.eu/eastern/index_en.htm).

is missing, agree that above all, the size of domestic adoption costs and the degree of consistency in the EU's rule promotion (be it through conditionality, socialisation and/or domestic empowerment mechanisms) are decisive for the EU's effectiveness in promoting democracy.[5] The international dimension of the EU's democracy promotion will be the main focus of this study.

One of the essential factors addressed in the literature for the effectiveness of the EU's democracy promotion is the consistency of democratic rule-transfer to target countries.[6] In order for the EU to be regarded as consistent, it has to be

> guided only by the democratic and human rights performance of the targeted countries and it ought not to discriminate against any country either positively or negatively on the basis of other considerations.[7]

For the EU that introduces itself as a normative power, the subordination of consistently promoting democracy to its other foreign policy concerns and interests not only undermines its policy credibility and legitimacy but also limits its potential positive impact on democratisation processes in target countries.[8] In referring to the EU's consistency, this chapter focuses particularly on the consistency between the EU's rhetoric regarding its normative ideas of democracy promotion and its assessment of the actual progress of target countries' democratic developments.

The reminder of the paper is structured as follows: I continue by outlining the EU's self-representation and motives behind its democracy promotion in its Eastern Neighbourhood (i.e., Armenia, Azerbaijan, Belarus, Georgia, Moldova, and Ukraine). Second, I discuss the notion of consistency and analyse how the EU's strategic interests and ability to promote democracy are linked to its (in)consistent democracy promotion

---

5   Schimmelfennig, Frank/ Sedelmeier, Ulrich (eds): The Europeanization of central and eastern Europe, Ithaca, NY: Cornell University Press, 2005; Schimmelfennig, Frank: EU political accession conditionality after the 2004 enlargement: consistency and effectiveness, in: Journal of European Public Policy, 2008 (vol. 15), no. 6, pp. 918–937.
6   Crawford, Gordon: The EU and democracy promotion in Africa: high on rhetoric, low on delivery, in: Mold, Andrew (ed.): EU development policy in a changing world: challenges for the 21st century, Amsterdam: Amsterdam University Press, 2007, pp. 169–197; Johnson, Juliet: The remains of conditionality: the faltering enlargement of the euro zone, in: Journal of European Public Policy, 2008 (vol. 15), no. 6, pp. 826–841; Schimmelfennig, Frank: EU political accession conditionality after the 2004 enlargement: consistency and effectiveness, in: Journal of European Public Policy, 2008 (vol. 15), no. 6, pp. 918–937; Jørgensen, Knud E./ Rosamond, Ben: Europe: regional laboratory for a global polity?, Coventry: University of Warwick, 2001 (CSGR Working Paper no. 71/01), http://wrap.warwick.ac.uk/2054/1/WRAP_Rosamond_wp7101.pdf.
7   Schimmelfennig, Frank: EU political accession conditionality after the 2004 enlargement: consistency and effectiveness, in: Journal of European Public Policy, 2008 (vol. 15), no. 6, pp. 918–937, here p. 920.
8   Crawford, Gordon: Foreign aid and political conditionality: issues of effectiveness and consistency, in: Democratization, 1997 (vol. 4), no. 3, pp. 69–108; Crawford, Gordon: The EU and democracy promotion in Africa: high on rhetoric, low on delivery, in: Mold, Andrew (ed.): EU development policy in a changing world: challenges for the 21st century, Amsterdam: Amsterdam University Press, 2007, pp. 169–197.

assessments. Next, I operationalise the EU's (in)consistency variable and show that there is cross-time and cross-country variation. Finally, using regression analysis, I suggest that there is a positive association between the degree of the EU's assessment consistency and the level of target countries' energy exports and export diversification.

9.2. The EU's Self-Representation as a Democracy Promoter

In the theoretical literature there is a large scholarly debate questioning the EU's external democracy promotion practices and its self-representation as an international actor in the globalised world.[9]

The idea of the EU representing a form of 'normative power' in its external relations holds significant appeal.[10] According to this approach, the EU's role in the development of liberal-democratic security communities is regarded as essentially different from those of other major powers. This approach was first explicitly formulated in the mid-1970s by Dûchene, who described Europe as a civilian power that is 'long on economic power and relatively short on armed forces'.[11] Manners' conception of normative power goes beyond this understanding of the EU's power. He notes that not only does the EU aim to achieve its goals without military power but also that the norms that it pursues are universal and anchored in a Kantian philosophy. Thus, the idea of a Normative Power Europe (NPE) implies that the EU constitutes 'a force for good' in world politics that binds itself to international norms, even if they are not in its interest, and acts accordingly. More importantly, being constructed on a normative basis, the EU is predisposed to act normatively in world politics.[12] Thus, in a nutshell, according to NPE, the EU is unique because of its ability to define what is 'normal' in world affairs and to promote those norms by non-military means through the attrac-

---

9   Manners, Ian: Normative power Europe: a contradiction in terms?, in: Journal of Common Market Studies, 2002 (vol. 40), no. 2, pp. 235–258; Youngs, Richard: Normative dynamics and strategic interests in the EU's external identity, in: Journal of Common Market Studies, 2004 (vol. 42), no. 2, pp. 415–435; Hyde-Price, Adrian: Normative power Europe: a realist critique, in: Journal of European Public Policy, 2006 (vol. 13), no. 2, pp. 217–234.
10 Whitman, Richard G.: From civilian power to superpower? The international identity of the EU, Basingstoke: Palgrave, 1998; Cerutti, Furio: Towards the political identity of the Europeans: an introduction, in: Cerutti, Furio/ Rudolph, Enno (eds): a Soul for Europe: on the political and cultural identity of the Europeans. a reader, vol. 1, Leuven: Peeters, 2001, pp. 1–31; Manners, Ian: Normative power Europe: a contradiction in terms?, in: Journal of Common Market Studies, 2002 (vol. 40), no. 2, pp. 235–258; Cederman, Lars-Erik: Nationalism and bounded integration: what it would take to construct a European demos, in: European Journal of International Relations, 2002 (vol. 7), no. 2, pp. 139–174.
11 Dûchene, François: The European Community and the uncertainties of interdependence, in: Kohnstamm Max/ Hager, Wolfgang (eds.): a nation writ large? Foreign-policy problems before the European Community, London: Macmillan, 1973, pp. 1–21, here p. 19.
12 Manners, Ian: Normative power Europe: a contradiction in terms?, in: Journal of Common Market Studies, 2002 (vol. 40), no. 2, pp. 235–258, here p. 252.

tiveness of its example, which reflects the EU's 'core' norms and principles of peace, liberty, democracy, respect for human rights, and the rule of law.[13]

At the same time, however, as the theoretical and empirical literature suggests, EU's self-portrayal as a normative power seeking to promote democracy beyond its borders is somewhat flawed. In reality, the EU's actions suggest a prioritisation of security and stability commitments rather than the promotion of normative values. From a neo-realist perspective, as Hyde-Price argues, the EU is a hegemonic power that imposes its norms on its Eastern Neighbours in a way that is beneficial for EU member states.[14] Hence, the EU's external democratic rules promotion is regarded merely as an instrument that serves its particular strategic interests and will always rank below national security interests.[15] In the same vein, Forsberg and Herd point to examples in which the EU's strategic agendas trump any concern for human rights.[16] On the other hand, as Chandler notes, because it is difficult to assess the effectiveness and impact of the EU's normative principles empirically, we are led to believe that in fact these principles are more 'rhetoric without responsibility'.[17]

Yet there is also another, relatively new perspective, according to which the EU is considered not as a wholly normative actor, but rather as an actor that mixes its normative positions with rationalist assumptions. This stream of thought introduces a degree of rationalism to the normative understanding of the EU.[18] As Youngs suggests, the EU's sort of self-representation is not only due to member states' importance in the EU's institutional structure but also because the EU operates in a predominantly realist international environment.[19]

---

13 Ibid.; Wood, Steve: The European Union: a normative or normal power?, in: European Foreign Affairs Review, 2009 (vol. 14), no. 1, pp. 113–128; Gerrits, André (ed.): Normative power Europe in a changing world: a discussion, The Hague: Netherlands Institute for International Relations Clingendael, 2009 (Clingendael European Papers no. 5), http://www.clingendael.nl/sites/default/files/20091200_cesp_paper_gerrits.pdf; Whitman, Richard G.: Normative power Europe: empirical and theoretical perspectives, Basingstoke: Palgrave Macmillan, 2011; Diez, Thomas: Normative power as hegemony. Paper for presentation at the 2011 EUSA Biennial International Conference, Boston, MA, 3–5 March 2011, http://euce.org/eusa/2011/papers/7l_thomas.pdf.
14 Hyde-Price, Adrian: Normative power Europe: a realist critique, in: Journal of European Public Policy, 2006 (vol. 13), no. 2, pp. 217–234, here p. 227.
15 Ibid.
16 Forsberg, Tuomas/ Herd, Graeme P.: The EU, human rights, and the Russo–Chechen conflict, in: Political Science Quarterly, 2005 (vol. 120), no. 3, pp. 455–478.
17 Chandler, David: Rhetoric without responsibility: the attraction of 'ethical' foreign policy, in: The British Journal of Politics & International Relations, 2003 (vol. 5), no. 3, pp. 295–316, here p. 303.
18 Youngs, Richard: Normative dynamics and strategic interests in the EU's external identity, in: Journal of Common Market Studies, 2004 (vol. 42), no. 2, pp. 415–435; Diez, Thomas: Constructing the self and changing others: reconsidering 'Normative Power Europe', in: Millennium, 2005 (vol. 33), no. 3, pp. 613–636; Pace, Michelle: The construction of EU normative power, in: Journal of Common Market Studies, 2007 (vol. 45), no. 5, pp. 1041–1064.
19 Youngs, Richard: Fusing security and development: just another Euro platitude?, in: European Integration, 2008 (vol. 30), no. 3, pp. 419–437.

9. The EU's Democracy Promotion in its Eastern Neighbourhood

To uncover the EU's pattern of (in)consistency, the issue will be situated within this third approach and contribute to its interpretation of the EU as a distinct international actor that acts based on its normative commitments, although it can be simultaneously embroiled in a process of rational calculations. This interpretation of the EU's image can potentially shed light on and help to understand the EU's democracy promotion practices in its neighbourhood and particularly in the EaP area.

9.3. The Concept of the EU's Consistency

The concept of the EU's consistency has been widely studied in the academic literature and is regarded as a crucial factor in the effectiveness of the EU's foreign policies.[20] The literature discusses three major dimensions of the EU's foreign policy, namely the vertical, horizontal, and external.

Vertical consistency refers to the coherence between member states' external policies and EU policies.[21] In particular, it determines to what extent member states' domestic policies line up with the EU's external policies, i.e., the ability of the EU to 'speak with one voice'. The horizontal approach to consistency determines compatibility between intergovernmental and communautaire aspects of the EU's external policies and 'the consistency of the individual policies formulated within these two dimensions'.[22] One particular subgroup of this dimension is institutional consistency, which denotes the coherence[23] of policy-making between the Council and Commission and of the different sub-structures within these institutions.[24]

---

20 Gauttier, Pascal: Horizontal coherence and the external competences of the European Union, in: European Law Journal, 2004 (vol. 10), no. 1, pp. 23–41; Nuttall, Simon: Coherence and consistency, in: Hill, Christopher/ Smith, Michael (eds): International relations and the European Union, Oxford: Oxford University Press, 2005, pp. 91–112; Gaspers, Jan: The quest for European foreign policy consistency and the Treaty of Lisbon, in: Humanitas Journal of European Studies, 2008 (vol. 2), no. 1, pp. 19–25; Cremona, Marise: Defining competence in EU external relations: lessons from the Treaty reform process, in: Dashwood, Alan/ Maresceau, Marc (eds): Law and practice of EU external relations: salient features of a changing landscape, Cambridge: Cambridge University Press, 2008, pp. 34–69; Smith, Karen E.: European Union foreign policy in a changing world, Cambridge: Polity, 2009.
21 Nuttall, Simon: Coherence and consistency, in: Hill, Christopher/ Smith, Michael (eds): International relations and the European Union, Oxford: Oxford University Press, 2005, pp. 91–112, here pp. 96–98.
22 Gaspers, Jan: The quest for European foreign policy consistency and the Treaty of Lisbon, in: Humanitas Journal of European Studies, 2008 (vol. 2), no. 1, pp. 19–25, here p. 21.
23 Following the argument proposed by Nuttall the terms 'coherence' and 'consistency' will be used interchangeably in the paper. As he puts 'attempts to distinguish between [the terms] risk ending up in linguistic pedantry' (Nuttall, Simon: Coherence and consistency, in: Hill, Christopher/ Smith, Michael (eds): International relations and the European Union, Oxford: Oxford University Press, 2005, pp. 91–112, here p. 93).
24 Ibid.

While the latter two dimensions refer to the internal features of the EU's foreign policy, the external dimension, which will be the main focus of this chapter, refers to the extent to which the EU's policies are consistent with respect to the norms it claims to advocate in target countries. In this regard, consistency means that the EU should treat all third countries similarly in spite of its foreign policy concerns and interests towards the countries. For instance, the EU's 'reaction to human right violations in any third country should be similar regardless of the fact in which country they take place in'.[25] Any inconsistencies raise doubts about the extent to which democracy promotion and human rights protection are genuine concerns in its foreign policy and thus undermine legitimacy, credibility and the effectiveness of its policies abroad.[26] Within this third approach, I examine a particular aspect of consistency, namely the consistency between the EU's rhetoric regarding its normative ideas of democracy promotion and its assessments of the actual progress of democratic development of target countries.

9.4. The EU's Interests and Ability to Promote Democratic Norms

While there are undoubtedly various factors behind the inconsistencies among the EU's democracy promotion assessments, two broad explanations can be identified. First, the EU's *interests* (security, strategic etc.) are expected to trump its democratic norms, i.e., when there is clash between the two strategies, the EU's interest is expected to prevail. The second reason inconsistencies emerge is the EU's *ability* to promote its norms.

According to the democratic peace proposition, democracies should have not only a normative commitment to spread democracy abroad but also strategic interests that enhance their own benefit such as security, trade, investment etc.[27]

Within the EU's neighbourhood policy, democracy promotion is tied to its security interests. This can be traced back to two key documents, the European Neighbourhood Policy (ENP)[28] strategy paper and the European Security Strategy (ESS). As posited in

---
25 Smith, Karen E.: European Union foreign policy in a changing world, Cambridge: Polity, 2009, p. 73.
26 Crawford, Gordon: The EU and democracy promotion in Africa: high on rhetoric, low on delivery, in: Mold, Andrew (ed.): EU development policy in a changing world: challenges for the 21st century, Amsterdam: Amsterdam University Press, 2007, pp. 169–197; Johnson, Juliet: The remains of conditionality: the faltering enlargement of the euro zone, in: Journal of European Public Policy, 2008 (vol. 15), no. 6, pp. 826–841; Schimmelfennig, Frank: EU political accession conditionality after the 2004 enlargement: consistency and effectiveness, in: Journal of European Public Policy, 2008 (vol. 15), no. 6, pp. 918–937; Smith, Karen E.: European Union foreign policy in a changing world, Cambridge: Polity, 2009.
27 Wolff, Jonas/ Wurm, Iris: Towards a theory of external democracy promotion: a proposal for theoretical classification, in: Security Dialogue, 2011 (vol. 42), no. 1, pp. 77–96.
28 Through its European Neighbourhood Policy (ENP), the EU works with its southern and eastern neighbours to develop political associations and economic integration. Partner countries agree on an ENP action plan demonstrating their commitment to democracy, human rights, the rule

9. The EU's Democracy Promotion in its Eastern Neighbourhood

the ENP strategy paper, which serves as a key document for the EU's policy making in its neighbourhood, the EU's task is to

> make a particular contribution to stability and good governance in [its] immediate neighbourhood [and] to promote a ring of well governed countries to the East of the European Union and on the borders of the Mediterranean with whom [EU] can enjoy close and cooperative relations.[29]

In the same vein, as is highlighted in the ESS, the overall objective of the EU is to ensure security, stability and prosperity in its immediate neighbourhood through the promotion of democracy, the rule of law, and human rights values.[30] Stated in this way, one can observe how the EU directly links the importance of democracy promotion with the assurance of security and stability in its neighbourhood.

Hence, it is expected that the EU generally promotes democracy in its neighbourhood, which prompts a legitimate question: when democracy and human right promotion policies sit at the top of the EU's neighbourhood agenda and are even regarded as the basis for the EU's stability and security, why then does the EU frequently eschew consistency in extending those norms to secure its interests? Why are the EU's democracy promotion assessments consistent for one country and flawed for others? To what extent and under what conditions is the EU (in)consistent in its EaP and what accounts for cross-country and cross-temporal variation?

Part of the explanation may lie in the rationalist understanding of the EU's norm promotion policies, as discussed above. According to this perspective, democracy is promoted as long as it is not detrimental to the EU's strategic interests.[31] In cases when a conflict emerges between its security interests and the policy of promoting democratic norms, however, the EU is inclined to give priority to the former.[32] Thus, the

---

of law, good governance, market economy principles, and sustainable development. Of the 16 ENP countries, 12 have currently agreed on ENP action plans (Armenia, Azerbaijan, Egypt, Georgia, Israel, Jordan, Lebanon, Moldova, Morocco, Palestine, Tunisia, Ukraine). While Algeria is currently negotiating an ENP action plan, Belarus, Libya and Syria remain outside most of the structures of the ENP (for more information, see http://eeas.europa.eu/enp/index_en.htm).

29 European Commission: European Neighbourhood Policy—strategy paper, COM/2004/373 final, 12 May 2004, p. 6, http://ec.europa.eu/world/enp/pdf/strategy/strategy_paper_en.pdf.
30 European Union: a secure Europe in a better world—European security strategy, 12 December 2003, http://www.consilium.europa.eu/uedocs/cmsUpload/78367.pdf.
31 Wolff, Jonas/ Wurm, Iris: Towards a theory of external democracy promotion: a proposal for theoretical classification, in: Security Dialogue, 2011 (vol. 42), no. 1, pp. 77–96.
32 Carothers, Thomas: Aiding democracy abroad: the learning curve, Wasington, DC: Carnegie Endowment for International Peace, 1999; Emerson, Michael/ Aydin, Senem/ Noutcheva, Gergana/ Tocci, Nathalie/ Vahi, Marius/ Youngs, Richard: The EU as a promoter of democracy in its neighbourhood, in: Emerson, Michael (ed.): Democratisation in the European neighbourhood, Brussels: Centre for European, Policy Research, 2005, pp. 169–230; Crawford, Gordon: The European Union and democracy promotion in Africa: the case of Ghana, in: European Journal of Development Research, 2005 (vol. 17), no. 4, pp. 571–600; Jünemann, Annette/ Knodt, Michèle (eds): Die externe Demokratieförderung der EU, Baden-Baden: Nomos, 2007.

underlying assumption of the EU's security-through-democracy nexus proves to be problematic in practice because democracy is given less policy attention.[33]

While this approach explains to a certain extent why the EU acts inconsistently in cases where there is a clash between normative values and interests, it is difficult to explain why inconsistent assessments occur when this clash is non-existent. In other words, it is possible to observe occasions when, although the EU has no strategic interests in a third country, its assessments of democracy promotion still reveal inconsistencies. One proposition that has been given less attention in the literature, and which I consider as another essential determinant to the EU's inconsistency, is the EU's ability to promote democracy. The latter idea is based on the assumption that the more intensive, diverse and strong the EU's interactions and co-operation with target countries are, the stronger the EU's leverage to push for reforms and, therefore, the more consistent the EU's assessments of democracy promotion will be.

The concept of ability is examined in the literature from three perspectives: external democracy promotion, or the 'leverage model' of intergovernmental co-operation; the 'linkage model' of transnational socialisation; and the 'governance model' of trans-governmental exchanges.[34] The leverage model represents a direct channel of democracy promotion through various instruments and strategies that are used intentionally to affect the democratisation process in target countries. In contrast, in the latter two models, an indirect and more nuanced mechanism of the EU's norm transfer is applied that, although they require no explicit policy to promote democracy, could exert positive effects on the democratisation processes.[35] Against this background, this chapter argues that as a consequence of intense direct and indirect interactions between the EU and target countries, the ability of the EU to push for reforms and to act more consistently in its democracy promotion assessments will increase.

9.5. Trade Structure and the EU's Consistency

The latest enlargements of the EU brought the EaP countries into its immediate neighbourhood, thus making the stability and security of these countries an object of direct concern and strategic value for the EU. Hence, the promotion of its normative principles

---

33　Youngs, Richard: The European Union and the promotion of democracy, New York: Oxford University Press, 2001; Gillespie, Richard/ Youngs, Richard: Themes in European democracy promotion, in: Democratization, 2002 (vol. 9), no. 1, pp. 1–16.

34　Freyburg, Tina/ Lavenex, Sandra/ Schimmelfennig, Frank/ Skripka, Tatiana/ Wetzel, Anne: Democracy promotion through functional cooperation? The case of the European Neighbourhood Policy, in: Democratization, 2011 (vol. 18), no. 4, pp. 1026–1054; Lavenex, Sandra/ Schimmelfennig, Frank: EU democracy promotion in the neighbourhood: from leverage to governance?, in: Democratization, 2011 (vol. 18), no. 4, pp. 885–909.

35　Freyburg, Tina: Transgovernmental networks as catalysts for democratic change? EU functional cooperation with Arab authoritarian regimes and socialization of involved state officials into democratic governance, in: Democratization, 2011 (vol. 18), no. 4, pp. 1001–1025.

9. The EU's Democracy Promotion in its Eastern Neighbourhood

of democracy, human rights, the rule of law, and the peaceful settlement of regional territorial conflicts became essential objectives for the preservation of sustainable peace and security in its neighbourhood.[36]

Another critical reason that enhanced the EU's interests in its Eastern Neighbourhood stability and security is linked to its growing concern for diversified energy supplies and alternative routes that bypass Russia and the Middle East. By 2030, the EU, which is the world's largest importer of oil and gas, is estimated to be reliant on imports for 70% of its energy needs. In this respect, the presence of Caspian energy resources and its neighbourhood location as a strategic energy transit corridor is clearly important and assumes special significance. Thus, the EaP area is vital to the supply of its energy needs. In this respect, and as the ESS explicitly states, 'greater diversification of fuels, sources of supply, and transit routes, is essential, as are good governance, respect for the rule of law and investment in source countries'. In support of these objectives, the EU calls for greater engagement in its neighbourhood.[37]

As discussed above, the EU may deviate from its consistency pattern in cases when its strategic objectives such as security or economic interests are prioritised over norms of democratic rule transfer.[38] In its Eastern Neighbourhood, the EU's energy supply security interests are regarded as one of the key concerns that can mitigate its attempts to promote democratic rule and good governance. a clear example of how the EU's security considerations can override its support for democracy and human rights reforms can be found in the case of Azerbaijan. Here, despite continuing violations of human rights and democratic principles, the EU has voiced very little criticism, which is mainly linked to strategic considerations.[39]

To test this line of argument empirically, I consider the proportion of energy in a target country's exports to the EU as an imperfect proxy to measure the degree of the EU's commercial interests in the target country.

---

36 Whitman, Richard G.: Normative power Europe: empirical and theoretical perspectives, Basingstoke: Palgrave Macmillan, 2011, p. 70.
37 European Union: a secure Europe in a better world—European security strategy, 12 December 2003, http://www.consilium.europa.eu/uedocs/cmsUpload/78367.pdf.
38 Crawford, Gordon: Foreign aid and political reform: a comparative analysis of political conditionality and democracy assistance, Basingstoke: Palgrave, 2001; Gillespie, Richard/ Youngs, Richard (eds): The European Union and democracy promotion: the case of North America. London: Routledge, 2002.
39 Stewart, Susan: The interplay of domestic contexts and external democracy promotion: lessons from Eastern Europe and the South Caucasus, in: Democratization, 2009 (vol. 16), no. 4, pp. 804–824; Stewart, Susan: Democracy promotion before and after the 'colour revolutions', in: Democratization, 2009 (vol. 16), no. 4, pp. 645–660; Börzel, Tanja/ Pamuk, Yasemin: Pathologies of Europeanisation: fighting corruption in the Southern Caucasus, in: West European Politics, 2012 (vol. 35), no. 1, pp. 79–97.

H1: Thus, the proportion of energy volume in target countries' exports to the EU is assumed to evoke more inconsistent EU democracy promotion assessments in target countries.

As a second step, I aim to test whether the EU's increased ability to promote democracy is associated with more consistent assessments. For the EU as a trade power, trade is a key source to disseminate its democratic rules.[40] Therefore, the structure[41] of target countries' exports, i.e., the extent of exports' concentration and diversification, is taken as a one possible proxy for the EU's ability to promote democracy.

While diversified trade enables the exchange of goods and services in many trade areas, as a virtuous circle, it will ease the EU's ability to diffuse its democratic rules, liberal institutions and democratic structures. Hence, under intensive trade links, successful and rather consistent EU assessments are presumed to follow. At the same time, super-concentrated trade based mostly on resources will impede the EU's ability to promote democracy.

H2: Thus, while the presence of many exporting sectors will facilitate more consistent EU assessments, the concentrated export structure of target countries will render the EU's assessment of democratic rule promotion less consistent.

Moldova and Azerbaijan represent two extreme cases of trade structure in the EaP area. While Azerbaijani exports to the EU comprised in 2012 mainly energy resources (99.1%), Moldova's trade balance was more diversified (with 33.8% miscellaneous manufactured articles accounted for the biggest share, followed by machinery and transport equipment with 16.8% and food and live animals with 15.8%).[42] Accordingly, it is expected that in the case of Azerbaijan, the EU will show less consistency in its assessment of democracy promotion, while in the case of Moldova, more consistent assessments should follow.

40　Youngs, Richard: The European Union and democracy promotion in the Mediterranean: a new or disingenuous strategy?, in: Democratization, 2002 (vol. 9), no. 1, pp. 40–62; Meunier, Sophie/ Nicolaidis, Kalypso: The European Union as a conflicted trade power, in: Journal of European Public Policy, 2006 (vol. 13), no. 6, pp. 906–925; Schimmelfennig, Frank/ Scholtz, Hanno: EU democracy promotion in the European Neighbourhood: political conditionality, economic development and transnational exchange, in: European Union Politics, 2008 (vol. 9), no. 2, pp. 187–215.
41　In the present study, export structure reflects target countries' product export portfolios to the EU. Export diversification means that the target country exports a wide range of products to the EU over time. Conversely, a country whose exports comprise a small number of products has a higher export concentration.
42　EU imports by SITC product grouping for 2012 according to EuroStat data, http://ec.europa.eu/trade/policy/countries-and-regions/statistics/index_en.htm (accessed 23 November 2013).

9.6. Data

The sample analysed in this chapter consists of five EaP countries—Armenia, Azerbaijan, Georgia, Ukraine, and Moldova—for the period from 2007 to 2012.[43] First, I will operationalise the dependent variable, which is the degree of consistency in the EU's assessment of the state of democracy in its Eastern Neighbourhood countries. Subsequently, I will use regression analysis to determine whether a high share of energy in the country's trade balance and export diversification leads to more consistent EU assessments. The data for energy trade volume and diversification are obtained from the EU's statistical database.[44]

9.6.1. Operationalisation of the EU's Assessment Consistency

The dependent variable measures the annual level of consistency in the EU's assessments of the state of democracy in its Eastern Neighbourhood. EU institutions carefully examine the process of democratisation in these countries and subsequently report on the improvements or drawbacks of democratic processes through the release of annual country progress reports, strategy papers, memoranda, statements, and so on. To measure the degree of assessment consistency, I compare Freedom House (FH)[45] datasets capturing the democratisation dynamics in the Eastern Neighbourhood with the EU's own assessments.[46] For this reason, I first take democracy measures reported by FH in its annual 'Nations in Transit' publication.[47] The latter gives a composite measure of the extent of democratisation in the post-Soviet countries and covers seven categories of democratisation, which are each ranked on a seven-point Likert scale with '1' representing the highest and '7' the lowest level of democratic progress; FH has used these categories since 2005 (see Table 9-1). The measure is counter-intuitive because the most democratic states receive the lowest score and vice versa.

---

43   Belarus, the sixth country from the Eastern Neighbourhood, is dropped from the study because thus far no EU progress reports have been conducted. Since the EU's progress reports for all five countries under study are available only from 2007, this year was chosen as a starting point to measure the EU's consistency.
44   Cf. EuroStat data, http://ec.europa.eu/trade/policy/countries-and-regions/statistics/index_en.htm (accessed 23 November 2013).
45   Freedom House, founded in 1941, is a US-based non-governmental organization that conducts global research and advocacy on democracy, political freedom and human rights (cf. http://www.freedomhouse.org/).
46   In a later stage of research, the study will embrace other international institutions' democracy measurements, such as those of the Human Right Watch, Transparency International, etc.
47   Cf. http://www.freedomhouse.org/report-types/nations-transit (accessed 21 November 2013).

Table 9-1: Freedom House Democratisation Categories

Democratisation Categories	Definition
1 National Democratic Governance	Considers the democratic character and stability of the governmental system; the independence, effectiveness, and accountability of legislative and executive branches; and the democratic oversight of military and security services.
2 Electoral Process	Examines national executive and legislative elections, electoral processes, the development of multiparty systems, and popular participation in the political process.
3 Civil Society	Assesses the growth of NGOs, their organisational capacity and financial sustainability, and the legal and political environment in which they function; the development of free trade unions; and interest group participation in the policy process.
4 Independent Media	Addresses the current state of press freedom, including libel laws, harassment of journalists, editorial independence, the emergence of a financially viable private press, and Internet access for private citizens.
5 Local Democratic Governance	Considers the decentralisation of power; the responsibilities, election, and capacity of local governmental bodies; and the transparency and accountability of local authorities.
6 Judicial Framework and Independence	Highlights constitutional reform, human rights protections, criminal code reform, judicial independence, the status of ethnic minority rights, guarantees of equality before the law, treatment of suspects and prisoners, and compliance with judicial decisions.
7 Corruption	Looks at public perceptions of corruption, the business interests of top policy makers, laws on financial disclosure and conflict of interest, and the efficacy of anticorruption initiatives.
Democracy Score	Straight average of the ratings for all categories

Source: Freedom House, Nations in Transit, http://www.freedomhouse.org/report/nations-transit-2013/nations-transit-2013-methodology (accessed 11 November 2013).

Subsequently, I employ Quantitative Content Analysis (QCA) to analyse the EU's annual progress reports, in which the democratisation processes of the EaP countries are individually addressed and evaluated. In this analysis, the EU's democratisation categories are adjusted to match FH's seven categories, namely national democratic governance; electoral process; civil society; independent media; local democratic governance; judicial framework and independence; and corruption. Since there are relatively few EU reports for the EaP countries, the indices of the EU's assessments of target countries' democratic developments are generated manually, and the following five text-driven codes are detected (see Table 9-2).

9. The EU's Democracy Promotion in its Eastern Neighbourhood

Table 9-2: EU Democratisation Assessment Units and Definitions

Democracy rating	Expressions used in the EU reports
-0.50	'Good progress'
-0.25	'Progress', 'positive steps are taken in improving', 'some progress', 'improved slightly', 'is a step forward'
0	'No progress', 'no improvement'
+0.25	'Raised concerns', 'slightly decreased'
+0.50	'Raised serious concerns'

Based on the available data, depending on the strengths/weaknesses of the EU's assessed progress/setback of democratisation processes, I have derived a scale of the EU's assessment of the democratisation dynamics in the target countries by building a straight average of the ratings for all categories per country per year. The measure is also counter-intuitive in that the most democratic states receive the lowest score.

Next, I compare FH's scores with those from the EU assessments. In this regard, to measure the deviation between FH's 'independent' scores and those of the EU, I focus on the change in FH's democracy scores from one year to the next rather than on their absolute level. This method of comparison allows for the possibility of tracking the processes of democratisation taking place in the target countries according to both assessments.

However, it is important to mention that while the FH index is the most widely used measurement of democracy in research and policy, the validity and reliability of its indices have been extensively criticised in the literature. Scholars mainly stress the index's methodological weaknesses,[48] arguing that by relying on expert-based evaluations, the methodology is likely to fall into measurement error.[49] Other grounds for criticism include the lack of specificity and rigour in the index's construction, the inadequate level of transparency and replicability of the scales and the ideological biases of its methodology.[50]

---

48 Landman, Todd/ Häusermann Julia: Map-making and analysis of the main international initiatives on developing indicators on democracy and good governance, Colchester: University of Essex, Human Rights Centre, 2003, http://www.oecd.org/development/governance-development/20755719.pdf.

49 Ibid.; Coppedge, Michael J.: Democratization and research methods, Cambridge: Cambridge University Press, 2012.

50 Bollen, Kenneth/ Paxton, Pamela: Subjective measures of liberal democracy, in: Comparative Political Studies, 2000 (vol. 33), no. 1, pp. 58–86; Diamond, Larry J.: Thinking about hybrid regimes, in: Journal of Democracy, 2002 (vol. 13), no. 2, pp. 21–35; Munck, Gerardo L./ Verkuilen, Jay: Conceptualizing and measuring democracy Evaluating alternative indices, in: Comparative Political Studies, 2002 (vol. 35), no. 1, pp. 5–34; Giannone, Diego: Political and ideological aspects in the measurement of democracy: the Freedom House case, in: Democratization, 2010 (vol. 17), no. 1, pp. 68–97.

However, although the objectivity of the FH indices is much criticised in the literature, what matters in my analysis is the 'independence' of FH's evaluations in comparison to those of the EU. In this respect, the FH index can be considered an imperfect benchmark against which to measure the consistency of the EU's assessments of the state of democracy in the target countries. The deviation between their scores, then, is regarded as a proxy for the degree of the EU assessments' consistency in two dimensions: cross-country (Armenia, Azerbaijan, Georgia, Ukraine, and Moldova) and cross-time (2007–2012). Thus, on the one hand, if the EU's overall democracy score is the same as the FH's democracy assessments, the EU will be considered to be consistent in its assessments. On the other hand, when there is a difference between the change of the overall democracy scores between the EU and FH, the former will be regarded as inconsistent in its assessments.

Graph 1 below shows democratisation developments in the EaP countries across time. The solid lines denote the EU's assessments (based on the content analysis estimates per country and year), while the dashed lines depict changes in the FH democracy scores in comparison to the previous year. The gap between the two curves shows the EU's deviation from the FH score. Thus, because these two institutions evaluate democratisation processes in target countries independently, we can treat the difference between two lines as a measure of the EU's assessment consistency. When the deviation between FH's scores and the EU's assessments is positive, i.e., the FH score is higher than that of EU, it implies that the EU was overly optimistic in its evaluations. In turn, when the bias is negative, i.e., the EU score was higher than the FH score, it denotes that the EU was overly pessimistic in its assessments.

For instance, in the case of Armenia, overall, we can observe that in 2008, FH's democracy rating deteriorated 0.18 units with no progress or setback reported in 2009, while in 2010, an improvement of only 0.04 units was estimated. In the case of the EU, a somewhat different picture emerges. Whereas in 2008 0.14 units of deterioration were reported, in 2009, an improvement of 0.04 units was indicated, an outlook that improved by an additional 0.07 units in 2010.

Figure 9-1: Democratic Change in EaP Countries According to Freedom House and the EU

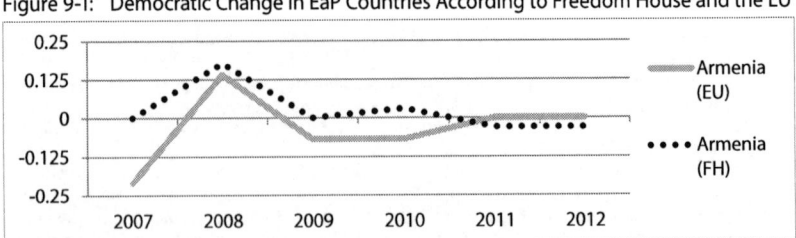

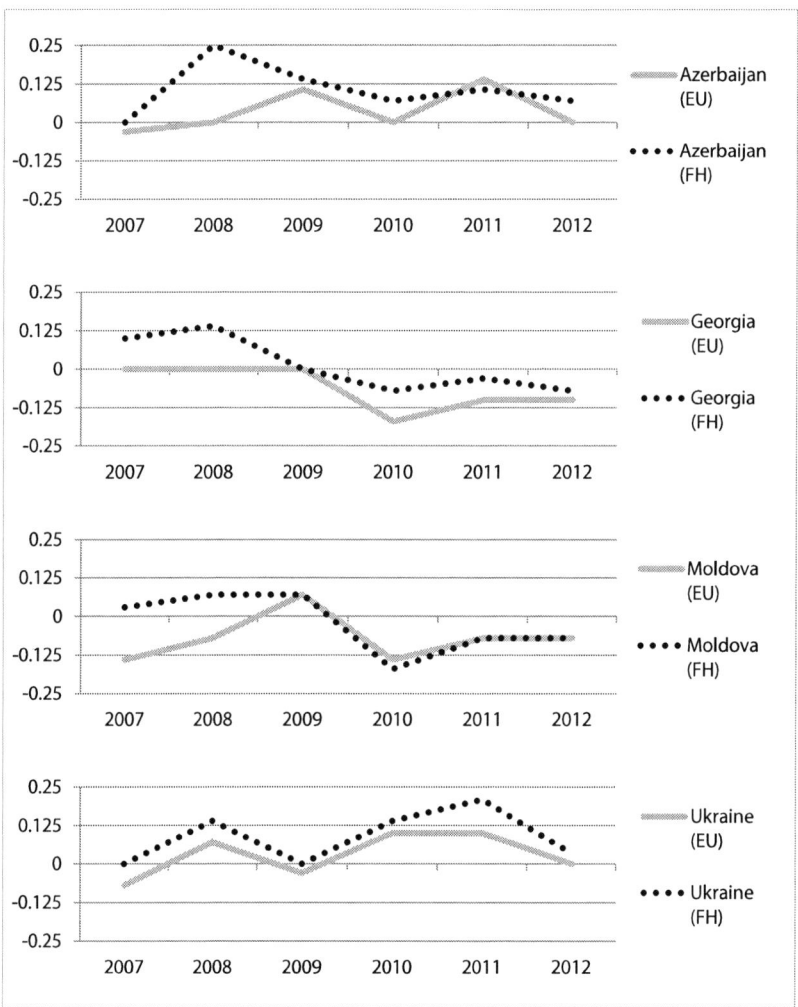

9.6.2. Identification

The study specifies two classes of models to empirically test hypotheses H1 and H2. The models are estimated using an Ordinary Least Squares (OLS) regression analysis with and without fixed effects of the following form:

(1) $EU\ Consistency_{i,t,h} = \alpha + \beta_1 TradeVolume_{i,t} + \beta_2 TradeConcentration_{i,t} + \beta_3 FragilityIndex_{i,t} + \beta_4 ElectionDummy_{i,t} + \beta_5 FHDemocracyIndex_{i,t,h} + \beta_6 GDP_{i,t} + \beta_7 Popolation_{i,t} + \beta_8 Energy\ Dummy_{i,t} + \mu_i + \varepsilon_{i,t,h}$

(2) $EU\ Consistency_{i,t,h} = \alpha + \beta_1 TradeVolume_{i,t} + \beta_2 TradeConcentration_{i,t} + \beta_3 FragilityIndex_{i,t} + \beta_4 ElectionDummy_{i,t} + \beta_5 FHDemocracyIndex_{i,t,h} + \beta_6 GDP_{i,t} + \beta_7 Popolation_{i,t} + \beta_8 Energy\ Dummy_{i,t} + \varepsilon_{i,t,h}$

The dependent variable 'EU Consistency' is the deviation of the EU's assessment of the democratic processes in target countries from that of the FH score per seven categories (h) for each country (i) and year (t). Thus, a positive (negative) number would indicate that the EU was overly optimistic (pessimistic) in evaluating a given country's progress from the previous year, while numbers closer to zero imply similar, thus consistent, behaviour between FH and the EU. On the right hand side, 'Trade Volume' is the (log) volume of bilateral trade (exports plus imports) between the EU and the EaP countries. a number of cross-time and cross-country variables enter into the regression as controls: 'Trade Concentration', which is the main variable of interest, is calculated as the sum of squares of the shares of different categories of traded goods and service in total imports and exports (i.e., 'Export Concentration' and 'Import Concentration'). 'Fragility Index' measures the effectiveness and legitimacy of state performance along four dimensions of state functions, namely the economic, political, security, and social spheres. The index ranges from '0', 'no fragility', to '25', 'extreme fragility'. 'FH Democracy Index' captures democracy ratings in levels per seven categories (h). The 'Energy Dummy' variable divides the countries into two groups in which the countries that possess or transport energy resources are coded '1' (Azerbaijan, Georgia, and Ukraine) and otherwise coded '0' (Armenia, Moldova). 'Elections Dummy' is coded '1' for the years when either parliamentary or presidential elections were held in the target countries and coded '0' when no elections were held. 'GDP' is the gross domestic product and 'Population' is the annual population, both are specified in logarithms. Alternative modifications of this baseline specification will additionally control for unobserved cross-country heterogeneity by including country fixed effects—μ_i. Finally, ε is the well-behaved error term, robust to heteroskedasticity and autocorrelation.

Regarding the model parameters, α is the constant term, while βs represent the coefficients capturing the effect of explanatory variables on the dependent variable. For example, β_1 will be the effect of a one per cent change in bilateral trade volume on the 'EU Consistency' score.

9.7. Results

Table 9-3 on pp. 248–249 presents the results from estimating equations (1) and (2). As discussed in the identification section above, the dependent variable is 'EU Consistency' and the main independent variable of interest is the level of target countries' 'Export Concentration'. The table consists of 4 panels: panel a (columns 1–2) is the baseline regression; panel B introduces country fixed effects to account for time-constant country-specific heterogeneity (columns 3–5); panel C (columns 6–8) is a robustness check for panel B with the same specification but with standard errors robust to heteroskedasticity and clustered at country and country-category levels; and finally, the last panel, D, tests the consistency of the results to the exclusion of energy-rich Azerbaijan (column 9).

Before interpreting the results, it should be re-noted that the sample consists of five EaP countries (Armenia, Azerbaijan, Georgia, Moldova, and Ukraine) for the time period from 2007 to 2012 for eight democracy categories (seven categories and their aggregated score); this results in a balanced hierarchical panel dataset of 240 (5 x 6 x 8) observations.

I will now discuss the interpretation of the empirical results. The first row of each independent variable represents the β coefficient of the models specified in the last section. a star denotes a significance level at 10%, two stars at 5%, and three stars denote 1% significance. Each number below the coefficient shows the standard error.

The 'Trade Volume' variable is positive, albeit insignificant in all models. However, in the last panel, D, which excludes the Azerbaijani case, the β coefficient reaches a conventional significance level. In the latter case, 'Trade Volume' shows significance at the 1% level and a positive sign, indicating that a higher trade volume is associated with more inconsistent EU assessments.

The different outcome of the last regression may suggest that Azerbaijan, with its abundance of natural resources, is an outlier that largely deviates from the rest of the countries in the sample.

The explanatory variable, 'Export Concentration', is positive, and in most of the estimations, it shows significance at the 1% level. It is also robust to the exclusion of Azerbaijan. For instance, in column 4 with country fixed effects (controlling for between-country variables), the 'Export Concentration' variable enters the regression with a coefficient of 0.776, which is significant at the 1% level. In other words, one unit change in the variable 'Export Concentration' produced a 0.776 unit increase in the 'EU Consistency' value.

At the same time, as expected, 'Import Concentration' is also positively correlated with 'EU Consistency', showing significance at the 5% level. 'Energy Dummy' is positively correlated with consistency and, in panel 2, shows significance at the 10% level.

In other words, in cases when the country transports or possesses energy resources ('Energy Dummy' equals 1), the EU becomes less consistent in its assessments.

The evidence of this chapter suggests that there is a negative association between 'Trade Concentration' and 'Energy Dummy' on the one hand and 'EU Consistency' on the other. At the same time, the findings for the 'Trade Concentration' variable are clear-cut: the target countries' export diversification is positively associated with more consistent EU assessments.Ordinary Least Squares (OLS) Results

9.8. Conclusion

The consistency of the EU's actions in general and of its assessments in particular is considered to be the key factor for the effectiveness of the EU's democracy promotion efforts in its Eastern Neighbourhood. This chapter examined the questions of to what extent and under what conditions the EU's assessments of democratic processes in the EaP countries are (in)consistent. To address these questions, this chapter proposed an original approach to measure the extent of the EU's assessment consistency in target countries, which can also be applied beyond the EaP area. In order to create a composite measure of the EU's consistency, the EU's annual progress reports on five EaP countries were first analysed using QCA. The results of the analysis showed how the EU's assessments of the democratisation processes vary across time and country. As a second step, the measurements were compared against the somewhat 'neutral' benchmark of FH's democratisation index. The gap between the change of FH's evaluations of democratisation in comparison to the previous year and the EU's assessments is considered to reveal the extent of the EU's assessment (in)consistency towards the EaP countries.

While the causes of the EU's assessment biases compared to the FH score undoubtedly vary, this chapter highlights two broad arguments. First, inconsistency is about the prioritisation of the EU's interests (in the case of EaP countries, particularly energy supply interests are considered) over its democratic norms. Thus, whenever there is clash between those two strategies, the EU's interests are expected to prevail over its norm promotion policies and thus affect the consistency of its assessments. Second, it is argued that (in)consistencies tend to emerge as a result of the EU's (in)ability to promote its norms. To test these arguments empirically, an OLS regression analysis is applied. In this respect, as an imperfect proxy for the explanatory variables, namely the EU's energy interests and ability, I use the share of energy and the trade structure of target countries' exports to the EU, respectively. While trade represents one of the key mechanisms to disseminate democratic norms to the EaP countries, this chapter argues that the more diverse the structure of trade, the greater the EU's ability to push for reforms in the target countries and, therefore, be more consistent in its assessments. The trade structure also reveals whether the target country exports energy resources to the EU. In this respect, three (Azerbaijan, Georgia, and Ukraine) out of five EaP countries either possess or transport energy resources to EU markets, while the remaining two (Armenia and Moldova) do not export energy resources. Thus, the 'Energy Dummy' variable is created and coded '1' when energy resources are present in the trade structure and '0' when the countries neither possess nor transfer such resources.

The results of the regression analysis suggest that there is positive and highly significant evidence of an association between diversified export trade structures and the consistency of the EU's assessments. The 'Energy Dummy' in its turn is also positively correlated with the 'EU Consistency' variable and is significant at the 10% level. Thus, it can be concluded that the EU's consistent assessments towards the target countries are strongly associated with more diversified trade structures and cases in which no energy interests are involved.

Table 9-3: Ordinary Least Squares (OLS) Results

Model	-1		-2		-3		-4		-5		-6		-7		-8		-9	
Sample	All		All		All		All		All		All		All		All		Excl. Aze	
VARIABLES	Consistency	Consistency	Consistency	Consistency	Consistency	Consistency	Consistency	Consistency	Consistency	Consistency	Consistency	Consistency	Consistency	Consistency	Consistency	Consistency	Consistency	Consistency
Log Trade Volume	0.0647		0.124		0.0554		0.109		0.0414		0.109		0.109		0.109		0.342***	
	-0.0828		-0.083		-0.0828		-0.0837		-0.0822		-0.081		-0.117		-0.081		-0.115	
Export Concentration			0.840***				0.776***				0.776**		0.776		0.776***		1.280***	
			-0.256				-0.28				-0.3		-0.441		-0.3		-0.356	
Import Concentration									2.194**									
									-0.974									
FH Democracy Score	0.00941		0.0098		0.0754		0.0699		0.129**		0.0699		0.0699		0.0699		0.0575	
Fragility Index	-0.0143		-0.014		-0.0613		-0.0603		-0.0651		-0.0608		-0.079		-0.0608		-0.0616	
	-0.0158		-0.0127		-0.0333		-0.0144		-0.03		-0.0144		-0.0144		-0.0144		-0.0116	
	-0.018		-0.0176		-0.0231		-0.0237		-0.0229		-0.0225		-0.0106		-0.0225		-0.0218	
Election Dummy	0.0124		-0.00069		-0.00377		-0.00284		0.00853		-0.00284		-0.00284		-0.00284		-0.0337	
	-0.0229		-0.0228		-0.0253		-0.0249		-0.0256		-0.028		-0.0343		-0.028		-0.029	
Log GDP	-0.599**		-0.689**		-0.268		-0.56		-0.311		-0.56		-0.56		-0.56		-1.267***	
	-0.296		-0.291		-0.34		-0.35		-0.337		-0.362		-0.487		-0.362		-0.455	
Log Population	0.523**		0.288		-2.361		-0.683		-0.917		-0.683		-0.683		-0.683		3.417*	
	-0.24		-0.246		-1.555		-1.644		-1.667		-1.399		-1.478		-1.399		-1.993	
Log Distance	0.858		0.737															
	-0.552		-0.542															
Log Genetic Distance	-0.219		-0.740*															
	-0.381		-0.406															

9. The EU's Democracy Promotion in its Eastern Neighbourhood

Energy Dummy	0.527	0.968*						
	-0.575	-0.579						
Constant	✓	✓	✓	✓	✓	✓	✓	✓
Country FE	-	-	✓	✓	✓	✓	✓	✓
Robust S.E.	-	-	-	-	-	✓	✓	✓
Cluster	-	-	-	-	-	-	Country	Count.&Cat.	Country
Observations	240	240	240	240	240	240	240	240	192
R-squared	0.027	0.071	0.046	0.083	0.071	0.083	0.083	0.083	0.138
F	0.707	1.74	1.567	2.488	2.096	2.081	.	2.081	2.953

Standard errors in parentheses
*** p<0.01, ** p<0.05, * p<0.1

Part IV. Ukraine's Energy Policy

Eric Pardo Sauvageot

10. Gas Disputes Between Russia and Ukraine: Patterns of Escalation and Russian Stakes (2006–2009)

10.1. Introduction

This research focuses on three gas disputes in January 2006, March 2008, and January 2009 in which Russia and Ukraine were at loggerheads. Ukraine is both a consumer of Russian gas and a transit country for its gas exports to Central and Western Europe. Approximately 70% of Russian gas that was exported to the rest of Europe during that period was transported through Ukraine.[1] For Russia, this complex triangular relation implies that any dispute with Ukraine can eventually disrupt its export market. This poses a serious dilemma for Russia, as it has to decide whether damage to its image as a reliable supplier is acceptable for the sake of prevailing in disputes with transit countries. The three gas disputes mentioned above are arguably the best cases to examine Russia's reactions to this dilemma because gas cut-offs, as opposed to oil cut-offs, have a greater impact on consumer countries.[2]

Each of the three disputes shows a different pattern of escalation. This leads to the question of how the pattern of escalation changed depending on different decisions made by either Russia or Ukraine in each of these disputes.[3] A short overview of the way that the disputes escalated shows these differences in both countries' decision making: on 1 January 2006, Russia cut off gas to Ukraine as a result of a pricing dispute, and thereafter, Ukraine started siphoning gas off the pipeline, which led Russia to re-establish their gas supply two days later; an agreement was reached on 4 January. In March 2008, Russia merely used partial gas cut-offs and continued negotiations on the debt dispute after having restored a normal gas supply. In January 2009, another pricing and debt dispute led to a similar pattern as three years before, but Russia also

---
[1] The remaining 30% were transported either through Belarus or shipped via the Finland Connector and the Blue Stream and Nord Stream pipelines directly to consumer countries bypassing Ukraine. The share of Russian gas transported through Ukraine has been declining since 2004; it fell from 78% in 2004 to 68% in 2006 and 64% in 2009 (IHS CERA, Ministry of Energy and Coal Industry of Ukraine: Natural gas and Ukraine's energy future, Washington, DC: IHS 2012, Chapter 2, p. 2–3, Figure 2-3, http://s05.static-shell.com/content/dam/shell-new/local/country/zaf/downloads/pdf/research-reports/Ukraine-Policy-Dialogue-report.pdf).
[2] Gas is more difficult and therefore expensive to transport and store than oil, and interconnections and alternative import routes for gas are less available in Europe than those for oil.
[3] It is assumed that a dispute begins when a cut-off has been applied, although there are many previous stages to a dispute in the complex pricing negotiations that take place in the weeks and months beforehand. I adopt the same concept of escalation that has been applied to the most common phenomenon in international relations so far, war.

decided to cut off the gas supply to Central and Western Europe to prevent Ukraine from siphoning gas off the pipeline; the gas supply for both Ukraine and the rest of Russia's consumers was re-established only on 20 January, after an agreement had been reached for the bilateral dispute. This chapter only focuses on Russian decision making in these disputes,[4] aiming to determine the extent to which the patterns of escalation in these three disputes changed owing to changes in Russia's valuation of the status quo. If a change in the patterns of escalation is found, whether factors strictly linked to the energy disputes or other political factors related to Russia–Ukraine relations affected Russian decision making and eventually its responses in each of these disputes will be analysed.

10.2. Literature Review: Russian Energy Policy

Russian energy policy is assessed differently depending on the interpretations provided.[5] Some authors focus (to different degrees) on Russia's political motivations and situation of power,[6] while another group of scholars highlight Russia's economic motivations or at least qualify Russia's political moves,[7] pointing mainly to Russian

---

[4] The reason for focusing on Russia is that in the gas disputes analysed here, Russia moves first, as it supplies gas to Ukraine, so the first consequence of a failure in negotiations on pricing is that Russia stops supplying gas given the absence of a contract.

[5] For a classification of Russia's motivations in its energy policy, see Orttung, Robert W./ Øverland, Indra: A limited toolbox: explaining the constraints on Russia's foreign energy policy, in: Journal of Eurasian Studies, 2011 (vol. 2), no. 1, pp. 74–85, here pp. 75–76.

[6] Goldman, Marshall I.: Petrostate: Putin, power, and the new Russia, Oxford: Oxford University Press, 2008; Lucas, Edward: The new Cold War: Putin's Russia and the threat to the West, New York: Palgrave Macmillan, 2009; Bugajski, Janusz: Dismantling the West: Russia's Atlantic agenda, Washington, D.C.: Potomac Books, 2009; Baran, Zeyno: EU energy security: time to end Russian leverage, in: Washington Quarterly, 2007 (vol. 30), no. 4, pp. 131–144; Nygren, Bertil: The rebuilding of greater Russia: Putin's foreign policy towards the CIS countries, London: Routledge 2008; Nygren, Bertil: Putin's use of natural gas to reintegrate the CIS region, in: Problems of Post-Communism, 2008 (vol. 55), no. 4, pp. 3–15; Socor, Vladimir: Gazprom, the prospects of a gas cartel, and Europe's energy security, in: Cornell, Svante E./ Nilsson, Niklas (eds.): Europe's energy security: Gazprom's dominance and Caspian supply alternatives, Stockholm/ Washington, DC: Central Asia–Caucasus Studies & Silk Road Studies Program, 2008, pp. 72–83; Walker, Martin: Russia v. Europe: the energy wars, in: World Policy Journal, 2007 (vol. 24), no. 1, pp. 1–8; Faber van der Meulen, Evert: Gas supply and EU–Russia relations, in: Europe-Asia Studies, 2009 (vol. 61), no. 5, pp. 833–856, here pp. 849–850; Closson, Stacy: a comparative analysis on energy subsidies in Soviet and Russian policy, in: Communist and Post-Communist Studies, 2011 (vol. 44), no. 4, pp. 343–356; Bruce, Chloë: Fraternal friction or fraternal fiction? The gas factor in Russian–Belarusian relations, Oxford: Oxford Institute for Energy Studies, 2005, http://www.oxfordenergy.org/wpcms/wp-content/uploads/2010/11/NG8-FraternalFrictionOrFraternalfictionTheGasFactorInRussianBelarusianRelations-ChloeBruce-2005.pdf.

[7] Balmaceda, Margarita M.: Belarus: oil, gas, transit pipelines and Russian foreign energy policy, London: GMB Publishing, 2006; Tsygankov, Andrei: New challenges for Putin's foreign policy, in: Orbis, 2006 (vol. 50), no. 1, pp. 153–165; Baev, Pavel K.: Russian energy policy and military power: Putin's quest for greatness, New York: Routledge, 2008; Orban, Anita: Power, energy, and the new Russian imperialism, Westport, CT: Praeger, 2008; Pirani, Simon (eds.): Russian and CIS gas

10. Gas Disputes Between Russia and Ukraine 255

interdependence with European consumers and the dilemmas involved therein.[8] Another strand of literature focuses on alternative factors, such as domestic politics, social constructions of identity that may introduce irrational actions,[9] and substate agents hijacking state agendas and imposing their particular interests. This line of research also provides an extensive overview of the energy sector,[10] by analysing

---

markets and their impact on Europe, Oxford: Oxford University Press, 2009; Pirani, Simon/ Stern, Jonathan/ Yafimava, Katja: The Russo–Ukrainian gas dispute of January 2009: a comprehensive assessment, Oxford: Oxford Institute for Energy Studies, 2009, http://www.oxfordenergy.org/wpcms/wp-content/uploads/2010/11/NG27-TheRussoUkrainianGasDisputeofJanuary2009AComprehensiveAssessment-JonathanSternSimonPiraniKatjaYafimava-2009.pdf; Poussenkova, Nina: The global expansion of Russia's energy giants, in: Journal of International Affairs, 2010 (vol. 63), no. 2, pp. 103–124; Yafimava, Katja: The June 2010 Russian–Belarusian gas transit dispute: a surprise that was to be expected, Oxford, Oxford Institute for Energy Studies, 2010, http://www.oxfordenergy.org/wpcms/wp-content/uploads/2010/11/NG43-TheJune2010RussianBelarusiaNGasTransitDisputeASurpriseThatWasToBeExpected-KatjaYafimava-2010.pdf; Orttung, Robert W./ Øverland, Indra: A limited toolbox: explaining the constraints on Russia's foreign energy policy, in: Journal of Eurasian Studies, 2011 (vol. 2), no. 1, pp. 74–85; Kivinen, Markku: Public and business actors in Russia's energy policy, in: Aalto, Pami (ed.): Russia's energy policies: national, interregional and global levels, Cheltenham: Edward Elgar, 2012, pp. 45–62.

8 Monaghan, Andrew: Russian oil and EU energy security, Watchfield: Conflict Studies Research Centre, 2005, www.da.mod.uk/CSRC/documents/Russian/05(65).pdf; Monaghan, Andrew: Russia–EU: an emerging energy security dilemma, in: Pro et Contra, 2006 (vol. 10), no. 2–3, pp. 1–13, http://www.carnegieendowment.org/files/EmergingDilemma1.pdf; Gölz, Roland: Russlands Erdgas und Europas Energiesicherheit, Berlin: Stiftung Wissenschaft und Politik, 2007, http://www.swp-berlin.org/fileadmin/contents/products/studien/2007_S21_gtz_ks.pdf; Quester, George H.: Energy dependence and political power: some paradoxes, in: Democratizatsya, 2007 (vol. 15), no. 4, pp. 445–454; Goldthau, Andreas: Rhetoric versus reality: Russian threats to European energy supply, in: Energy Policy, 2008 (vol. 36), no. 2, pp. 686–692; Jaffe, Amy M./ Soligo, Ronald: Militarization of energy: geopolitical threats to the global energy system, Houston, TX: Rice University, James A. Baker III Institute for Public Policy, 2008, http://bakerinstitute.org/files/621/; Ericson, Richard E.: Eurasian natural gas pipelines: the political economy of network interdependence, in: Eurasian Geography and Economics, 2009 (vol. 50), no. 1, pp. 28–57; Kropatcheva, Elena: Playing both ends against the middle: Russia's geopolitical energy games with the EU and Ukraine, in: Geopolitics, 2011 (vol. 16), no. 3, pp. 553–573; Schmidt-Felzmann, Anke: EU member states' energy relations with Russia: conflicting approaches to securing natural gas supplies, in: Geopolitics, 2011 (vol. 16), no. 3, pp 574–599.

9 Sherr, James: Europe, Russia, Ukraine and energy: the last warning, in: The World Today, 2009 (vol. 65), no. 2, pp. 14–17; Saari, Sinikukka: Strategies for nurturing trust in the wider Black Sea region, in: Southeast European and Black Sea Studies, 2011 (vol. 11), no. 3, pp. 333–343.

10 Guillet, for example, shows that in the Russia–Belarus gas crisis of January 2007, Russia unexpectedly renounced the favourable transit fee that it had offered to Minsk. The author suggests that bureaucratic interests were to blame, although this suggestion should be regarded as mere conjecture (Guillet, Jérôme: Gazprom as a predictable partner: another reading of the Russian–Ukrainian and Russian–Belarusian energy crises, Paris: Institut Français des Relations Internationales, 2007 (Russia.Nei.Visions, no. 18), http://www.ifri.org/?page=contribution-detail&id=4770&id_provenance=97). In general, siloviki are divided among those dealing with domestic politics and those dealing with international affairs. Greater cohesion exists among the former, although this does not necessarily suggest that relations among the groups are competitive (Kryshtanovskaya, Olga/ White, Stephen: Inside the Putin court: a research note, in: Europe-Asia Studies, 2009 (vol. 57), no. 7, pp. 1065–1075, here pp. 1070–1071). Relying on news information, however, Tsyganov does find that there exists a split between these groups

deficiencies in the functioning of institutions;[11] competing rationality frames where state, business, and personal rent-seeking interests intermingle;[12] the symbiotic relation between the Russian government and Gazprom and its balance between politics and business;[13] or the role of intermediaries, either as vehicles of self-enrichment or as means to enforce strategic objectives (e.g., asset acquisition in transit countries, assertion of explicitly political or geopolitical objectives or 'strategic buffers').[14]

Regarding energy disputes, Orttung and Øverland identify 31 Russian energy disputes with more than 20 countries from 2000 to 2010,[15] to which we could add those that occurred in the 1990s.[16] The two transit countries Belarus and Ukraine in particular have witnessed many disputes where,[17] as several case studies suggest, energy has

(Tsygankov, Andrei P.: Russia's international assertiveness: what does it mean for the West?, in: Problems of Post-Communism, 2008 (vol. 55), no. 2, pp. 38–55, here p. 45, footnote 19).

11 Aalto, Pami: Introduction: Russian energy, old and new, in Aalto, Pami (ed.): Russia's energy policies: national, interregional and global levels, Cheltenham: Edward Elgar, 2012, pp. 1–19.

12 Balmaceda, Margarita M.: Russia's central and eastern European energy transit corridor: Ukraine and Belarus, in Aalto, Pami (ed.) Russia's energy policies: national, interregional and global levels, Cheltenham: Edward Elgar, 2012, pp. 136–155.

13 Campbell, Kurt A./ Price, Jonathan: The global politics of energy, New York: The Aspen Institute, 2008; Stulberg, Adam: Well-oiled diplomacy: strategic manipulation and Russia's energy statecraft in Eurasia, Albany, NY: State University of New York Press, 2007; Balmaceda, Margarita M.: Russia's central and eastern European energy transit corridor: Ukraine and Belarus, in Aalto, Pami (ed.) Russia's energy policies: national, interregional and global levels, Cheltenham: Edward Elgar, 2012, pp. 136–155; see also Kalman, Kalotay: Russian transnationals and international investment paradigms, in: Research in International Business and Finance, 2008 (vol. 22), no. 2, pp. 85–107; Poussenkova, Nina: The global expansion of Russia's energy giants, in: Journal of International Affairs, 2010 (vol. 63), no. 2, pp. 103–124; Bilgin, Mert: Energy security and Russia's gas strategy: the symbiotic relationship between the state and firms, in: Communist and Post-Communist Studies, 2011 (vol. 44), no. 2, pp. 119–127; Casier, Tom: The rise of energy to the top of the EU–Russia agenda: from interdependence to dependence?, in: Geopolitics, 2011 (vol. 15), no. 3, pp. 237–248; Balmaceda, Margarita M.: Russian energy companies in the new Eastern Europe, in: Perovic, Jeronim/ Orttung, Robert W./ Wenger, Andreas (eds): Russian business power: the role of business power in foreign and security relations, London: Routledge, 2006, pp. 67–87.

14 Closson, Stacy: a comparative analysis on energy subsidies in Soviet and Russian policy, in: Communist and Post-Communist Studies, 2011 (vol. 44), no. 4, pp. 343–356, here p. 352; Ševce, Peter: The role of intermediary companies in Russian–Ukrainian relations and implications for the countries of Central, in: Despite Borders, Central and Eastern European Watch, 11 January 2006, p. 3, http://www.despiteborders.com/clanky/data/upimages/sprostredkovadelske_spoloc nosti_ang.pdf; Balmaceda, Margarita M.: Corruption, intermediary companies, and energy security: Lithuania's lessons for Central and Eastern Europe, in: Problems of Post-Communism, 2008 (vol. 55), no. 4, pp. 16–28, here pp. 18–19, 20, 24–25.

15 Orttung, Robert W./ Øverland, Indra: A limited toolbox: explaining the constraints on Russia's foreign energy policy, in: Journal of Eurasian Studies, 2011 (vol. 2), no. 1, pp. 74–85.

16 Stern, Jonathan P: The Russian natural gas 'bubble': consequences for European gas markets, London: Royal Institute of International Affairs, 1995; Mabro, Robert/ Wybrew-Bond, Ian (eds): Gas to Europe: the strategies of four major suppliers, Oxford: Oxford University Press, 1999; Pirani, Simon: Ukraine's energy sector, Oxford: Oxford Institute for Energy Studies, 2007, http://www.oxforden ergy.org/wpcms/wp-content/uploads/2010/11/NG21-UkrainesGasSector-SimonPirani-2007.pdf.

17 From 1993 to 2004, there were only three years in which a dispute did not occur (Bruce, Chloë: Fraternal friction or fraternal fiction? The gas factor in Russian–Belarusian relations, Oxford:

10. Gas Disputes Between Russia and Ukraine   257

been used as either an economic or a political tool for coercion. In the case of Ukraine, where disputes were very common before Russia established a policy of price convergence between most of the European countries and the former Soviet Union (FSU), as officially approved in November 2006,[18] we focus on the three main disputes that occurred in January 2006, March 2008, and January 2009. The existing literature has interpreted these disputes as follows:

Regarding the January 2006 dispute,[19] analysts disagree about whether economic or political motivations were underlying the dispute[20] and whether Russia backed off as a result of European criticism;[21] another line of argument concerns whether Russia had been ready to accept a lesser deal having miscalculated Ukraine's response[22] or whether, in fact, Russia had managed to gain more than what was lost[23] (the expanded role of the intermediary RosUkrEnergo[24] raised many questions as to whether it was

---

Oxford Institute for Energy Studies, 2005, pp. 2, 8, http://www.oxfordenergy.org/wpcms/wp-content/uploads/2010/11/NG8-FraternalFrictionOrFraternalfictionTheGasFactorInRussianBelarusianRelations-ChloeBruce-2005.pdf.

18   Cf. Tomberg, Igor: Russia to raise gas prices for CIS states, in: RIA Novosti, 13 November 2006, http://en.ria.ru/analysis/20061113/55572753.html; for a review of the process of revising prices and the reasons that Russia proceeded with this convergence, see: Svoboda, Karel: Business as usual? Gazprom's pricing policy toward the Commonwealth of Independent States, in: Problems of Post-Communism, 2011 (vol. 58), no. 6, pp. 21–35.

19   The first general accounts of the dispute and the road leading to the dispute itself can be found in: Stern, Jonathan P.: The Russian–Ukrainian gas crisis of January 2006, Oxford: Oxford Institute for Energy Studies, 2006, http://www.oxfordenergy.org/wpcms/wp-content/uploads/2011/01/Jan2006-RussiaUkraineGasCrisis-JonathanStern.pdf; Stern, Jonathan P.: Natural gas security problems in Europe: the Russian–Ukrainian crisis of 2006, in: Asia-Pacific Review, 2006 (vol. 13), no. 1, pp. 32–59; Dubien, Arnaud: L'Énergie, vulnérabilité stratégique persistante de l'Ukraine, in: Revue d'Etudes comparatives Est-Ouest, 2006 (vol. 37), no. 4, pp. 165–180.

20   Dubien, Arnaud: The opacity of Russian–Ukrainian energy relations, Paris: Institut Français des Relations Internationales, 2007, p. 10 (Russia.Nei.Visions, no. 19), http://www.ifri.org/?page=contribution-detail&id=4800&id_provenance=97; Campbell, Kurt A./ Price, Jonathan: The global politics of energy, New York: The Aspen Institute, 2008, pp. 80–81.

21   Chow, Edward/ Elkind, Jonathan: Where East meets West: European gas and Ukrainian reality, in: Washington Quarterly, 2009 (vol. 32), no. 1, pp. 77–92, here p. 82.

22   Moshes, Arkady: Change in the air in Ukraine, in: Russia in Global Affairs, 2006, no. 2, pp. 142–155; Pirani, Simon/ Stern, Jonathan/ Yafimava, Katja: The Russo–Ukrainian gas dispute of January 2009: a comprehensive assessment, Oxford: Oxford Institute for Energy Studies, 2009, http://www.oxfordenergy.org/wpcms/wp-content/uploads/2010/11/NG27-TheRussoUkrainianGasDisputeofJanuary2009AComprehensiveAssessment-JonathanSternSimonPiraniKatjaYafimava-2009.pdf; Baev, Pavel: From west to south to north: Russia engages and challenges its neighbours, in: International Journal, 2008 (vol. 63), no. 2, pp. 291–305, here p. 298.

23   Abdelal, Rawi: The profits of power: commercial Realpolitik in Europe and Eurasia, Cambridge, MA: Harvard Business School, 2010 (Working Paper 11-028), p. 22, http://www.hbs.edu/faculty/Publication%20Files/11-028.pdf.

24   Cf. Global Witness: It's a gas: funny business in the Turkmen–Ukraine gas trade, Washington, DC: Global Witness, 2006, http://www.globalwitness.org/library/its-gas-funny-business-turkmen-ukraine-gas-trade.

designed to bankrupt Naftogaz,[25] whether it was Gazprom's ally,[26] or whether its latest links to Viktor Yushchenko, as reported by the press some months before, made it rather a 'Ukrainian' creation[27]). The conflict of March 2008 did not reach the headlines, as no disruptions affected Europe, so few relevant analyses exist. However, the January 2009 dispute,[28] the longest dispute to date and one that affected customers in the rest of Europe to a greater extent than that in January 2006, attracted considerable attention, and many different interpretations of the motivations underlying the dispute were raised. Among many political interpretations, the dispute is explained to have resulted from Russia's interest in provoking a conflict for strategic reasons;[29] in this sense, some authors highlight Russia's alleged strategy of destabilising Ukraine with periodical price renegotiations,[30] and others note Russia's reduced threshold for tolerance since the leaders of the 'Orange Revolution' came to power.[31] By contrast,

---

25 Dubien, Arnaud: L'Énergie, vulnérabilité stratégique persistante de l'Ukraine, in: Revue d'Etudes comparatives Est-Ouest, 2006 (vol. 37), no. 4, pp. 165–180.
26 Pirani, Simon: a gas dependent state, in: Pirani, Simon (ed.): Russian and CIS gas markets and their impact on Europe, Oxford: Oxford University Press, 2009, pp. 93–132, here p. 101.
27 Obscure payments by RosUkrEnergo in favour of PetroGas (allegedly belonging to Viktor Yushchenko's brother) were reported on the eve of the agreement (Kiselev, Sergei: Oranzhevii boomerang, in: Moskovskie Novosti, 27 January 2006; Gevorkian, Natalia: Tselikom mutnaia istoria, in: Gazeta.ru, 27 April 2006, http://wap.gazeta.ru/column/gevorkyan/626707.shtml); indirect relations between Viktor Yushchenko and Ivan Fursin (owner of 10% of Centragas and 50% of RosUkrEnergo) were also reported, and there were rumours that part of UkrGazEnergo (JV between Naftogaz and RosUkrEnergo) benefits accrued to the President through PetroGas (Grib, Natal'ya/ Solov'ev, Vladimir/ Gavrish, Oleg: Viktora Yushchenko travyat gazom, in: Kommersant, 16 January 2008).
28 Perovic, Jeronim: Farce um Gas: Russland, die Ukraine und die EU-Energiepolitik, in: Osteuropa, 2009 (vol. 59), no. 1, pp. 19–35, here pp. 25–26; Pirani, Simon/ Stern, Jonathan/ Yafimava, Katja: The Russo–Ukrainian gas dispute of January 2009: a comprehensive assessment, Oxford: Oxford Institute for Energy Studies, 2009, http://www.oxfordenergy.org/wpcms/wp-content/uploads/2010/11/NG27-TheRussoUkrainianGasDisputeofJanuary2009AComprehensiveAssessment-JonathanSternSimonPiraniKatjaYafimava-2009.pdf.
29 Ebel, Robert E.: The geopolitics of Russian energy: looking back, looking forward, Washington, DC: Center for Strategic and International Studies, 2009, pp. 9–13, http://csis.org/files/publication/090708_Ebel_RussianEnergy_Web.pdf; Pirani, Simon/ Stern, Jonathan/ Yafimava, Katja: The Russo–Ukrainian gas dispute of January 2009: a comprehensive assessment, Oxford: Oxford Institute for Energy Studies, 2009, http://www.oxfordenergy.org/wpcms/wp-content/uploads/2010/11/NG27-TheRussoUkrainianGasDisputeofJanuary2009AComprehensiveAssessment-JonathanSternSimonPiraniKatjaYafimava-2009.pdf. We can find this strategy mentioned among possible strategies, while Smith Stegen supports the idea that Russia was targeting Europe in order to convincingly demonstrate Ukraine's unreliability (Smith Stegen, Karen: Deconstructing the 'energy weapon': Russia's threat to Europe as case study, in: Energy Policy, 2011 (vol. 39), no. 10, pp. 6505–6513, here p. 6509).
30 Smith, Keith C.: Bringing Energy Security to East Central Europe, Washington, DC: Center for Strategic and International Studies, 2010, http://csis.org/files/publication/100402_Smith_BringingEnergySecurity_Web.pdf; Milov, Vladimir: Gazprom 'harms Russian interests', Ukraine refuses to compromise, in: RFE/RL, 8 January 2009, http://www.rferl.org/content/Interview_Gazprom_Harms_Russian_Interests_/1367968.html.
31 Pirani, Simon: Am Tropf. Die Ukraine, Russland und das Erdgas, in: Osteuropa, 2010 (vol. 60), no. 2, pp. 237–256; Lough, John: Russia's energy diplomacy, London: Chatham House, 2011, p. 9

other authors argue for Russia's economic reasons for increasing gas prices in starting the dispute.[32] Apart from these political and economic interpretations, other authors focus on the escalation of Russia's actions independent of its motivations and consider Russia's desperation rather than confidence,[33] or Russia's conscious decision to escalate the conflict out of proportion with Ukraine's appropriation of gas.[34]

10.3. Theoretic Framework and Research Design

Economic rationales, strategic and political calculations, and free riding by sub-state actors are presented in the literature review as explanations for how these three disputes developed. While the literature review highlights economic and political motivations for these disputes, many questions regarding the role of intermediaries, such as RosUkrEnergo, are left unanswered; thus, only a detailed analysis of the disputes may provide more authoritative answers. To identify the changing pattern of escalation in each of the disputes, we need a theoretical framework to clarify the logic underlying such patterns.

The model of escalation proposed by Zagare and Kilgour provides a useful tool to follow the stages through which the three disputes under study evolved.[35] Within this model, different actors are depicted according to their decisions (to either escalate or de-escalate) at each node of a dispute. Insights from the debate regarding the deterrence model versus the spiral model and contributions from cognitive psychology are also applied to understand Russia's decision making in these disputes and its underlying reasons for being either a risk-acceptant or a risk-taking actor.

(Briefing Paper), http://www.chathamhouse.org/sites/default/files/19352_0511bp_lough.pdf.
32 Pirani, Simon/ Stern, Jonathan/ Yafimava, Katja: The Russo–Ukrainian gas dispute of January 2009: a comprehensive assessment, Oxford: Oxford Institute for Energy Studies, 2009, http://www.oxfordenergy.org/wpcms/wp-content/uploads/2010/11/NG27-TheRussoUkrainianGasDisputeofJanuary2009AComprehensiveAssessment-JonathanSternSimonPiraniKatjaYafimava-2009.pdf; Perovic, Jeronim: Farce ums Gas: Russland, die Ukraine und die EU-Energiepolitik, in: Osteuropa, 2009 (vol. 59), no. 1, pp. 19–35; Rutland, Peter: Gasmail, in: Transitions Online, 12 January 2009, http://prutland.web.wesleyan.edu/Documents/GAS%20MAIL.pdf.
33 Shleifer, Andrei/ Treisman, Daniel: Why Moscow says No: a question of Russian interests, not psychology, in: Foreign Affairs, 2011 (vol. 90), no. 1, pp. 122–138, here p. 127.
34 Pirani, Simon/ Stern, Jonathan/ Yafimava, Katja: The Russo–Ukrainian gas dispute of January 2009: a comprehensive assessment, Oxford: Oxford Institute for Energy Studies, 2009, pp. 60–62, http://www.oxfordenergy.org/wpcms/wp-content/uploads/2010/11/NG27-TheRussoUkrainianGasDisputeofJanuary2009AComprehensiveAssessment-JonathanSternSimonPiraniKatjaYafimava-2009.pdf.
35 Zagare, Frank C./ Kilgour, Marc D: Perfect deterrence, New York: Cambridge University Press, 2000; Zagare, Frank C: The games of July: explaining the Great War, Ann Arbor, MI: University of Michigan Press, 2011; Zagare, Frank C: The dynamics of deterrence, Chicago: University of Chicago Press, 1987.

Figure 10-1: Decision Tree by Zagare and Kilgour[36]

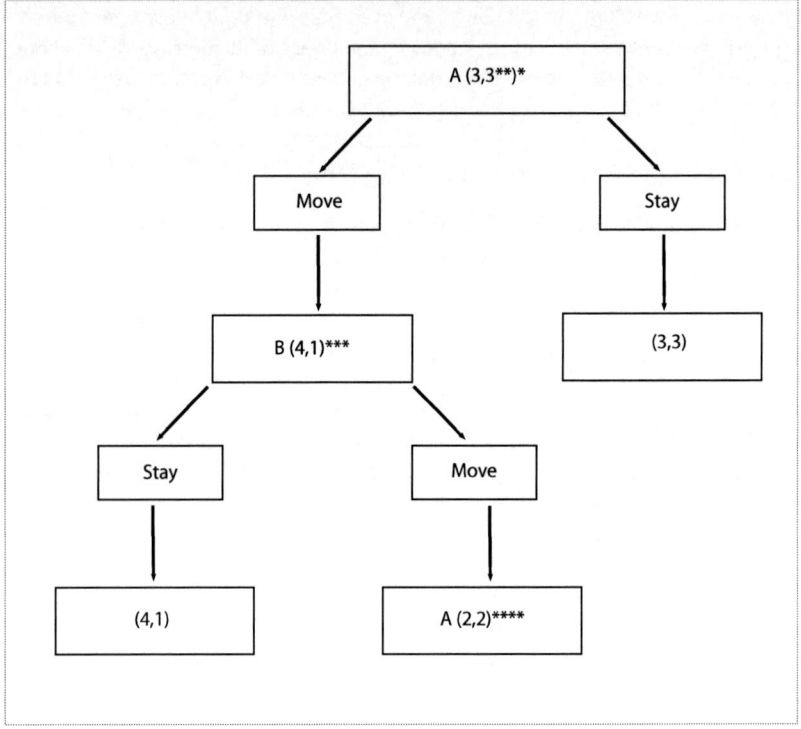

* Where the first number represents A and the second B. Whenever A or B is represented, it means that this is the situation (status quo) when the actor in question must take a decision (either 'move' or 'stay').
** Both A and B are located at the value (3,3) independent of whichever both 'objective' or 'subjective' values may be. They are equal for both actors as this value will be taken as a reference (status quo at the beginning) to value the end of the game where both losses and gains for each actor will be registered.
*** A 'wins' (4) by challenging B who is 'losing' (1), as its status quo worsens as a result of being challenged.
**** B challenges (moves) A so it re-balances the situation in its favour; compared to the first box, peace no longer prevails down the spiral of conflict, so re-balancing does not lead back to the value (3,3), but to the lesser value (2,2)

In their model of escalation, Kilgour and Zagare distinguish among several nodes of decision trees[37] where state a and state B have to either accept the status quo or esca-

---

36 For the decision-tree representation, see Zagare, Frank C.: The dynamics of escalation, in: Information and Decision Technologies, 1990 (vol. 16), no. 3, pp. 249–261, here p. 253.
37 The 'nodes of the decision tree' refer to the moments in which any of the actors involved have to make a decision that will either escalate or de-escalate a conflict. In our case, the nodes of

10. Gas Disputes Between Russia and Ukraine 261

late the dispute in order to challenge that status quo. Thus, we find different kinds of states depending on their acceptance of the status quo. States can thus be reduced in their decision-tree representation to 'hard states', which prefer war to capitulation, and 'soft states' which prefer capitulation to war.[38] This model can be extended to describe each actor according to its response at each node of the decision tree in any pattern of dispute[39] as well as Russia's response at each node of the decision tree for the three analysed disputes.

Within this model of escalation, there are two different models underlying the inner mechanism of escalation. The deterrence model assumes that an actor will refrain from escalating a conflict when it faces credible threats. What has not been fully clarified is the logic underlying the deterrence model: the model of deterrence (which, according to Huth, 'can be defined as the use of threats by one party to convince another party to refrain from initiating some course of action')[40] is based on expected utility models of rationality. However, as Jervis notes, there seems to be circumstances where deterrence fails and leads to situations where reprisals do not deter but, on the contrary, further fuel conflict. This opposite model is the spiral model, which shows that, in certain instances, an actor will escalate a conflict when faced with resistance.[41] In fact, Schelling's insights that deterrence is actually more easily achieved than compellence, since the latter is an offensive goal and thus liable to cause more resistance from the target, go precisely in the same direction.[42] As an alternative, to merge both models, prospect theory offers interesting insights from the field of cognitive psychology.

---

the decision tree apply to each of the stages of the dispute: the first node of the decision tree is Russia's reaction to the absence of a contract with Ukraine after negotiations fail, whether it pursues negotiations or applies a gas cut-off; the second node of the decision tree is Ukraine's reaction to this cut-off, in which case escalation would entail Ukraine's disruption of the flow of gas to the rest of Europe through its own territory (we do not consider here whether taking gas is necessary for technical reasons or whether it is done intentionally as a measure of retaliation); the final node of the decision tree concerns Russia's decision to either de-escalate to avoid damages to its reputation due to a disruption in the gas transit towards its end consumers or signal its resolve to Ukraine by stopping all gas supplies in order to deny Ukraine any possibility of siphon gas off the pipeline.

38 Zagare, Frank C./ Kilgour, Mark D.: Asymmetric deterrence, in: International Studies Quarterly, 1993 (vol. 37), no. 1, pp. 1–27, here p. 8.
39 Cf. Carlson, Lisa J./ Dacey, Raymond: Sequential decision analysis of the traditional deterrence game, in: Peace Economics, Peace Science and Public Policy, 2004 (vol. 10), no. 3, pp. 1–14; Simon, Michael W.: Rogue state response to BDM: the regional context, in: Defense and Security Analysis, 2002 (vol. 18), no. 3, pp. 271–292.
40 Huth, Paul K.: Deterrence and international conflict: empirical findings and theoretical debates, in: Annual Review of Political Science, 1991 (vol. 2), pp. 25–48, here p. 26.
41 Jervis, Robert: Perception and misperception in international politics, Princeton, NJ: Princeton University Press, 1976; Jervis, Robert: Cooperation under the security dilemma, in: World Politics, 1978 (vol. 30), no. 2, pp. 167–214.
42 Schelling, Thomas: Arms and influence, New Haven, CT: Yale University Press, 1966.

Prospect theory,[43] in challenging expected utility theory,[44] suggests that actors tend to be risk averse regarding gains and risk acceptant regarding losses. Perception in either domain is determined by the point of reference that the actor chooses in order to frame and then value the status quo. In the context of international relations, states would not risk their status quo for prospective gains if they are rather satisfied with that status quo, while they would take risks to prevent losses if they consider their status quo to be unsatisfactory, so deterrence may be effective with gain-seeking states, whereas appeasement may be effective for loss-averse states.[45] The expectations through which an actor frames the status quo as satisfactory or unsatisfactory and draws a particular baseline for expectations are determined by the actor's perceptions and the point of reference chosen.[46] Given the complex triangular relation in the energy disputes studied, it is worth considering how Russia's framing of its relations with Ukraine could explain its changing disposition towards bearing responsibility in affecting European gas consumers as an externality. Comparing the escalation in relation to the interests involved, I consider the explanatory power of a possible change in Russia's valuation of the status quo in explaining the different patterns of escalation.

---

43 Tversky, Amos/ Kahneman, Daniel: Prospect theory: an analysis of decision under risk, in: Econometrica, 1979 (vol. 47), no. 2, pp. 263–291; Jervis, Robert/ Lebow, Richard N./ Stein, Janice Gross: Psychology and deterrence, Baltimore, MD: Johns Hopkins University Press, 1989; Stein, Janice Gross/ Pauly, Louis W.: Choosing to co-operate: how states avoid loss, Baltimore, MD: Johns Hopkins University Press, 1993; Davis, James W.: Threats and promises: the pursuit of international influence, Baltimore, MD: Johns Hopkins University Press, 2000; Jervis, Robert: Prospect theory, human nature and values, in: Political Psychology, 2004 (vol. 35), no. 2, pp. 163–176; Taliaferro, Jeffrey W.: Balancing risks: great power intervention in the periphery, Ithaca, NY: Cornell University Press, 2004.
44 Expected utility theory has been the basis for most rational choice models. The theory very simply supposes that actors will choose actions that will yield the greatest return and avoid those that, on the contrary, will not bring benefits. Prospect theory follows a similar assumption of rational actors but introduces a risk factor, which is absent in expected utility theory.
45 Levy, Jack S.: An introduction to prospect theory, in: Political Psychology, 1992 (vol. 13), no. 2, pp. 171–186.
46 Prospect theory so far has generally been tested in extreme cases short of war: Jervis, Robert/ Snyder, Jack L.: Dominoes and bandwagons: strategic beliefs and great power competition in the Eurasian rimland, New York: Oxford University Press, 1991; Jervis, Robert: Political implications of loss aversion, in: Political Psychology, 1992 (vol. 13), no. 2, pp. 187–204.; Berejikian, Jeffrey: Revolutionary collective action and the agent-structure problem, in: The American Political Science Review, 1992 (vol. 86), no. 3, pp. 647–657; Levi, Ariel S./ Whyte, Glen: a cross-cultural exploration of the reference dependence of crucial group decisions under risk: Japan's 1941 decision for war, in: Journal of Conflict Resolution, 1997 (vol. 41), no. 6, pp. 792–813; Haas, Mark L: Prospect theory and the Cuban missile crisis, in: International Studies Quarterly, 2001 (vol. 45), no. 2, pp. 241–270; Jeffrey, Berejikian: a cognitive theory of deterrence, in: Journal of Peace Research, 2002 (vol. 39), no. 2, pp. 165–183; Cha, Victor: Hawk engagement and preventive defense on the Korean peninsula, in: International Security, 2002 (vol. 27), no. 1, pp. 40–78; Tessman, Brock. T./ Chan, Steve: Power cycles, risk propensity, and great-power deterrence, in: Journal of Conflict Resolution, 2004 (vol. 48), no. 2, pp. 131–153.

The dependent variable is the set of responses that Russia made in response to Ukrainian decisions leading to the genesis of and different patterns of escalation in the three energy disputes between Russia and Ukraine in January 2006, March 2008, and January 2009. The independent variable is the rationale (either economic or political) for de/escalating in each of the three disputes. The independent variable assumes that there was a convergence in interests between Gazprom and the Russian government and that the former was subordinated to the latter. Pending clear evidence that this supposition is correct and in the absence of clear evidence that Gazprom free-rode against the state in the context of an increasing influence of the state as a stakeholder since the arrival of Alexei Miller in succeeding Rem Viakhirev in 2001,[47] I make this assumption for the sake of parsimony even if this could reduce the explanatory power of the variable.

10.4. Analysis of the Three Energy Disputes in January 2006, March 2008, and January 2009

The main question to answer is whether Russia's changing valuation of the status quo was the reason for the different patterns of escalation in the three disputes, namely, whether Russia's decisions were related to the changing stakes involved in each of the disputes and to the different pay-offs for escalating or de-escalating the conflicts. This question will be combined with Ukraine's own decisions and its influence on the patterns of escalation, as Russia's stakes might have remained constant while Ukraine's stakes, on the contrary, changed.

10.4.1. The January 2006 dispute

The agreement reached on 4 January 2006 has to be considered in the light of the background of the previous year and two important elements: Russian intentions to end the highly disadvantageous subsidising of Ukraine's gas supply and to enforce market prices instead and Turkmenistan's parallel plans to introduce higher prices for its gas supply. Ukraine imported a total volume of approximately 57.4bcm of gas: 37bcm of gas came from Turkmenistan[48] (13bcm was supposedly kept by the interme-

---

47 The Russian government managed to take full control as the majority stakeholder in 2004. Sorbello, Paolo/ Grandi, Ludovico: From concentration to competition: the struggle for power between the Kremlin and Gazprom through the study of TNK-BP and South-Stream, in: Irish Slavonic Studies, 2013 (vol. 25), pp. 106–119, here pp. 109–110.
48 Pirani, Simon: Ukraine's energy sector, Oxford: Oxford Institute for Energy Studies, 2007, p. 28, Table 4.1, http://www.oxfordenergy.org/wpcms/wp-content/uploads/2010/11/NG21-UkrainesGasSector-SimonPirani-2007.pdf.

diary RosUkrEnergo as payment for its services),[49] while the rest, amounting to 24bcm, was reserved for Ukrainian consumption. Moreover, 20.4bcm of Russian gas came in form of transit fees for Russian gas exported to Central and Western Europe.[50] It was calculated that Russian gas was priced at US$50 per 1,000cm, and Naftogaz imposed a transit fee of US$1.09 per 1,000cm/100km, a generous price given that gas was sold to Europe at as much as three times that price.[51]

As a result of this situation, at the end of 2005, Ukraine faced a set of demands from its two gas suppliers: Russia demanded a price of US$230 instead of US$50 for its gas (Gazprom had offered a lower price of US$160 earlier), while Turkmenistan, which had been negotiating prices with both Gazprom and the Ukrainian state company Naftogaz, insisted on raising prices from US$44 per 1,000cm to US$60 per 1,000cm. Gazprom eventually won out by agreeing to buy all Turkmen exports at US$65 per 1,000cm, meaning that Ukraine would surely have to accept prices for Turkmen gas ranging from US$85 to US$95 per 1,000cm at its border with Russia. As no agreement was reached on 1 January, Gazprom decided to stop the gas supply to Ukraine. After supply disruptions were reported by Russia's European customers, the gas supply was restored on 3 January, and a final agreement was reached a day later.[52]

---

49 The figure is provided by: Pavel, Ferdinand: The Ukrainian–Russian gas agreement: analysis and alternatives, presentation at the Harvard Ukrainian Research Institute, 5 February 2006, https://www.youtube.com/watch?v=iXuzSx-QbeM. This was already reported in 2004, when the agreements were signed; it was also stated that, while 3–5bcm would re-exported to Eastern Europe by RosUkrEnergo, an unspecified amount of gas would be sold to Naftogaz ('Gazovoe primirenie', in: Neftegazavaia Vertikal', 27 August 2004).
50 Pirani, Simon: Ukraine's energy sector, Oxford: Oxford Institute for Energy Studies, 2007, p. 28, Table 4.1, http://www.oxfordenergy.org/wpcms/wp-content/uploads/2010/11/NG21-UkrainesGasSector-SimonPirani-2007.pdf; Pavel provides a higher figure of 23bcm (Pavel, Ferdinand: The Ukrainian-Russian gas agreement: analysis and alternatives, presentation at the Harvard Ukrainian Research Institute, 5 February 2006, https://www.youtube.com/watch?v=iXuzSx-QbeM). However, the lower than expected figure for Russian gas stems from an agreement reached in the summer of 2004. For the 2005–2009 period, Naftogaz would receive 21bcm instead of 26bcm to recover from a debt accumulated for the period 1997–2000 (Stern, Jonathan P.: Natural gas security problems in Europe: the Russian–Ukrainian crisis of 2006, in: Asia-Pacific Review, 2006 (vol. 13), no. 1, pp. 32–59, here pp. 37–38); how exactly Ukraine closed up the gas deficit was never clarified, although it seems that additional volumes of gas from Central Asia were imported, as foreseen.
51 Krasnians'kii, Mikhailo: Kto i kogo shantazhiruet gazom, in: Ukrainska Pravda, 8 December 2005, http://www.pravda.com.ua/articles/2005/12/8/3030037/; Kurdel'chuk, Daniil/ Malinovskii Aleksandr/ Novak, Inna: Ukraina – Rossiia: evropeiskii podkhod k resheniyu gazovoi dilemmy na osnove verkhovenstva prava, in: Zerkalo Nedeli, 17 December 2005, http://gazeta.zn.ua/ARCHIVE/ukrainarossiya_evropeyskiy_podhod_k_resheniyu_gazovoy_dilemmy_na_osnove_verhovenstva_prava.html; Korchemkin, Mikhail: Economics of Russian–Ukrainian gas conflict: summary for beginners, in: East European Gas Analysis, 30 December 2005, http://www.eegas.com/ukrtran3.htm.
52 Stern, Jonathan P.: Natural gas security problems in Europe: the Russian–Ukrainian crisis of 2006, in: Asia-Pacific Review, 2006 (vol. 13), no. 1, pp. 32–59.

Figure 10-2: The Gas Conflict in 2006

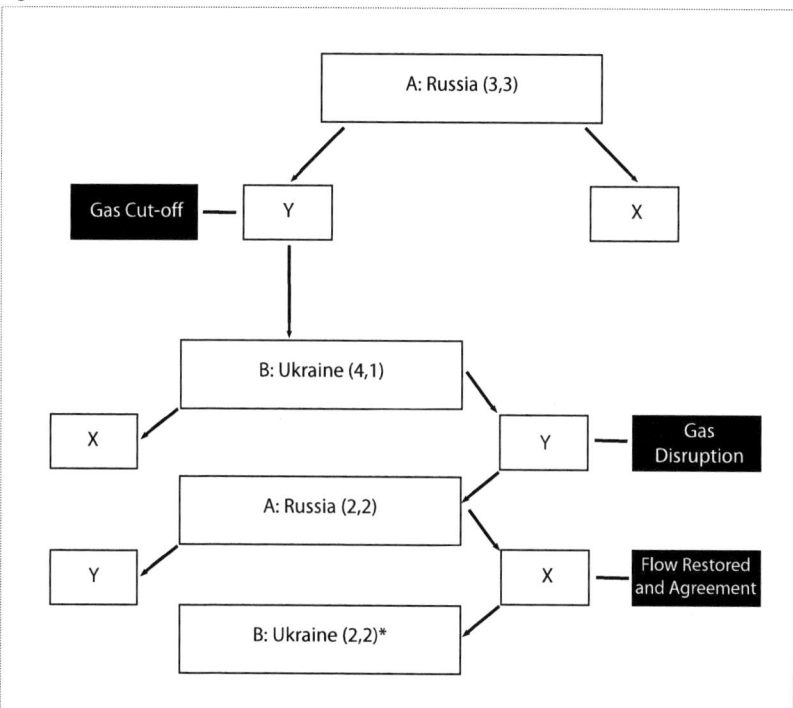

* The new status quo as represented here (2,2) conveys the idea that neither side obtained a satisfactory result, as opposed to the 'true winner' of the dispute, RosUkrEnergo.

The agreement on 4 January was very confusing and difficult to interpret.[53] It initially appeared to be a victory for Gazprom: cash-based payments to Ukraine for Russian gas transit were established, while the transit tariff was raised from US$1.09 per 1,000cm/100km to US$1.60 per 1,000cm/100km. Russian gas was set to a price of US$230 per 1,000cm, exactly what Gazprom had been demanding. However, closer examination of some of the provisions of the agreement, many of them only clarified in the following weeks, reveals a much more complex picture.

First, it must be highlighted that somehow, confusingly, the agreed price for Ukraine was US$95 per 1,000cm, which at first seemed to be a relative victory for Ukraine in the dispute, as it only exceeded the price of US$80 per 1,000cm that had been set as a maximum by US$15. This, however, was mostly misleading, as the price

---
53 Cf. ibid.

resulted from a calculation of the median price derived from mixing Central Asian and Russian gas.

Second, the breakdown of the gas share that Ukraine would be importing changed: 41bcm of gas was to come from Turkmenistan, and a volume of 17bcm of gas would be from Russia. The decrease in the volume of Russian gas was a consequence of changes to the transit payment system that now favoured Russia compared to the previous subsidies. However, if these figures roughly corresponded to Ukraine's gas needs, there was confusion regarding a third volume of gas, comprising a maximum of 7bcm of gas from Uzbekistan and 8bcm of gas from Kazakhstan. Adding all of the figures, Ukraine would end up importing well beyond its gas needs. The explanation was that, in fact, little if any of the Russian gas would go to Ukraine, which would import 100% of its gas needs from Central Asia instead.[54] It seems safe to surmise that Russia had insisted on this agreement as a means of saving face. Russian gas would instead be sold to RosUkrEnergo (50% of which was owned by Gazprom itself and the rest of which was owned by two Ukrainian businessmen, Dmitro Firtash and Ivan Fursin), the intermediary trader that sold Turkmen gas to Naftogaz in 2005, which would henceforth monopolise all gas imports for Ukraine.

Many considered RosUkrEnergo's expanded role to be a victory for Russia, as it enabled Gazprom to penetrate and control Ukraine's internal gas import market through the creation of the joint venture (JV) UkrGazEnergo between RosUkrEnergo and Naftogaz. RosUkrEnergo's expanded role seemed logical, as the share of Central Asian gas being imported to Ukraine, for which the company already had a monopoly, would increase. In addition, the 17bcm of Russian gas that was to be sold to RosUkrEnergo to be exported to Europe may be understood in the light of the percentage of gas that intermediaries generally claim as part of their tariff for serving as traders. As much of this gas was resold earlier to Naftogaz for re-export, RosUkrEnergo's role was beneficial for Gazprom, as it fully replaced Naftogaz's role as a competitor in Central Europe.[55] On the negative side for Gazprom, the new agreement diverted more Turkmen, Uzbek, and Kazakh gas, which was previously sold by Gazprom, towards the

---

54 Korchemkin provides calculations assuming that none of the Russian gas was used for Ukraine's gas needs (Korchemkin, Mikhail: RosUkrEnergo wins, Gazprom and NAK lose, in: East European Gas Analysis, 6 March 2006, http://www.eegas.com/ukrtran6.htm). Pirani, following official statistics, finds that 9.1bcm of Russian gas was supplied to Ukraine in 2006; at least partly supporting Korchemkin's assessment, he assumes that part of the 9.1bcm of gas was physically swapped with gas from Uzbekistan and Kazakhstan, as Russian gas was symbolic and RosUkrEnergo possibly assumed the extra costs instead of Naftogaz (Pirani, Simon: Ukraine's energy sector, Oxford: Oxford Institute for Energy Studies, 2007, p. 28, Table 4.1, p. 37, footnote 58, http://www.oxfordenergy.org/wpcms/wp-content/uploads/2010/11/NG21-UkrainesGasSector-SimonPirani-2007.pdf).

55 Before the new agreement, Naftogaz re-exported certain volumes at market prices to gain profits, which were reinvested to subsidise gas prices within Ukraine. Through the agreement that ended the 2006 dispute, Russia required Ukraine to henceforth lose this lucrative market; in 2005, gas that Naftogaz would buy from RosUkrEnergo as part of the gas RosUkrEnergo received for its transit services, which was then resold in Central Europe, would now remain in the intermediary's hands. The final destination of this gas remained within Central Europe, and

10. Gas Disputes Between Russia and Ukraine

Ukrainian market. Gazprom's losses resulting from the agreement were calculated to range from US$1 billion to US$741 million, and the agreement closed the door for Gazprom to progressively introduce market prices and freely control gas from Central Asia.[56] Many assumed that Gazprom consciously accepted the losses and established RosUkrEnergo as a strategic ally to penetrate the Ukrainian market and/or bankrupt Naftogaz. Gazprom reached an agreement that seemed to fail to meet its initial objectives and that was arguably suboptimal, unless it preferred RosUkrEnergo as a strategic vehicle.

Regarding the January 2006 dispute and the agreement on 4 January, the complexity of the agreement poses a challenge to analysts aiming to determine which country's interests were best secured. Taking into account the pricing agreements, Gazprom seems to have accepted a defeat, unless it placed greater value on the strategic role of RosUkrEnergo. Although the June 2004 agreement with the administration of President Leonid Kuchma foresaw that Ukraine would eventually import only Central Asian gas after 2007,[57] it was assumed that Turkmenistan would increase its production. In addition, in 2002, Ukraine and Russia had agreed to an international consortium for Ukraine's transportation system.[58] The situation radically changed with the election of Viktor Yushchenko, who rejected previous plans that Russia favoured. Given that Gazprom's strategy was to either end financially unprofitable subsidisation in the FSU or to maintain such subsidies in exchange for energy assets, Gazprom does not seem to have de-escalated the dispute in January 2006 after Ukraine had accepted a deal that was considered optimal for the Russian monopoly. Instead, Gazprom arguably accepted a lesser deal to restore its reputation in the face of considerable criticism from European countries that were collaterally affected by the dispute. RosUkrEnergo replaced Naftogaz's exports and facilitated Gazprom's penetration into the Ukrainian internal gas market. The intermediary could be considered either a strategic buffer that 'sweetened' a financially unprofitable deal or a genuine strategic vehicle. In the case of the latter, Gazprom would have traded off financial benefits and control of most of the Central Asian gas supply for much less than an international consortium and without having yet reaped the expected production increases from Turkmenistan. In any case, Gazprom's would have failed to meet its initial objectives either compared to Gazprom's negotiating position in 2005 or compared to long-term projects agreed upon during Kuchma's tenure. As prospect theory shows, an actor will not take risks for the sake of gains if such risks could lead to further losses and will thus de-escalate

---

it was sold either through Gazprom Export or directly to end consumers. Thus, RosUkrEnergo assumed Naftogaz previous role.
56 Korchemkin, Mikhail: RosUkrEnergo wins, Gazprom and NAK lose, in: East European Gas Analysis, 6 March 2006, http://www.eegas.com/ukrtran6.htm.
57 'Gazovoe primirenie', in: Neftegazavaia Vertikal', 27 August 2004.
58 'Ukraina i Rossiskii Gaz', in: Neft' i Kapital, 25 October 2004.

in the face of such risks. Gazprom's response to the dispute in January 2006 fits such a pattern, and Gazprom's eventual acceptance of losses as the outcome of the dispute could be explained by the fact that it did not want to further risk its reputation among customers in Central and Western Europe. Gazprom's rhetoric from the second day of the dispute to make up for the reduced gas supply and the decision to restore the supply on the third day, before an agreement had been reached, are strong indications of the Russian side's risk avoidance. Ukraine's acceptance of a scheme in which Naftogaz also lost (in January 2009, Timoshenko's government would not have accepted any deal involving RosUkrEnergo) implies that the Ukrainian side also showed flexibility, which arguably helped an agreement be reached after four days.

10.4.2. The March 2008 Dispute

A dispute arose in January 2008 regarding a debt amounting to US$1.5 billion and unpaid Russian gas sold to make up for the reduced gas supply from Central Asia.[59] As agreed on 12 February with President Viktor Yushchenko, Ukraine would begin paying off its debts corresponding to all gas consumed in the last two months of the previous year. In addition, it was agreed that RosUkrEnergo would eventually be removed as an intermediary and that UkrGazEnergo would be replaced by another Naftogaz/Gazprom JV.[60] However, because of Yulia Timoshenko's reluctance to honour the agreement, Gazprom threatened to cut off the gas supply and eventually did so on 3 March: Gazprom started by reducing 25% of the supply, with 25% more added the next day. Given the Russian gas cut-offs, the Ukrainian government warned that European gas supply could not be guaranteed if the situation persisted. The gas supply finally started being restored on 5 March, and further negotiations led to a new agreement on 14 March. According to the new agreement, the outstanding debt would be repaid and Russian gas would be replaced with Central Asian gas. The eventual removal of RosUkrEnergo as an intermediary was confirmed, but instead of the new JV in place of UkrGazEnergo, Gazprom would be granted direct access to the Ukrainian market through a new supplier, Gazprom Sbyt Ukraina, which was granted approximately 15% of the import market share.[61]

Regarding March 2008, the 14 March agreement did guarantee Russia's minimum objective, which was the repayment of debt, but it is doubtful that this agreement improved the terms that had been agreed upon on 12 February. The fact that partial cut-offs were progressively applied instead of a total cut-off could be explained by the fact that the conflict concerned debt, not pricing; however, Gazprom's decision not to escalate to a full supply cut-off coincided with Ukraine's warnings of possible supply

---

59   Reznik, Irina/ Kashin, Vasilii: Dym nad vodoi, in: Vedomosti, 8 February 2008.
60   Grib, Natal'ya/ Gavrish, Oleg: 'Gazprom' dal Ukraine poltora mesiatsa, in: Kommersant, 15 February 2008.
61   Sergei Kulikov: Gazprom sdalsia Ukraine, in: Nezavisimaia Gazeta, 14 March 2008.

10. Gas Disputes Between Russia and Ukraine

problems for the rest of Europe if Ukraine's energy security was affected, as Russia likely wanted to avoid repeating what occurred in January 2006. In a similar fashion as in January 2006, Gazprom ended up not only accepting a compromise to ensure uninterrupted gas supply through Ukraine, ensuring that debt would be recovered, but also accepting loss some of the gains delineated in the 12 February agreement. The timing of the dispute, where the gas supply was restored on 5 March but an agreement was only reached on 14 March, as in January 2006, indicates Gazprom's unwillingness to risk its reputation as a reliable supplier. Using terminology from prospect theory, we find that Gazprom again accepted a deal that avoided losses but stopped before it recovered at least all of the gains achieved one month earlier.

Figure 10-3: The Gas Conflict in 2008

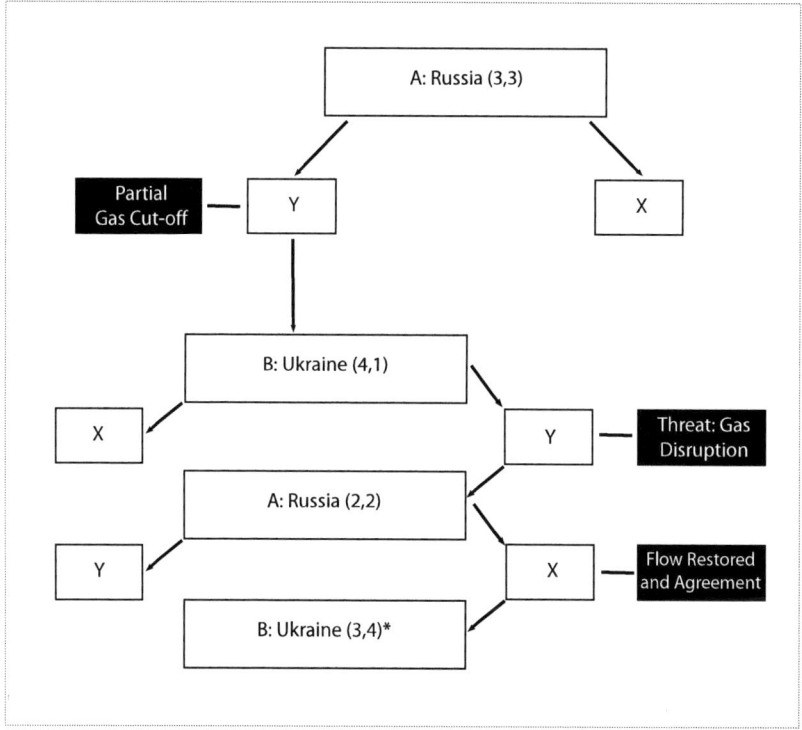

* The new status quo as represented here (3,4) conveys the idea that Russia has solved the problem that forced it to move, but has lost some of the gains initially made, whereas Ukraine, although having decided to repay the debt to Russia's satisfaction, has nevertheless moved forward against intermediaries. Even if Ukraine has 'lost' repaying the debt, not paying a debt is not considered a realistic goal; thus, the reference is what is gained in exchange of repaying that debt.

10.4.3. The January 2009 Dispute[62]

In March 2008, Central Asian countries had declared that they would raise gas prices,[63] with some estimates indicating that Ukraine would end up paying about US$320 per 1,000cm in 2009 compared to US$179.5 per 1,000cm in 2008.[64] This was a more substantial price increase than that of previous years, as prices increased to US$95 in 2006 and to US$130 in 2007, although the price was still short of actual market prices, assuming as a benchmark prices paid by Western countries such as Germany, which were close to US$500 in 2008. This pricing dispute was coupled with a new debt dispute concerning about US$2 billion.[65] However, the situation did not look as bleak as to nullify any prospects for an agreement. At the latest stage of negotiations, Russia was offering US$250 per 1,000cm, while Ukraine balked and insisted on US$235 per 1,000cm along an increase in the transit fee to US$1.7 or US$1.8 per 1,000cm/100km. The fact that debts were only partially paid at the last moment on 31 December reduced the margin for any agreement. When an agreement could not finally be reached, negotiations broke down, and Gazprom reverted to its threats of imposing market prices of US$450 per 1,000cm.

The dispute started like that on 1 January 2006 and led to a total cut-off of the gas supply to Ukraine, which began siphoning off gas being transported to Central and Western Europe. Russia then decided on 4 January to reduce the flow of gas being transported through Ukraine by the amount of gas being appropriated by Ukraine. Ukraine kept appropriating gas, and by 7 January, consumers in Central and Western Europe were left without any gas, resulting in a serious energy crisis in the most dependent countries, such as Bulgaria and Serbia, which did not have the means to replace the absent Russian gas. The EU intermediated in the dispute and, on 11 January, sent a team to oversee the process of resuming the gas supply. However, the gas supply could not resume, as according to Ukraine, resuming the gas supply through their country would force them to stop the emergency supply in reverse mode to its Eastern industrial regions. This new spat led to further days of interruption until finally on 19 January, an agreement was reached between Russia and Ukraine that ended the gas cut-offs for both Ukraine and other customers in Central and Western Europe.[66]

---

62 The most detailed general account to date on the dispute is: Pirani, Simon/ Stern, Jonathan/ Yafimava, Katja: The Russo–Ukrainian gas dispute of January 2009: a comprehensive assessment, Oxford: Oxford Institute for Energy Studies, 2009, http://www.oxfordenergy.org/wpcms/wp-content/uploads/2010/11/NG27-TheRussoUkrainianGasDisputeofJanuary2009AComprehensiveAssessment-JonathanSternSimonPiraniKatjaYafimava-2009.pdf.
63 Mazneva, Elena: Aziia trebuet doplatit', in: Vedomosti, 12 March 2008.
64 Grib, Natal'ya/ Gavrish, Oleg/ Konstantinov, Aleksandr: Natsenka po povedeniiu, in: Kommersant, 12 March 2008.
65 Grib, Natal'ya/ Gavrish, Oleg: Ukraina zaplatit za Evropu, in: Kommersant, 11 November 2008.
66 Pirani, Simon/ Stern, Jonathan/ Yafimava, Katja: The Russo–Ukrainian gas dispute of January 2009: a comprehensive assessment, Oxford: Oxford Institute for Energy Studies, 2009, pp. 19–26,

10. Gas Disputes Between Russia and Ukraine

Gas supplies were fully restored on the 20 January. According to the new agreement, Ukraine having rejected the latest Russian offer of US$250 per 1,000cm, which could (at least partly) have covered the price increases from Central Asia, would now have to accept Russian gas instead, which would be priced at US$418 per 1,000cm for the first quarter but be reduced to US$360 owing to the application of a 20% discount. RosUkrEnergo would be eliminated as an intermediary, and gas would come directly from Gazprom.[67]

Figure 10-4: The Gas Conflict in 2009

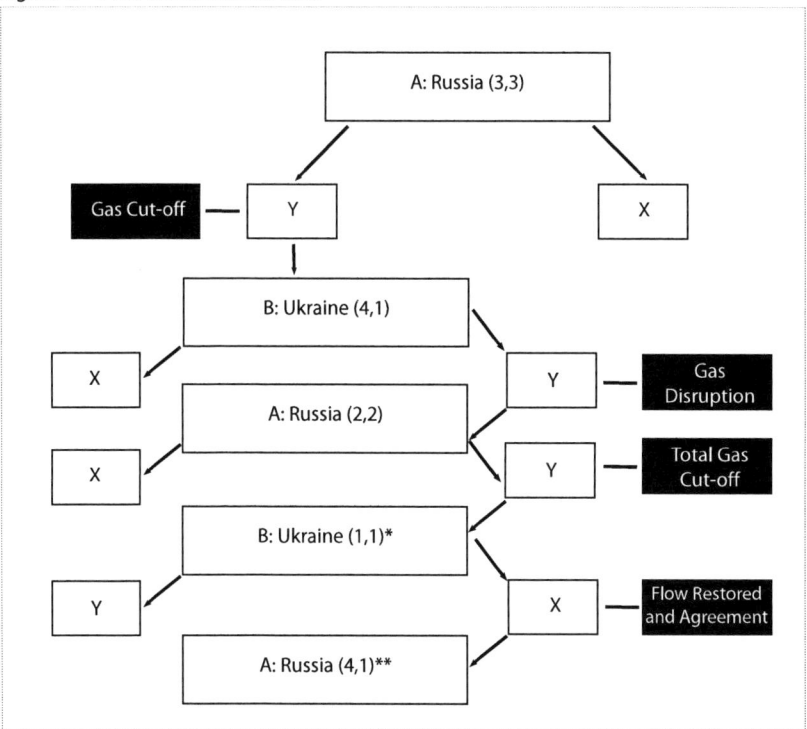

* As opposed to the (2,2), the dispute has further worsened for both actors, so the value is even lower (1,1). We also consider that Russia's move puts both Ukraine and itself in a worse position, as it deepens the dilemma of effects on consumers who rely on Russia as a reliable supplier.

** As the game ends with Russia's victory and Ukraine's defeat, the value (4,1) is the same as that after Russian's first move against Ukraine (4,1).

---

http://www.oxfordenergy.org/wpcms/wp-content/uploads/2010/11/NG27-TheRussoUkrainian GasDisputeofJanuary2009AComprehensiveAssessment-JonathanSternSimonPiraniKatjaYafim ava-2009.pdf.
67 Ibid., pp. 26–31.

The January 2009 dispute stands in large contrast to the two previous disputes. Russia secured what was in question in the days before the dispute broke out, and went beyond forcing Ukraine to accept Central Asian gas at a price sufficient to cover the costs of the new price increases, as it had done in the price revisions for 2006, 2007, and 2008. Rather, Russia forced Ukraine to accept Russian gas at market prices (US$418 per 1,000cm) similar to those paid by customers in other countries in Central and Western Europe, with Ukraine being barely compensated with only a 20% discount (US$360 per 1,000cm) and a new transit tariff at US$1.90 per 1,000cm/100km, also with 20% discount (US$1.70 per 1,000cm/100km).[68] Using the tenets of prospect theory to explain Russia's motivations, we face a more complex picture here if we want to identify a particular factor that may explain the different outcome. Although the divergence between the price of US$250 before the dispute and the final price of US$360 may be explained by a delicate and unstable balance between Gazprom's and the Russian government's positions, which was finally resolved in favour of the former, strong candidates for the trigger of this final switch may lie in both the general economic situation, with Russia facing both capital flights after the August war against Georgia and free-falling oil prices,[69] and the hardening of negotiations towards Ukraine since the return of the second Timoshenko government in response to its new revisionist projects in the energy sector.[70] If Russia framed the general situation as one of losses, which made it imperative to impose market prices on Ukraine, this would explain why reputation losses derived from maintaining a protracted cut-off affecting all European customers became acceptable and why it had become unacceptable

---

68   For the further arrangements on transit fees see, ibid., p. 27.
69   Oil prices were in free fall, from a peak in July of US$147 per barrel to only US$86 in October, to close to US$40 at the beginning of 2009 (Rutland, Peter: The Impact of the Global Financial Crisis on Russia, in: Russian Analytical Digest, 17 October 2008 (no. 48), pp. 2–14, http://www.isn.ethz.ch/Digital-Library/Publications/Detail/?ots591=0c54e3b3-1e9c-be1e-2c24-a6a8c7060233&lng=en&id=92848; Khan, Mohsin S.: The 2008 oil price 'bubble', in: Peterson Institute for International Economics, Poliy Brief PB19-09, August 2009, http://www.iie.com/publications/pb/pb09-19.pdf).
70   When her second government came into power, Yulia Timoshenko's plans to exclude RosUkrEnergo from the scheme of energy relations between Russia and Ukraine (as the trader had become the political nemesis for the Ukrainian politician, its removal from the arrangement became one of her main priorities) seemed to have scared Gazprom: before she was nominated as Prime Minister in December 2007, Gazprom made an unusual move when a new dispute emerged regarding debt accumulated by Naftogaz, threatening gas supply cuts if the debt was not recovered (the debt was eventually recovered). Apparently, Gazprom wanted to recover the debt before the new government was formed and Timoshenko started putting her plans into practice. Then, in the March 2008 dispute, a new debt dispute elicited a harsher than usual response by Gazprom, which in the face of the Ukrainian government's refusal to return the newly accumulated debt and to use it as an argument to eliminate RosUkrEnergo did put earlier threats into practice. Finally, Gazprom established a linkage when new debt problems started emerging in mid-2008, stating that there would be no agreement on gas prices before existing debts were recovered.

to forgo the opportunity costs of not selling Russian gas at market prices to Ukraine. As conclusive evidence cannot be found for either explanation, there is also ground to suspect that other more strictly political factors might have been influential, as the August 2008 war with Georgia had occurred months before the dispute and Ukraine had showed support for Georgia in the face of Russia's attack. The fact that the Russian government suddenly reversed its priorities and forced higher prices instead of prices high enough to absorb the increased prices of Central Asian gas, as Gazprom seemed to have favoured, could be an indication in itself of a prioritisation of politics over economics, although as for this latest set of reasons, it is difficult to find a connection between Ukraine's political intentions and the debt and price negotiations. The general context nevertheless seems to show that whether for strictly economic or more 'political' reasons, the status quo worsened for Russia; thus, consistent with the tenets of prospect theory, Russia's response to the dispute clearly reflects its increased risk-acceptance of reputational costs.

10.5. Conclusion

The empirical results of this research, even if they may not provide as conclusive evidence as would be desirable, in particular in the analysis of the third and the most complex dispute of January 2009, show a relation between Russia's changing decisions and a changing status quo. The greatest divergence seems to affect the last dispute, which shows a different pattern of escalation, with Russia facing a more critical situation for different reasons, as already noted above. Instead of Ukraine's decisions, it seems that the best explanation for the different patterns of escalation are decisions made in nodes of decision trees on the Russian side, with Gazprom assuming losses to end the dispute in January 2006, reverting initial partial gas cut-offs in March 2008 before an agreement was reached, but pushing its initial demands in January 2009 when faced with Ukraine's resistance. As for the reasons for Russia's changing position, Russia's perception of a changing status quo is likely a strong factor accounting for its responses reflected in the different patterns of escalation.

If we are to consider insights from cognitive psychology provided by prospect theory and how it may explain the success or failure of deterrence and the appearance of a spiral effect, the analysis of these three disputes provides interesting findings. Prospect theory focuses on how actors rationally (even if the selection of a point of reference may not be rational) behave in the face of risks and how their responses may differ depending on whether they value the status quo positively or negatively. In the analysis of these three disputes, it seems obvious to consider that the main risk factor for Gazprom, and thus for Russia, was the effect of gas cut-offs to Ukraine on its reputation as a reliable supplier. The triangular relation between Russia, Ukraine, and gas consuming countries in the rest of Europe is complex: Russia supplies gas both for

Ukraine and for the rest of Europe; Ukraine consumes either Russian gas or Central Asian gas transported through Russia and is the main transit corridor for Russian gas further West; lastly, European countries depend on not only Russia for gas but also Ukraine as a guarantor of safe transit. Due to this complexity, the 'energy weapon' used by Russia against Ukraine in escalating these disputes can either intentionally or unintentionally, due to Ukraine's conscious decisions, affect end consumers in other countries. The dilemma that this supposes for Russia implies a risk in making certain decisions. This situation arguably forced Russia to assess the stakes in each of the disputes and to consider whether risks were worth taking to prevail over Ukraine in energy disputes.

The empirical analysis has shown, after having tried to clarify the intricacies of some of the highly opaque agreements reached, that the changes and derived divergences in the patterns of escalation were mostly due to decisions made not by the Ukrainian side but by the Russian side. Changes resulted from different decisions in the same nodes of the decision tree, something, as noted above, that becomes more apparent in the case of the January 2009 dispute. To use the terminology of deterrence theory, after Ukraine responded, Gazprom/Russia was 'deterred' by the risks in reputational costs in January 2006 and was 'deterred' to an even greater degree in March 2008 but was not 'deterred' in January 2009. Prospect theory shows that depending on the valuation of the status quo, actors will behave in different ways: they will be deterred if prospective gains imply a risk of worsening a status quo framed as positive, but they will assume risks if they need to improve a status quo framed as negative. In this respect, considering the stakes in each of the disputes shows a picture that conforms to the tenets of prospect theory. Determining exactly how Russian decision makers both in Gazprom and the government framed the status quo is arguably challenging, but the fact that these disputes involved money makes it much easier to determine 'what was in it' for Russia, even if we still cannot make absolute conclusions.

Thus, we can assume that having accepted a less than optimal deal, Russia preferred not to assume many risks in reputational costs in January 2006. Regarding March 2008, contrary to January 2006, the dispute only involved debts instead of pricing agreements, and the stakes were lower, which could explain why Russia de-escalated earlier and was even less willing to assume risks. Regarding January 2009, Russia faced a pricing dispute with Ukraine similar to the dispute in January 2006. Following a strictly positivistic assessment, it would be difficult to explain the difference in Russia's responses to these disputes. Here, the picture becomes even more complex, as the factors resulting in a worsened status quo, accounting for the spiral effect that prevailed in the more than two weeks that the dispute lasted, are multiple. Not only the economic factor of the financial crisis but also the worsened geostrategic situation and worsened Russian–Ukrainian relations translating into more tense energy negotiations since the second Timoshenko government came into power should be taken into account. Unfortunately, the explanatory power has its limits, and the empirical findings cannot show exactly which of these factors accounts for Russia's response to the dispute that took place in January 2009. However, we find worth considering the fact that, as predicted by prospect theory, the outcome of the January 2009 dispute was consistent with the spiral model when the status quo clearly worsened for Russia.

Katerina Malygina

11. The Struggle Over Ukraine's Gas Transit Pipeline Network Through the Lenses of Securitisation Theory

11.1. Introduction

For more than twenty years, Russia has attempted to obtain control over Ukraine's gas transit system (GTS). In Russia's discourse, Ukraine has consistently been presented as an unreliable partner, either as unable to pay for its gas debts or as siphoning off Russian gas bound for European customers. Although such a discourse corresponded to reality in the 1990s, this was no longer the case in the 2000s. Nevertheless, Russia continued to uphold the situation artificially, bringing the dispute, usually over gas prices, to a state of crisis. Two such acute gas crises took place in winter 2006 and 2009, arousing considerable interest among researchers.

Since this time, a great amount of academic literature has been written on Russia's 'energy weapon', or Russia's alleged efforts to manipulate other countries' dependency on its energy resources to obtain economic and political gains.[1] Yet, the dependency between Russia and Ukraine is mutual; Ukraine depends on Russia for gas supplies, and Russia depends on Ukraine for the transit of its gas to Europe. However, the question of gas transit management by Ukraine has been largely overlooked, and most studies have concentrated either on the geopolitics of Russia's bypass projects[2] or on Ukraine's dependency on gas imports from Russia.[3] The focus on the other perspective—Ukraine's perspective on gas transit—has been limited.

---

[1] Cf. e.g., Hedenskog, Jakob/ Larsson, Robert L.: Russian leverage on the CIS and the Baltic states, Stockholm: Swedish Defence Research Agency, 2007; Baran, Zeyno: EU energy security: time to end Russian leverage, in: The Washington Quarterly, 2007 (vol. 30), no. 4, pp. 131–144; Smith, Keith C.: Russia and European energy security: divide and dominate, Washington, DC: Center for Strategic & International Studies, 2008; Orttung, Robert W./ Øverland, Indra: A limited toolbox: explaining the constraints on Russia's foreign energy policy, in: Journal of Eurasian Studies, 2011 (vol. 2),no. 1, pp. 74–85.

[2] Cf. e.g., Victor, Nadejda M./ Victor, David G.: Bypassing Ukraine: exporting Russian gas to Poland and Germany, in: Victor, David G./ Jaffe, Amy M./ Hayes, Mark H. (eds): Natural gas and geopolitics: from 1970 to 2040, Cambridge: Cambridge University Press, 2006, pp. 122–168; Gonchar, Michael/ Martynyuk, Vitalii/ Chubyk, Andriy: The impact of Nord Stream, South Stream on the gas transit via Ukraine and security of gas supplies to Ukraine and the EU, in: Liuhto, Kari (ed.): EU–Russia gas connection: pipes, politics and problems, Turku: Pan-European Institute, 2009, pp. 49–69.

[3] Cf. e.g., Balmaceda, Margarita M.: Gas, Oil and the linkages between domestic and foreign policies: the case of Ukraine, in: Europe-Asia Studies, 1998 (vol. 50), no. 2, pp. 257–286; Balmaceda, Margarita M.: Explaining the management of energy dependency in Ukraine. Possibilities and limits of a domestic-centered perspective, Mannheim: Mannheimer Zentrum für Europäische Sozialforschung, 2004, http://www.uni-mannheim.de/fkks/fkks30.pdf.

This article seeks to fill this gap and presents a case study of the midstream gas sector that involves gas transit via Ukraine. Since independence in 1991, the Ukrainian gas transport network has remained owned by the state. This is in great contrast to many other industrial sectors captured by various oligarchs, including the gas imports sector. The fact that Russia was not able to purchase Ukraine's pipeline operator despite the similar policy and success across many other post-Soviet states[4] makes this case study interesting.

The main question of the case study is how Ukraine managed to avoid handing over control of its GTS to Russia[5] despite the constant pressure from the latter. Thus, the focus is the logic that shaped the actions of various actors at different points in time. For this purpose, I utilise the securitisation theory that helps to situate the case of gas transit in the wider context of power relations in Ukraine. My hypothesis is that when the Ukrainian government was negotiating on GTS status with Russia, the opposition publicly securitised the issue, thus diminishing the government's room for manoeuvring, on the one hand, and damaging the government's image, on the other. Moreover, I argue that such an instrumentalisation of discourses as exemplified in the case of Ukraine's GTS should not be regarded as a special case of energy relations in the post-Soviet region but as a typical pattern in Ukraine's decision-making process. As a result, this article aims to contribute not only to the 'energy weapon' literature but also to the research on Ukraine's governance practices.

The article is structured as follows. First, I consider different hypotheses that may explain the research question and show their shortcomings. In the subsequent sections, I present my analytical framework based on the securitisation theory, which I contextualise for Ukraine's political environment. I then describe the debate over GTS privatisation from the 1990s to 2010. Lastly, I examine the struggle over Ukraine's GTS through the lens of securitisation theory and draw conclusions.

11.2. Theoretical Considerations

Since the breakup of the Soviet Union, Russia has consistently pursued the policy of transit avoidance and transit control. The methods that it has used are well documented in the academic literature. These include subsidies, cut-offs, and the construction of alternative

---

4   For instance, Gazprom purchased the shares of Moldova's and Armenian's GTS in exchange for their gas debts in the 1990s. It also bought 50% of Beltransgaz, the gas transit operator in Belarus, in 2007, and since 2012 has fully owned it. Mitrova, Tatiana/ Pirani, Simon/ Stern, Jonathan P.: Russia, the CIS and Europe: gas trade and transit, in: Pirani, Simon (ed.): Russian and CIS gas markets and their impact on Europe, Oxford: Oxford University Press, 2009, pp. 395–441, here p. 411.
5   I further refer to it as 'GTS privatization blockade'. By privatization attempts, I mean all attempts to change the status of GTS as a public entity, whether it is a concession, privatization in the proper sense, creation of a consortium, etc.

11. The Struggle Over Ukraine's Gas Transit Pipeline Network

transit pipelines that would bypass the transit countries.[6] Such a policy was generally, but not always, successful, as Ukraine's case suggests. This situation gives some authors reason to doubt the potency of Russia's energy leverage. Smith Stegen, for instance, argues that that the deployment of an 'energy weapon' could be regarded as successful only when the targeted state modifies its behaviour in a case of threats or continues its support in case of rewards.[7] Thus, for Smith Stegen, the limits of the Russian energy leverage should be sought outside Russia. Unfortunately, she does not develop the idea further and does not explain how some targeted states were able to resist Russia's pressure.

This perspective is central to this analysis. The aim of this article is not to find the reason for Ukraine's GTS privatisation blockade but to understand the process of 'how' it is happening and the logic behind it. From the theoretical literature, three primary hypotheses for explaining the GTS privatisation blockade can be proposed: the geopolitical leverage thesis, the rent-seeking thesis, and the economic nationalism thesis. In the following, I will consider each of the hypotheses individually.

11.2.1. Hypothesis 1: Geopolitical Leverage Thesis

For more than a century, classical geopolitics assigned Ukraine to be part of the so-called 'heartland', the control of which was essential for worldwide dominance.[8] In the same vein, Brzezinski defined the contemporary role of Ukraine on a geopolitical chessboard as one of the 'geopolitical pivots'

> whose importance is derived not from their power and motivation but rather from their sensitive location and from the consequences of their potentially vulnerable condition for the behaviour of geostrategic players.[9]

The geopolitical leverage thesis was formulated for Ukraine largely by external actors[10] and was not fully internalised at home. The Ukrainian actors never formulated any foreign policy doctrine based on this view and never endeavoured to rationalise the country's energy sector to serve geopolitical purposes, as Russia did. This is despite the fact that due to its geographical position, Ukraine's GTS has significant geopolitical importance for Russia as a large gas supplier and the EU as a gas customer.[11] What

---

6 Cf. e.g., Orttung, Robert W./ Øverland, Indra: A limited toolbox: explaining the constraints on Russia's foreign energy policy, in: Journal of Eurasian Studies, 2011 (vol. 2),no. 1, pp. 74–85.
7 Smith Stegen, Karen: Deconstructing the 'energy weapon': Russia's threat to Europe as case study, in: Energy Policy, 2011 (vol. 39), no. 10, pp. 6505–6513, here p. 6510.
8 Mackinder, Halford John: Democratic ideals and reality: a study in the politics of reconstruction, London: Constable, 1919.
9 Brzezinski, Zbigniew: The Grand chessboard: American primacy and its geostrategic imperatives, New York: Basic books, 1997, p. 41.
10 I refer here to Zbigniew Brzezinski and his famous thesis that without independent Ukraine, Russia would never become an empire again. Ibid., p. 46.
11 For more details on Ukraine's gas transit state and policy, see Malyhina, Kateryna: Die Erdgasversorgung der EU unter Berücksichtigung der Ukraine als Transitland,

Ukraine attempted to do many times was to put pressure on Russia in issues of transit volumes and fees before or during gas trade negotiations. However, the logic behind it was fully economic, emerging from the desire to improve the terms of trade rather than to gain some geopolitical weight. The geopolitical importance of Ukraine as a gas transit state was therefore a by-product of this policy, not its primary motivation. Otherwise, how can we explain the fact that Ukraine was ready to hand over its GTS to Russia several times in the last twenty years, as I will show later in this article?

Furthermore, unlike Russia, Ukraine never developed a strategic vision for its gas transit status despite the fact that its significance is widely acknowledged in Ukraine. Until 2006, Ukraine had no single gas transit policy because gas import from Russia and gas transit via Ukraine were bundled in one package, which also speaks in favour of the economic underpinning of Ukraine's actions, as argued above. When the 'Energy Strategy of Ukraine to 2030' was finally adopted in 2006, it only described Ukraine's gas transport capacity and did not foresee any measures on its further development apart from its modernisation needs.[12] As a result, both opportunities and risks for Ukraine's transit have been poorly evaluated. For instance, Ukraine never considered the EU's energy strategy aimed at lowering its energy consumption,[13] a move that would have direct consequences for Ukraine's importance as a gas transit state. Likewise, the possibility of Russia's declining gas production, widely discussed by the policy advisors in the EU,[14] was never even problematised in Ukraine. Furthermore, Ukraine attempted only once to coordinate its transit policy with another transit country, Belarus, but to no avail. Lastly, Ukraine never seriously cared about the creation of an attractive regime for investments into its gas pipelines despite their urgent need for modernisation.

This lack of vision surprisingly contradicts Ukraine's successful experience of combating Russia's attempt to gain control over its GTS. Therefore, there should be some other mechanism in place. The geopolitical leverage hypothesis can be discarded for the purposes of this chapter.

Bremen: Forschungsstelle Osteuropa, 2009, http://www.forschungsstelle.uni-bremen.de/UserFiles/file/06-Publikationen/Arbeitspapiere/fsoap105.pdf.
12 See Enerhetychna stratehiya Ukrainy na period do 2030 roku, 15 March 2006, http://zakon4.rada.gov.ua/laws/file/text/8/f197393n9.zip.
13 For an overview of the EU energy policy, Malyhina, Kateryna: Die Erdgasversorgung der EU unter besonderer Berücksichtigung der Ukraine als Transitland, Bremen: Forschungsstelle Osteuropa, 2009, pp. 14–17, http://www.forschungsstelle.uni-bremen.de/UserFiles/file/06-Publikationen/Arbeitspapiere/fsoap105.pdf.
14 Cf. e.g., Orttung, Robert W./ Perović, Jeronim/ Pleines, Heiko/ Schröder, Hans-Henning (eds): Russia's energy sector between politics and business. Bremen: Research Centre for East European Studies, 2008, http://www.laender-analysen.de/pages/arbeitspapiere/fsoAP92.pdf; Nies, Susanne: Oil and gas delivery to Europe: an overview of existing and planned infrastructures, Paris: IFRI, 2008, p. 61; Ahrend, Rüdiger/ Tompson, William: Unnatural monopoly: the endless wait for gas sector reform in Russia, in: Europe-Asia Studies, 2005 (vol. 57), no. 6, pp. 801–821, here p. 804.

11.2.2. Hypothesis 2: Rent-Seeking Thesis

The second hypothesis that may help us to understand how Ukraine managed to avoid handing over its GTS to Russia is the rent-seeking thesis, whose main proponent with regard to Ukraine's energy policy is Balmaceda.[15] Unlike many other scholars who treat the energy dependency of post-Soviet states on Russia as dependent variables (i.e., as a product of Russia's economic pressure), she shifted the research focus to the internal political processes in Ukraine and viewed the Ukrainian gas policy as an independent variable in her analysis of Ukrainian–Russian energy relations. Balmaceda argues that Ukraine is unable or unwilling to decrease its high energy dependency on Russia because the gas trade creates lucrative rents for Ukrainian political and economic elites. For this reason, the Ukrainian actors are interested in maintaining the status quo. The desire to increase these 'rents of energy dependency'[16] guides political behaviour, resulting in an incoherent and inconsistent gas policy that undermines Ukraine's energy security. According to Balmaceda, the rent-seeking behaviour of the Ukrainian policy makers results in Ukraine's perpetual energy dependency and leads to the success of Russia's 'energy weapon' doctrine in Ukraine.

Balmaceda's work falls within the broader literature on 'state capture', which generally presents a modification of the classical 'regulatory capture' theory with application to the post-communist context.[17] In 1998, Hellman wrote an influential article seeking to explain the slow reform progress in post-Soviet states. He advanced a simple idea: 'The most serious political obstacles to the process of economic reform have come not from the short-term losers but from the short-term winners'.[18] Thus, the immediate winners of the initial reform phase managed to obstruct further reform implementation because they threatened to 'eliminate the special advantages and market distortions upon which their own early reform gains were based'.[19] The situation led

---

15 Balmaceda, Margarita M.: Gas, Oil and the linkages between domestic and foreign policies: the case of Ukraine, in: Europe-Asia Studies, 1998 (vol. 50), no. 2, pp. 257–286; Balmaceda, Margarita M.: Explaining the management of energy dependency in Ukraine. Possibilities and limits of a domestic-centered perspective, Mannheim: Mannheimer Zentrum für Europäische Sozialforschung, 2004, http://www.uni-mannheim.de/fkks/fkks30.pdf; Balmaceda, Margarita M.: Ukraine's persistent energy crisis, in: Problems of Post-Communism, 2004 (vol. 51), no. 4, pp. 40–50; Balmaceda, Margarita M.: Energy dependency, politics and corruption in the former Soviet Union: Russia's power, oligarchs' profits and Ukraine's missing energy policy, 1995–2006, London: Routledge, 2008.
16 Balmaceda, Margarita M.: Energy dependency, politics and corruption in the former Soviet Union: Russia's power, oligarchs' profits and Ukraine's missing energy policy, 1995–2006, London: Routledge, 2008, p. 8.
17 For a literature review, see Dal Bo, Ernesto: Regulatory capture: a review, in: Oxford Review of Economic Policy, 2006 (vol. 22), no. 2, pp. 203–225.
18 Hellman, Joel S.: Winners take all: the politics of partial reform in postcommunist transitions, in: World Politics, 1998 (vol. 50), no. 2, pp. 203–234, here p. 203.
19 Ibid. p. 204.

to a partial reform equilibrium, which slowly transformed into a 'capture economy'[20] or 'capitalism for the few'.[21]

In general, the literature on regulatory capture, including the Hellman thesis for the post-socialist states, analyses the influence of the powerful economic groups on the policy formation process, or the direct interference of powerful interest groups into politics by means of formal and informal lobbying. In the case of Ukraine, the proponents of the 'state capture' concept have been investigating strategies of influence exercised by the so-called 'oligarchs' to achieve preferential treatment from politicians.[22]

The rent-seeking thesis could well explain why GTS was not privatised. It is, however, poorly suited to answer the 'how' question formulated above. First, the GTS has always been under the state's control, and private actors' access to it was rather limited. Thus, the GTS case clearly shows the limits of 'state capture' in Ukraine.[23] Moreover, the rent-seeking thesis does not capture the dynamics that this issue has experienced since Ukraine's independence. The question of GTS privatisation is highly contested in Ukraine's domestic politics. Several times, Ukraine was ready to hand over its GTS to Russia, which is against the logic of rent seeking because the transfer of control to the external player would virtually mean the renunciation of future rents. Thus, if the rent-seeking thesis were valid, no Ukrainian actor would be interested in the GTS lease or sale. Therefore, the logic of the rent-seeking thesis is not sufficient to explain the dynamics of the GTS privatisation blockade.

11.2.3. Hypothesis 3: Economic Nationalism Thesis

The economic nationalism thesis could be regarded as a counter-argument against asymmetric economic interdependence in general and Russia's 'energy weapon' as its concrete form. The idea that unequal economic interdependence might be a source

---

20   Hellman, Joel S./ Kaufmann, Daniel: Confronting the challenge of state capture in transition economies, in: Finance & Development, 2001 (vol. 38), no. 3, pp. 31–35.
21   Havrylyshyn, Oleh: Divergent paths in post-communist transformation: capitalism for all or capitalism for the few?, Basingstoke: Palgrave Macmillan, 2006.
22   Pleines, Heiko: Ukrainische Seilschaften. Informelle Einflussnahme in der ukrainischen Wirtschaftspolitik 1992–2004, Münster: LIT, 2005; Pleines, Heiko: The political role of the oligarchs, in: Besters-Dilger, Juliane (ed.): Ukraine on its way to Europe: interim results of the orange revolution, Frankfurt/ Main: Peter Lang, 2009, pp. 103–120; Pleines, Heiko: Demokratisierung ohne Demokraten: Die Oligarchen in der ukrainischen Politik, in: Osteuropa, 2010 (vol. 60), no. 2–4, pp. 123–134; Puglisi, Rosaria: The rise of the Ukrainian oligarchs, in: Democratization, 2003 (vol. 10), no. 3, pp. 99–123; Puglisi, Rosaria: a window to the world? Oligarchs and foreign policy in Ukraine, in: Fischer, Sabine (ed.): Ukraine. Quo vadis?, Paris: Institute for Security Studies, 2008, pp. 55–86.
23   Here, I define 'state capture' in narrow terms and refer to the 'captors' from the private sector seeking rents from the state. It is, however, possible that state actors themselves might capture state institutions to promote their own financial interests. For conceptual clarification of the 'state capture', see World Bank: a contribution to the policy debate, Washington, DC: World Bank, 2000, p. 3.

of power was articulated by Hirschman in 1945.[24] With regard to Ukraine, Abdelal utilised this approach to explain Ukraine's foreign policy change in the early 1990s, when it reversed its nationalist stance and moved closer to Russia, following Russia's energy cut-offs.[25]

In the 2000s, however, a number of scholars representing constructivist economic nationalism questioned the potency of the asymmetric interdependence approach.[26] These researchers sought to explain the variety of economic policies adopted by Soviet successor states towards Russia despite the fact that all of them were heavily dependent on Russia economically after the Soviet Union's break-up. According to the economic nationalism thesis, the varied outcomes could be explained by different perceptions of the state's economic dependence on Russia. Thus, although the Baltic states and Ukraine viewed it as a national threat, Belarus saw it as a benefit. As a result, varying identity narratives played a crucial role in determining the states' foreign economic preferences. Accordingly, Abdelal formulated the limits of Russia's 'energy weapon' as follows: 'Russia's attempts to exploit the coercive power inherent in Ukraine's energy dependence failed, primarily because the Ukrainian government interpreted Russia as a security threat'.[27]

However, the economic nationalism thesis does not explain why Ukraine signed a lease agreement with Russia on the Black Sea Fleet in 1997 but not on the GTS, although the former was perceived as a threat in a similar way as the latter. Therefore, the identity factor alone cannot explain the complexity of Ukrainian political decision-making.

---

24 Hirschman, Albert O.: National power and the structure of foreign trade, Berkeley, CA: University of California Press, 1980.
25 Abdelal, Rawi/ Kirshner, Jonathan: Strategy, economic relations, and the definition of national interests, in: Security Studies, 1999 (vol. 9), no. 1–2, pp. 119–156; Abdelal, Rawi: Interpreting interdependence: national security and the energy trade of Russia, Ukraine, and Belarus, in: Legvold, Robert/ Wallander, Celeste A. (eds): Swords and sustenance: the economics of security in Belarus and Ukraine, Cambridge, MA: MIT Press, 2004, pp. 101–127.
26 Abdelal, Rawi: National purpose in the world economy: post-Soviet states in comparative perspective, Ithaca, NY: Cornell University Press, 2005; Abdelal, Rawi: Nationalism and international political economy in Eurasia, in: Helleiner, Eric/ Pickel, Andreas (eds): Economic nationalism in a globalizing world, Ithaca, NY: Cornell University Press, 2005, pp. 21–43; Eichler, Maya: Explaining post-communist transformations: economic nationalism in Ukraine and Russia, in: Helleiner, Eric/ Pickel, Andreas (eds): Economic nationalism in a globalizing world, Ithaca, NY: Cornell University Press, 2005, pp. 69–87; Shulman, Stephen: Nationalist sources of international economic integration, in: International Studies Quarterly, 2000 (vol. 44), no. 3, pp. 365–390; Tsygankov, Andrei P.: Pathways after empire: national identity and foreign economic policy in the post-Soviet world, Lanham, MD: Rowman & Littlefield, 2001; Tsygankov, Andrei P.: The return to Eurasia: Russia's identity and geoeconomic choices in the post-Soviet world, in: Helleiner, Eric/ Pickel, Andreas (eds): Economic nationalism in a globalizing world, Ithaca, NY: Cornell University Press, 2005, pp. 44–68.
27 Abdelal, Rawi: Interpreting interdependence: national security and the energy trade of Russia, Ukraine, and Belarus, in: Legvold, Robert/ Wallander, Celeste A. (eds): Swords and sustenance: the economics of security in Belarus and Ukraine, Cambridge, MA: MIT Press, 2004, pp. 101–127, here p. 108.

11.2.4. Hypothesis 4: Securitisation Thesis

The main problem of the above-mentioned hypotheses is their static conceptualisation of the gas transit issue that downplays the dynamics of the GTS privatisation debate. Some scholars have attempted to combine different hypotheses to better capture Russian–Ukrainian relations. For instance, D'Anieri (1999) combined the asymmetric interdependence and economic nationalism theses to explain Ukrainian–Russian tensions in the 1990s in terms of 'Ukraine's dilemma', a trade-off between Ukraine's quest for complete political independence from Russia and the need to take into account the deep economic interdependence between the two states.[28]

Although the proposed reading is still relevant, I would like to suggest another interpretation of Ukraine's relationship with Russia from the perspective of internal power struggles, seen through the lens of the securitisation theory. As noted by Guzzini, security analysis can explain why actors take certain moves, 'why some of them may find a receptive audience [...], and indeed why certain action-complexes can then follow'.[29] My hypothesis is that when the Ukrainian government was negotiating GTS status with Russia, the opposition securitised the issue, thus diminishing the government's room for manoeuvring, on the one hand, and damaging the government's image, on the other. I now turn to presenting the securitisation theory as an analytical framework.

11.3. Securitisation Theory as Analytical Framework

The securitisation theory was developed by Buzan, Wæver, and their collaborators (dubbed the Copenhagen School). This theory presents a completely novel approach to looking at security not as objectively given but as socially constructed in political practice. Security is thus understood as a speech act that presents a particular development as a threat to a 'referent object' and therefore moves it to the security agenda, which has a specific rationality or logic implied by security language.[30] Ultimately, securitisation serves to legitimate the use of extraordinary measures that would not have been applicable under 'normal' conditions of everyday politics.[31] Thus, securitisation is an explicitly political act involving a decision by an actor to securitise and is therefore comparable with politicisation, but the two concepts are not identical: whereas

---

28 D'Anieri, Paul: Economic interdependence in Ukrainian–Russian relations, Albany, NY: State University of New York Press, 1999.
29 Guzzini, Stephano: Securitization as a causal mechanism, in: Security Dialogue, 2011 (vol. 42), no. 4–5, pp. 329–341, here p. 338.
30 Cf. Wæver, Ole: Securitization and desecuritization, in: Lipschutz, Ronnie D. (ed.): On security, New York: Columbia University Press, 1995, pp. 46–86, here p. 55.
31 Buzan, Barry/ Wæver, Ole/ de Wilde, Jaap: Security: a new framework for analysis, Boulder, CO: Lynne Rienner, 1998, pp. 23–24.

11. The Struggle Over Ukraine's Gas Transit Pipeline Network 285

politicisation is a matter of choice and open debate, securitisation is a matter of urgency.[32] Such a distinction clearly shows that political and security issues underlie different handling policies. Through successful securitisation, one can legitimately bend and/or break the established decision-making rules.[33]

Yet, not every issue that is presented as a threat ('securitising move')[34] could lead to the intended consequence of rule bending/breaking. The existential threats and emergency action may be necessary but not sufficient conditions for securitisation success. These conditions refer to what Stritzel calls the 'internalist' centre of gravity in the securitisation theory, according to which new meaning is established via a speech act outside context by the very act of its utterance. This internalist/ poststructural perspective on security concentrates on the performative power of speech acts[35] and is temporarily limited to the event of their utterance.[36] There is, however, also the 'externalist' or contextual grounding of the Copenhagen School, namely the understanding of security as a 'structured field' where 'some actors are placed in positions of power' over the others.[37] According to Stritzel, the externalist elements of the securitisation theory are reflected in the conceptualisation of 'facilitating conditions' and the speaker–audience relationship.[38] These points more precisely define the possibility of the success of securitisation.

The Copenhagen School defines 'facilitating conditions' as 'the conditions under which the speech act works, in contrast to cases in which the act misfires or is abused'.[39] These conditions are subdivided into internal and external facilitating conditions. Whereas the former present linguistic-grammatical conditions, the latter involve securitisation actors and constructed threats.

The internal facilitating conditions present the 'grammar of security', 'a plot that includes existential threat, point of no return, and a possible way out [...] plus the particular dialects of the different sectors'.[40] As explicated by other authors, security lan-

---
32   Ibid., p. 29.
33   Ibid., p. 24.
34   Ibid., p. 25.
35   More concrete, performative means that '[b]y saying the words, something is done' (ibid., p. 26).
36   Stritzel, Holger: Towards a theory of securitization: Copenhagen and beyond, in: European Journal of International Relations, 2007 (vol. 13), no. 3, pp. 357–383, here p. 359.
37   Ibid., p. 365; Buzan, Barry/ Wæver, Ole/ de Wilde, Jaap: Security: a new framework for analysis, Boulder, CO: Lynne Rienner, 1998, pp. 31–33.
38   Stritzel, Holger: Towards a theory of securitization: Copenhagen and beyond, in: European Journal of International Relations, 2007 (vol. 13), no. 3, pp. 357–383, here p. 374.
39   Buzan, Barry/ Wæver, Ole/ de Wilde, Jaap: Security: a new framework for analysis, Boulder, CO: Lynne Rienner, 1998, p. 32.
40   Ibid., p. 33. The scholars give the following examples of the sector-specific dialects: identity in the societal sector, recognition and sovereignty in the political sector, sustainability in the environmental sector.

guage has several universal features. First, it promotes zero-sum, 'us vs. them' thinking.[41] Second, the security frames[42] can become petrified and inert.[43]

With regard to the external facilitating conditions, the Copenhagen School assumes that the power position or authority of the securitising actor, which could be either formal or informal, is directly related to the likelihood of audience consent for the securitising move. The Copenhagen School also argues that certain features of the threatened (referent) objects might change the degree of plausibility of the securitisation claims. Thus, the chances of successful securitisation are higher when the securitising actors refer in their speech to a state's sovereignty or a nation's identity or when the referent objects can be linked to the state, which 'can claim a natural right to survive, to defend its existence'.[44] In this respect, Stritzel goes even further by proposing that the success of securitisation depends upon its resonance with existing discourses.[45] In their later work, Buzan and Wæver acknowledged the role of such discourses—macrosecuritisations (e.g., world religions, political ideologies, the Cold War) that have the power to structure security constellations and therefore to influence the lower-level securitisations.[46]

Regarding the speaker–audience relationship, the Copenhagen School claims that securitising moves become securitisations only after the audience accepts them as such. As Buzan et al. put it, 'Successful securitization is not decided by the securitizer but by the audience of the security speech act: Does the audience accept that something is an existential threat to a shared value?'.[47] This claim is, however, one of the weakest and most highly criticised points in securitisation theory.[48] As Stritzel rightly recognised, this part of securitisation theory is at odds with the internalist part. On the one hand, the introduction of the 'audience' undermines the internalist perspective

---

41 Cf. Williams, Michael C.: Words, images, enemies: securitization and international politics, in: International Studies Quarterly, 2003 (vol. 47), no. 4, pp. 511–531, here pp. 519–520.
42 Frames are defined here as conscious constructions that intentionally omit or emphasise different aspects of social life and are a product of a socio-constructivist process called framing (cf. Benford, Robert D./ Snow, David A.: Framing processes and social movements: an overview and assessment, in: Annual Review of Sociology, 2000 (vol. 26), no. 1, pp. 611–639).
43 Williams, Michael C.: Words, images, enemies: securitization and international politics, in: International Studies Quarterly, 2003 (vol. 47), no. 4, pp. 511–531, here p. 519.
44 Buzan, Barry/ Wæver, Ole/ de Wilde, Jaap: Security: a new framework for analysis, Boulder, CO: Lynne Rienner, 1998, p. 38.
45 Stritzel, Holger: Towards a theory of securitization: Copenhagen and beyond, in: European Journal of International Relations, 2007 (vol. 13), no. 3, pp. 357–383, here pp. 370.
46 Buzan, Barry/ Wæver, Ole: Macrosecuritisation and security constellations: reconsidering scale in securitisation theory, in: Review of International Studies, 2009 (vol. 35), no. 2, pp. 253–276.
47 Buzan, Barry/ Wæver, Ole/ de Wilde, Jaap: Security: a new framework for analysis, Boulder, CO: Lynne Rienner, 1998, p. 31.
48 Stritzel, Holger: Towards a theory of securitization: Copenhagen and beyond, in: European Journal of International Relations, 2007 (vol. 13), no. 3, pp. 357–383; Balzacq, Thierry: The three faces of securitization: political agency, audience and context, in: European Journal of International Relations, 2005 (vol. 11), no. 2, pp. 171–201; McDonald, Matt: Securitization and the construction of security, in: European Journal of International Relations, 2008 (vol. 14), no. 4, pp. 563–587.

with its reliance on the context-free speech act performing security by its mere articulation.[49] Without a consent-giving 'audience', on the other hand, there would be no checks and balances against rhetorical manipulations such as, for example, a fascist speech act.[50] Such a differentiation between speaker and audience acquires a special meaning. As argued by Williams, 'Casting securitization as a speech act places that act within a framework of communicative action and legitimation' that requires 'a process of argument, the provision of reasons, presentation of evidence, and commitment to convincing others of the validity of one's position'.[51]

Placing securitisation in the realm of communicative action, however, restricts the applicability of the original securitisation theory to the liberal democratic context for which deliberation is characteristic. By the same token, the differentiation between 'normal' and 'special' politics is not really appropriate for non-democratic political systems with fluid decision procedures, where extraordinary measures and rule breaking could be quite an ordinary practice.[52] As a result, there is a certain 'democratic bias' in the securitisation theory[53] that should be accounted for when it is utilised for non-Western contexts. These cultural limits of the original securitisation theory have recently been characterised as an emerging debate in the Copenhagen School.[54] However, the applications remain very limited.

11.4. Contextualising the Securitisation Framework to Hybrid Regimes

In summary, the analytical framework of securitisation consists of such elements as 'who securitises, on what issues (threats), for whom (referent object), why, with what results, and, not least, under what conditions (i.e., what explains when securitisation is successful)'.[55] However, there is a need to contextualise the securitisation framework for the Ukrainian political environment.

Despite more than twenty years of transition, the Ukrainian political regime is still not consolidated. In the last decade, Ukraine experienced two major shifts in its

---

49   McDonald, Matt: Securitization and the construction of security, in: European Journal of International Relations, 2008 (vol. 14), no. 4, pp. 563–587, here p. 572.
50   Williams, Michael C.: Words, images, enemies: securitization and international politics, in: International Studies Quarterly, 2003 (vol. 47), no. 4, pp. 511–531, here pp. 522–523.
51   Ibid., p. 522.
52   Vuori, Juha A.: Illocutionary logic and strands of securitization: applying the theory of securitization to the study of non-democratic political orders, in: European Journal of International Relations, 2008 (vol. 14), no. 1, pp. 65–99, here p. 69.
53   Ibid., p. 64; see also Wilkinson, C.: The Copenhagen School on tour in Kyrgyzstan: is securitization theory useable outside Europe?, in: Security Dialogue, 2007 (vol. 38), no. 1, pp. 5–25.
54   Wæver, Ole: Politics, security, theory, in: Security Dialogue, 2011 (vol. 42), no. 4–5, pp. 465–480.
55   Buzan, Barry/ Wæver, Ole/ de Wilde, Jaap: Security: a new framework for analysis, Boulder, CO: Lynne Rienner, 1998, p. 32.

politics, from Leonid Kuchma's authoritarian rulership to the more democratic leadership of Viktor Yushchenko in 2004 and back toward authoritarianism under Viktor Yanukovich after the presidential elections in 2010. Paraphrasing what Grzymala-Busse and Jones Luong once said about the post-communist state in general, the state in Ukraine is still not a unitary, intentional actor as 'no one single agent has uniform influence or authority across all state sectors, and state action is neither centralized nor coherent'.[56] The widespread rent-seeking and corruption, the presence of powerful economic actors beyond the state level (oligarchs), and the East–West division of Ukraine all prevent the Ukrainian state from consolidating. As a result, the competition over state financial and administrative resources is particularly fierce, and the rules themselves become a powerful resource in the power struggles. In the literature on political regimes, such a state is defined as 'competitive authoritarianism',[57] a type of hybrid regime between democracy and authoritarianism in which elections are not a mere façade, as in authoritarian regimes, but are meaningful for access to power. However, unlike in democracies, the playing field is systematically skewed in favour of the incumbents.

In such a political context, rule bending/breaking (formulated as the sole logic behind securitisation in the original framework) does not need special legitimisation, as opposed to the consolidated liberal democracies. Yet, at the same time, due to a weak formal setting and 'multiple centers of authority-building',[58] the country is particularly vulnerable to any rhetorical manipulation. Just like the rules, discourses become the means of power as every action or non-action can be appropriately framed to either build support or to discredit. Previous researchers have explored only the structural side of these practices, namely who controls the media and how it is manipulated for the incumbent's interests, most prominently by means of censorship.[59] The ideational side, namely the power of different frames and especially the securitisation of one of them, has not been analysed.

When contextualising the securitisation framework to the case of Ukraine, it should be clear what part of the securitisation theory requires refinement. As presented in the previous section, the security logic or rationality presents the internalist part of the theory and has a claim to be universal and therefore context invariant. The logic of securitisation,

---

56 Grzymala-Busse, Anna/ Jones Luong, Paula: Reconceptualizing the state: lessons from post-communism, in: Politics & Society, 2002 (vol. 30), no. 4, pp. 529–554, here pp. 532–533.
57 Levitsky, Steven/ Way, Lucan: Competitive authoritarianism: hybrid regimes in the post-Cold War era, Cambridge: Cambridge University Press, 2010.
58 Grzymala-Busse, Anna/ Jones Luong, Paula: Reconceptualizing the state: lessons from post-communism, in: Politics & Society, 2002 (vol. 30), no. 4, pp. 529–554, here p. 533.
59 Dyczok, Marta: Was Kuchma's censorship effective? Mass media in Ukraine before 2004, in: Europe-Asia Studies, 2006 (vol. 58), no. 2, pp. 215–238; Ryabinska, Natalya: The media market and media ownership in post-communist Ukraine, in: Problems of Post-Communism, 2011 (vol. 58), no. 6, pp. 3–20.

on the contrary, has to do with intentions of securitising actors that might want to instrumentalise the security language for their own purposes. Accordingly, it falls within the externalist part of the theory and should be considered in each particular case.

In general, the logic of securitisation depends highly on who the securitising actors are and how they are embedded in the overall 'structured field'. Although the original securitisation theory defines the securitising actors quite broadly on the basis of general social capital, which does not necessarily entail the possession of formal authority, the logic of bending and breaking the rules implies that potential securitisation actors are more likely to be found among policy-makers as 'it is difficult to see how un-authorized agents can break the normal practices of democratic politics in any meaningful way'.[60] Indeed, most applications of the securitisation framework focus on the securitisation moves conducted by decision-makers (governments or international organisations) producing meaningful effects on specific policies in democratic contexts.[61] In closed authoritarianisms, on the contrary, the incumbent's control over the media ensures that the alternative actors have little opportunity for securitisation. In hybrid regimes, due to their multiple and competing loci of authority, the spectrum of possible securitising actors is much broader than in democratic or authoritarian regimes. As shown by Wilkinson, a securitisation speech in a hybrid regime could be uttered not only by state officials but also by the opposition.[62]

The assumption that there could be various securitising actors necessitates the correction of the analytical framework with regard to securitisation logic. As argued by Vuori, 'Security speech can be utilized for other purposes than legitimating the breaking of rules. Security can be used to reproduce the political order, for renewing discipline, and for controlling society and the political order, for example'.[63] In his analysis of Chinese politics, Vuori explicated five possible purposes of securitisation:
- raising an issue on the agenda,
- legitimating future acts,
- deterrence,

---

60 Aradau, Claudia: Security and the democratic scene: desecuritization and emancipation, in: Journal of International Relations and Development, 2004 (vol. 7), no. 4, pp. 388–413, here p. 395.
61 Cf. e.g., Abrahamsen, Rita: Blair's Africa: the politics of securitization and fear, in: Alternatives, 2005 (vol. 30), no. 1, pp. 55–80; Arifianto, Alexander R.: The securitization of transnational labor migration: the case of Malaysia and Indonesia, in: Asian Politics & Policy, 2009 (vol. 1), no. 4, pp. 613–630; Hudson, Natalie F.: Securitizing women's rights and gender equality, in: Journal of Human Rights, 2009 (vol. 8), no. 1, pp. 53–70; Kamradt-Scott, Adam/ McInnes, Colin: The securitisation of pandemic influenza: framing, security and public policy, in: Global Public Health, 2012 (vol. 7), Supplement 2, pp. S95–S110; Youde, Jeremy: Who's afraid of a chicken? Securitization and avian flu, in: Democracy and Security, 2008 (vol. 4), no. 2, pp. 148–169.
62 Wilkinson, C.: The Copenhagen School on tour in Kyrgyzstan: is securitization theory useable outside Europe?, in: Security Dialogue, 2007 (vol. 38), no. 1, pp. 5–25.
63 Vuori, Juha A.: Illocutionary logic and strands of securitization: applying the theory of securitization to the study of non-democratic political orders, in: European Journal of International Relations, 2008 (vol. 14), no. 1, pp. 65–99, here p. 69.

- legitimating past acts or reproducing a security status,
- control.[64]

In all of these cases, the securitising actors apply the security language, but they differ in their intentions. As a result, the consequences of securitisation also vary. When the task is to raise public attention on a particular issue, the securitisation process will not necessarily end with the adoption of extraordinary measures. Similarly, securitisation for deterrence purposes is aimed at non-action rather than extraordinary measures. In a similar vein, the logic of securitisation affects the construction of audiences. Vuori argues that 'who has to be convinced of the necessity of security action changes with the cultural and political systems in which the securitization is taking place'.[65] It does not have to be the general public; it could also be the power elite, for instance. It is important that the audience is in a position 'to provide the securitizing actor with whatever s/he is seeking to accomplish with the securitization'.[66]

The last point I want to discuss here is the securitisation process. The Copenhagen School describes securitisation as a singular speaker-audience interaction. There is, however, growing concern about this oversimplification of the securitisation process. Empirical studies show that the securitisation process is not linear and involves a multitude of actors.[67] Moreover, the securitisation process might not only trigger certain effects (e.g., policy change) but could also be triggered by something else.[68] That means that the speech act may not only precede action but the opposite is also valid, and the process may start at any point.[69]

In conclusion, the securitisation process is much more complex than the original securitisation theory suggests. There may be multiple securitising moves by various actors that may have different goals in mind but apply the security language. There may also be different variations of the security frame tuned to different audiences,[70]

---

64 Ibid., p. 76. The type of securitization introduced by Wæver is in Vuori's typology the one for legitimating future acts.
65 Ibid., p. 72.
66 Ibid.
67 Wilkinson, C.: The Copenhagen School on tour in Kyrgyzstan: is securitization theory useable outside Europe?, in: Security Dialogue, 2007 (vol. 38), no. 1, pp. 5–25; Salter, Mark B./ Piché, Geneviève: The securitization of the US–Canada border in American political discourse, in: Canadian Journal of Political Science, 2011 (vol. 44), no. 4, pp. 929–951; Roe, Paul: Actor, audience(s) and emergency measures: securitization and the UK's decision to invade Iraq, in: Security Dialogue, 2008 (vol. 39), no. 6, pp. 615–635; McInnes, Colin/ Rushton, Simon: HIV/AIDS and securitization theory, in: European Journal of International Relations, 2013 (vol. 19), no. 1, pp. 115–138; McDonald, Matt: The failed securitization of climate change in Australia, in: Australian Journal of Political Science, 2012 (vol. 47), no. 4, pp. 579–592.
68 Guzzini, Stephano: Securitization as a causal mechanism, in: Security Dialogue, 2011 (vol. 42), no. 4–5, pp. 329–341, here p. 337.
69 Wilkinson, C.: The Copenhagen School on tour in Kyrgyzstan: is securitization theory useable outside Europe?, in: Security Dialogue, 2007 (vol. 38), no. 1, pp. 5–25.
70 McInnes, Colin/ Rushton, Simon: HIV/AIDS and securitization theory, in: European Journal of International Relations, 2013 (vol. 19), no. 1, pp. 115–138.

and the audience itself differs in each particular case. The security frames may also compete with other frames of the same issue, serving as an 'argumentative weapon' against other statements.[71] All this should be accounted for when adopting the securitisation framework beyond its original context of Western democracies and especially for analysing relations in hybrid regimes where power struggles between multiple actors are not seldom.

11.5. Case Study: Ukraine's GTS Privatisation Blockade

Before analysing the GTS privatisation blockade through the lens of securitisation, I want to explicate the political context and show the debate as a process to better understand the logic behind securitisation. In this section, I therefore analyse the main events around Ukraine's GTS privatisation debate from the 1990s until 2010. Because I am interested in the internal dynamics of Ukraine's politics with regard to GTS, Russia's actions and discourse are explicated only narrowly. Similarly, I do not consider the discourse around bypass pipelines. To reconstruct the relevant events in their chronological order, I use process tracing as a method and rely on a number of sources: the legislation database of the Verkhovna Rada, the Integrum database, which provides electronic access to Russian and mass media in the former Soviet Union, and various secondary sources.

Ukraine's gas transit system is the second-largest in Europe after Russia's. Approximately 110 billion cubic metres (bcm) of gas are pumped through it each year from Russia to the EU, which makes up approximately 80% of the Russian gas exports to Europe, or 25% of the EU's gas consumption.[72] In addition to the extensive pipeline network, Ukraine has one of the largest gas storage facilities in the world. The GTS valuation ranges from US$9 billion to US$25 billion and has always been a subject of political and commercial dispute.[73] Most pipelines were built in the 1970s and 1980s and now urgently need to be modernised, although the system has worked reliably and with no major technical disruptions.

The gas transit has always performed more than just a technical function in Ukraine. Until 2006, the transit fees for Russian gas were linked to gas prices for Ukraine, guaranteeing the latter relatively inexpensive gas. However, even after the removal of this mechanism from the bilateral trade, Ukraine received advance payments for transit

---

71 Cf. Rothe, Delf: Security as a weapon: how cataclysm discourses frame international climate negotiations, in: Scheffran, Jürgen (ed.): Climate change, human security and violent conflict: challenges for societal stability, Berlin: Springer, 2012, pp. 243–258.
72 Malyhina, Kateryna: Die Erdgasversorgung der EU unter besonderer Berücksichtigung der Ukraine als Transitland, Bremen: Forschungsstelle Osteuropa, 2009, http://www.forschungsstelle.uni-bremen.de/UserFiles/file/06-Publikationen/Arbeitspapiere/fsoap105.pdf.
73 Pirani, Simon: Ukraine's gas sector, Oxford: Oxford Institute for Energy Studies, 2007, p. 74, http://www.oxfordenergy.org/pdfs/NG21.pdf.

from Russia to pay back gas imports. Moreover, gas transit revenues have been used to defray the cost of low gas tariffs for households and heating companies, thus taking a social function. Such subsidies became possible due to the corporate structure of the GTS operator Ukrtransgaz, a subsidiary of Ukraine's national oil and gas stock company Naftogaz Ukrainy, which has been directly subordinated to the government since its creation in 1998.

Despite the operational control, the Ukrainian government does not have unrestricted competency over the GTS ownership status. As early as 1995, competency over GTS privatisation was transferred to the parliament. As a result, the issue of GTS privatisation soared to a prominent place on the political agenda in Ukraine and has been occasionally contested due to continuous attempts from Russia's side to obtain control over it. In general, there were several stages of the GTS privatisation debate, culminating in 1995, 2000, 2002, and 2007.

11.5.1. The GTS Privatisation Debate in the 1990s

The Russian state-owned gas company Gazprom developed the idea to obtain control over the gas infrastructure in Ukraine as a logical consequence of the deplorable situation in the early 1990s. During that time, the company could neither recover gas debts from Ukraine nor obtain guarantees from Kyiv of their future repayment. Gazprom's strategy to eliminate the debt by cutting off the gas supplies worked only partially. Although Ukraine did return some of its debt to resume gas deliveries, it illegally diverted some of the gas to Europe for its own use.[74]

Consequently, the first debate over GTS privatisation took place in 1995. While signing the gas restructuring agreement with Ukraine in March 1995,[75] Gazprom made it conditional upon the creation of a joint venture, Gaztransit, to be founded on a parity basis with Ukraine's largest underground gas storage facilities as its authorised capital.[76] The gas debt deal was facilitated and mediated by the International Monetary Fund (IMF), which made it a condition for granting stand-by credits both to Russia and Ukraine.[77] Running short of funds, the government supported Gaztransit's creation.

---

74 For Russian–Ukrainian gas relations in the 1990s see, Smolansky, Oles M.: Ukraine's quest for independence: the fuel factor, in: Europe-Asia Studies, 1995 (vol. 47), no. 1, pp. 67–90; Balmaceda, Margarita M.: Gas, Oil and the linkages between domestic and foreign policies: the case of Ukraine, in: Europe-Asia Studies, 1998 (vol. 50), no. 2, pp. 257–286.
75 'Uhoda mizh Uryadom Ukrainy i Rosijs'kym akcionernym tovarystvom Hazprom pro pryncypy vrehulyuvannya zaborhovanosti Ukrainy za postavku rosijs'koho pryrodnoho hazu v 1994 roci i zabezpechennya potochnykh platezhiv za postavku pryrodnoho hazu v 1995 roci', 18 March 1995, http://zakon4.rada.gov.ua/laws/show/643_312.
76 The two facilities were Bilche-Volitsko-Uhers'ke and Bohorodchany underground reservoirs, which make up 70% of the total capacity of Ukraine's gas storage sites.
77 'Ukrainskie obligatsii: prodavets uzhe est', ostalos' nayti pokupatelya', in: Kommersant Vlast, no. 31, 29 August 1995, http://www.kommersant.ru/doc/11426; see also Balmaceda, Margarita M.: Gas, Oil and the linkages between domestic and foreign policies: the case of Ukraine, in:

11. The Struggle Over Ukraine's Gas Transit Pipeline Network

However, it met fierce opposition in the parliament. In September 1995, the parliament added pipeline maintenance companies to the list of enterprises that were not eligible for privatisation.[78] It also adopted the law 'On pipeline transport' that explicitly prohibited the privatisation or any other type of transfer of the Ukrainian gas transit network to foreign states.[79] By the end of 1995, the establishment of Gaztransit in its initial form was completely blocked.

The first attempts to privatise Ukrainian GTS in 1995 failed due to the hard-line position of the parliament. After the parliamentary elections in 1994, communists managed to take advantage of their relative majority and occupy the key leadership posts in the parliament.[80] Paradoxically at first glance, the communists counteracted the attempts to create a joint gas company with Russia that would recreate the old Soviet ties. The project, however, fell victim to the power struggle between the president and the parliament, which had radically different views on the speed and form of economic reforms. For instance, in 1995, President Kuchma began the privatisation policy, which included the privatisation of GTS. By promoting a liberal agenda, Kuchma took a strategically advantageous position. On the one hand, the West promised financial backing to a cash-strapped Ukraine in exchange for neo-liberal reforms. Partial monetary stabilisation with the help of the IMF credit not only brought Kuchma popularity among the people but also improved Ukraine's negotiating position with regard to Russia. On the other hand, the reform discourse helped to legitimise the demands for increased presidential powers at the expense of the parliament, which, under leftist domination, pursued a reactionary economic policy and opposed all private ownership. In such a power struggle, the parliament vetoed the president's decree on financial-industrial groups (FIGs) and adopted its own version of the law on FIGs, which promoted the interests of large state-owned enterprises. In line with this general policy, the parliament also blocked the creation of Gaztransit in 1995 and the privatisation of other gas state companies, such as Ukrgazprom and Ukrnafta.

Europe-Asia Studies, 1998 (vol. 50), no. 2, pp. 257–286, here p. 262. The clause on Gaztransit was Gazprom's own initiative and was not co-ordinated with the IMF.

78 Postanova Verkhovnoii Rady Ukrainy no. 334a/95-VR 'Pro vnesennya zmin i dopovnen' do pereliku ob'yektiv, yaki ne pidlyahayut' pryvatyzaciyi u zv'yazku z yikh zahal'noderzhavnym znachennyam, zatverdzhenoho Postanovoiu Verkhovnoyi Rady Ukrainy vid 3 bereznya 1995 roku', 19 September 1995, http://zakon4.rada.gov.ua/laws/show/334%D0%B0/95-%D0%B2%D1%80/ed19950919.

79 Zakon Ukrainy no. 192/96-VR 'Pro truboprovidnyj transport', 15 May 1996, http://zakon4.rada.gov.ua/laws/show/192/96-%D0%B2%D1%80/ed19960515.

80 Whitmore, Sarah: State building in the Ukraine: the Ukrainian parliament, 1990–2003, London: Routledge Curzon, 2004, pp. 71–72.

11.5.2. The GTS Privatisation Debate in the Early 2000s

The debates around GTS privatisation began anew in 2000. By that time, there was a broad consensus among the Ukrainian political elite on GTS de-nationalisation and the need for third-party involvement in its operation apart from Russia and Ukraine. The controversy concerned the concrete form of the private sector involvement. Although the Yushchenko government advocated for concession,[81] which was also lobbied by Western stakeholders,[82] President Kuchma favoured its partial privatisation, which was more favourable for Russia as the most likely buyer.[83] The only parliamentary group that objected to both privatisation and concession was the Socialist Party, which, together with other oppositional but non-parliamentary parties, expressed deep concern in a statement.[84] Even Yulia Timoshenko, who occupied the post of Vice-Minister of Energy at that time, preferred not to speak about GTS in public.

In September 2000, a group of pro-presidential deputies introduced a bill that proposed to privatise 49% of GTS.[85] The initiative was supported by President Kuchma. Meeting with Russian President Vladimir Putin in Sochi in October 2000, he agreed to sell Russia a stake in its GTS in exchange for gas debt restructuring and the transit of Turkmen gas via Russian territory.[86] However, the plans were never realised due to the 'Kuchmagate' scandal.[87] In this new political situation, President Putin decided to support his Ukrainian colleague. In December 2000, Kuchma signed an agreement with Russia on conditions favourable for Ukraine. Not only did Russia give consent to restructure Ukrainian gas debts and to transit the Turkmen gas through its territory, but it also no longer pressed for the GTS handover.[88]

---

81   'Yushhenko negativno otsenivaet vozmozhnost' privatizatsii gazotransportnoy sistemy Ukrainy', in: UNIAN, 31 October 2000.
82   Thus, Shell was lobbying for Ukraine's GTS concession for some time already. Furthermore, the German advisors to the Ukrainian government argued in favour of concession. See Opitz, Petra/ von Hirschhausen, Christian: Ukraine as the gas bridge to Europe? Economic and geopolitical considerations, Kiev: Institute for Economic Research and Policy Consulting, 2000, http://www.ier.com.ua/files/publications/WP/2000/wp3_eng.pdf.
83   Gerodot, Nikita: Yushhenko, princ Datskij, in: Ukrainskaya pravda, 31 October 2000, http://www.pravda.com.ua/rus/news/2000/10/31/4360499/.
84   'Ob ugroze ehkonomicheskoj bezopasnosti Ukrainy v svyazi s dejstviyami vlasti', in: Tovarishch no. 44, 3 November 2000.
85   Proekt Zakonu pro osoblyvosti pryvatyzacii ob'yektiv hazotransportnoyi systemy Ukrayiny no. 6040, 7 September 2000, http://w1.c1.rada.gov.ua/pls/zweb2/webproc4_1?pf3511=8915.
86   'Kuchma comes to terms with Putin on partial takeover of gas system', in: Eurasia Daily Monitor, vol. 6, Issue 194, 18 October 2000, http://www.jamestown.org/single/?no_cache=1&tx_ttnews%5Btt_news%5D=22467.
87   In November 2000, the Socialist leader, Oleksandr Moroz, disclosed tapes recorded by the presidential bodyguard, Mykola Melnychenko, at Kuchma's office. The recordings, among other things, revealed that Kuchma was involved in the murder of the journalist Georgii Gongadze, who disappeared in September 2000 and was later found beheaded.
88   There is reason to believe that the Kremlin's plans to oust Rem Vyakhirev, Gazprom's powerful CEO, were the main reason for Russia's backdown in December 2000, when the talks about

11. The Struggle Over Ukraine's Gas Transit Pipeline Network 295

As in 1995, Kuchma favoured GTS privatisation in 2000 and sought active contact with Russia at a time of changing internal and geopolitical situations. In autumn 2000, the political situation in Ukraine was quickly deteriorating. In September, the first Ukrainian parliamentary majority formed to support the reform programme of the Prime Minister Victor Yushchenko began to fall apart as Yushchenko's economic policy caused increasing tensions with the oligarchs.[89] In October, the political crisis became an open confrontation between Yushchenko and Kuchma,[90] which ultimately led to the aforementioned Sochi agreements on Russia's participation in the GTS sale.

These changes were not only internal; the geopolitical situation had also been rapidly changing. Ukraine's image in the West was spoiled after the disappearance of the opposition journalist Geogrii Gongadze in September 2000. From the end of 1999, Ukraine was in a trade dispute with Russia, which demanded that Ukraine stop siphoning its gas and repays the debts of gas traders accumulated in the late 1990s. The resumed talks with the IMF did not progress, and the EU–Ukraine summit in September 2000 showed that Ukraine missed the opportunity to become an EU accession candidate. Russia's relations with the EU, on the contrary, were improving. In October 2000, Gazprom and four European gas companies signed the protocol of intentions to build the gas pipeline through Belarus, Poland, and Slovakia, thus bypassing Ukraine.[91] a few days later, Russia began a comprehensive 'Energy Dialogue' with the EU. These developments put enormous pressure on Ukraine's negotiating position.

The talks on GTS did not resume until 2002. On 9 June 2002 at the St. Petersburg Economic Forum, Presidents Putin and Kuchma and German Chancellor Gerhard Schröder agreed to form an international consortium on the operation and management of the Ukrainian GTS (referred to as the St. Petersburg Declaration).[92] On 7 October 2002, the Russian and Ukrainian governments signed an agreement on the creation

---

the GTS sale were buried for some time. In fact, in 2000, Russia had not yet consolidated its state control over its domestic gas sector. After Vladimir Putin's rise to power, the position of Gazprom's CEO, the powerful Rem Vyakhirev, began to weaken (Goldman, Marshall I.: Petrostate: Putin, power, and the new Russia, Oxford: Oxford University Press, 2008, pp. 100–105). In May 2001, Rem Vyakhirev stepped down from the post that he had occupied for almost ten years.

89 For more on Yushchenko's reforms as Prime Minister in 2000–2001, see Åslund, Anders: How Ukraine became a market economy and democracy, Washington, DC: Peterson Institute for International Economics, 2009, pp. 133–140.

90 The crisis was sparked by Timoshenko's positive account of the energy market situation at the Government Day in the parliament. At that time, Timoshenko was the Vice-Prime Minister of Energy. Her statement was like a bombshell because it ran counter to the dominant discourse that emphasised the deplorable situation in the energy sector. In this situation, Kuchma took the anti-government stance and took the initiative in gas negotiations with Russia and Turkmenistan away from the government.

91 Bushueva, Yuliya/ Trudolyubov, Maksim/ Gavrish, Oleg: Obkhodnoj konsorcium, in: Vedomosti, 19 October 2000.

92 'Deputaty kommentiruyut sozdanie mezhdunarodnogo gazovogo konsorciuma', in: UNIAN, 10 June 2002.

of the GTS consortium on a parity basis.[93] However, the decision was merely a political one, and many questions about its operation, such as the consortium's structure, specific activities, and even economic viability, were left unanswered.[94] As a result, the beginning of its work was postponed many times. Just before the Orange Revolution in August 2004, Ukraine and Russia reached the final intergovernmental agreement that marked the beginning of the consortium's work.[95] The parties agreed that the first investment project of the consortium would be the construction of a Bogorodchany–Uzhgorod pipeline in Ukraine that promised to increase Ukraine's transit capacity by up to 30bcm annually.[96] This meant that the consortium did not take over Ukraine's GTS but would work to build new pipelines.

Ukraine's rapprochement with Russia in 2002 was again pre-programmed by the geopolitical and internal situations. In 2002, Kuchma's external scope of actions was limited as his image (ruined by 'Kuchmagate') had worsened even more following the Kolchuga scandal,[97] freezing relations with the North Atlantic Treaty Organization (NATO) and the United States (US). Moreover, after the terrorist attacks in September 2001, the US shifted its priorities away from Ukraine toward Russia as an ally in the US-led 'war on terror'.[98] In an attempt to regain his credibility, Kuchma renewed his pro-Western policy. In addition to Kuchma's 'European Choice' programme and the first official statement on Ukraine's aspirations to join NATO, the St. Petersburg Declaration is considered one such step.

At the same time, following the St. Petersburg Declaration, the internal political situation in Ukraine changed dramatically. Despite the fact that the opposition received a considerable share of votes following the parliamentary elections in March 2002, the pro-presidential 'United Ukraine' faction began growing rapidly thanks to deputies elected in single-mandate constituencies.[99] The process ended with the formation of the pro-presidential majority the day after the signing of the GTS consortium's agreement. The majority in the parliament was only nominally pro-Kuchma; its core formed the rival

---

93 Uhoda mizh Kabinetom Ministriv Ukrainy ta Uryadom Rosijs'koyi Federaciyi pro stratehichne spivrobitnyctvo v hazovij haluzi, 7 October 2002, http://zakon4.rada.gov.ua/laws/show/643_790.
94 Saprykin, Volodymyr: Konsorcium yak vidobrazhennya stanu Ukrainy, 28 February 2002, http://www.razumkov.org.ua/ukr/article.php?news_id=248.
95 Uhoda mizh Kabinetom Ministriv Ukrainy ta Uryadom Rosijs'koyi Federaciyi pro zakhody shchodo zabezpechennya stratehichnoho spivrobitnyctva v hazovij haluzi, 18 August 2004, http://zakon4.rada.gov.ua/laws/show/643_283.
96 'Nachata realizatsiya pervogo etapa sozdaniya gazotransportnogo konsortsiuma', in: LIGA online, 11 February 2004.
97 According to the tape recordings, Kuchma was involved in illegal sales of Kolchuga radar systems to Iraq, in violation of UN sanctions. The scandal began in spring 2002 and reached its peak in October 2002 as the US froze financial aid to Kyiv.
98 Kuzio, Taras: Neither east nor west: Ukraine's security policy Under Kuchma, in: Problems of Post-Communism, 2005 (vol. 52), no. 5, pp. 59–68, here p. 64.
99 Whitmore, Sarah: State building in the Ukraine: the Ukrainian parliament, 1990–2003, London: Routledge Curzon, 2004, p. 47.

11. The Struggle Over Ukraine's Gas Transit Pipeline Network

centrist parties that were non-programmatic in their nature and represented the interests of oligarchs allied around the president as long as he acted as their benefactor. Thus, they did not form a stable power base for the latter. The centrist coalition in the parliament was strengthened in November 2002 by the appointment of Viktor Yanukovich as the new Prime Minister, who openly promoted the interests of the oligarchs. The latter advocated closer relations with Russia and the creation of the GTS consortium.

Following the signing of the St. Petersburg Declaration, the opposition, led by Yulia Timoshenko, started the anti-consortium campaign, which was not effective in 2002–2004. There are at least three reasons for this. First, Timoshenko's offensive was still occasional rather than systematic. Second, in 2002 and in the years to follow, the presidential administration had intensified its control over the media through government-issued media directives, known as 'temniki'.[100] State officials were forbidden to criticise the GTS consortium in public on pain of dismissal.[101] Third, the pro-presidential parliamentary majority created in October 2002 exercised extensive control over legislative activities. All attempts by the opposition to put the question of the consortium's creation progress on the agenda were to no avail. By declaring the intergovernmental agreement on the GTS consortium valid upon signature, the government managed to bypass its ratification process in the parliament.[102] The applications filed by the opposition to the Constitutional Court on that matter in 2002 were left unconsidered for a long time and were finally turned down in December 2006, when Viktor Yanukovich took the Prime Minister's office once again.[103]

11.5.3. The GTS Privatisation Debate in 2005–2009

After the Orange Revolution and the change of power in Kyiv, the new government changed its position with regard to the GTS consortium.[104] Although it did not stop the consortium's operation, it unilaterally reviewed its statute and excluded the paragraph on the joint GTS operation. Simultaneously, Ukraine resumed talks with the German companies, starting with E.ON Ruhrgas, for their participation in the consortium.[105]

---

100 Dyczok, Marta: Was Kuchma's censorship effective? Mass media in Ukraine before 2004, in: Europe-Asia Studies, 2006 (vol. 58), no. 2, pp. 215–238.
101 In September 2002, President Kuchma fired a deputy secretary of the Ministry of Finance who announced that Ukraine might lose approximately US$1 billion from the GTS consortium's creation. In December 2003, the President also fired the Vice Prime Minister of Energy and the head of the working group on GTS, Vitalii Haiduk, who also happened to criticise the GTS consortium.
102 Syrovatka, Sergej: Fors-mazhor dlya konsorciuma, in: Galickie kontrakty, no. 17, 25 April 2004, http://archive.kontrakty.ua/gc/2004/17/31-fors-mazhor-dlya-konsorciuma.html.
103 Ukhvala Konstytucijnoho Sudu Ukrainy no. 1-up/2006,12 December 2006, http://zakon4.rada.gov.ua/laws/show/v001u710-06.
104 'Yushhenko soglasen s mneniem o netselesoobraznosti sushhestvovaniya gazovogo konsortsiuma mezhdu Rossiey i Ukrainoy na dvustoronney osnove', in: UNIAN, 25 January 2005.
105 Butrin, Dmitrii: E.ON vvyazalsya v gazovyy konflikt mezhdu Rossiey i Turkmeniey na storone Ukrainy, in: Kommersant Vlast, 11 March 2005, http://www.kommersant.ru/doc/553577.

However, despite the rumours, the GTS consortium was not liquidated and continued working on the new pipeline even after the gas conflict with Russia in January 2006.[106] However, narrowing the work of the consortium to just one pipeline was not in Russia's interests. By the end of 2005, the gas relations between Ukraine and Russia deteriorated drastically as the latter demanded higher prices for its gas. In such a situation, Gazprom began talking about the GTS consortium in its 'complete' form that had to operate the entire GTS, not only a small pipeline that had yet to be constructed.[107] On 14 December 2005, Gazprom's chief executive officer (CEO), Alexei Miller, announced that a stake in Ukraine's GTS was the only possible compromise on the eve of the 'gas war'.[108] However, Ukraine rejected Miller's ultimatum. As a result, the negotiations were stalled completely, and a gas crisis followed.

The prospects for Russia obtaining control over Ukraine's GTS improved in summer 2006 following the victorious comeback of Viktor Yanukovich as Prime Minister. In 2006, however, the political situation changed drastically compared to 2002, when Yanukovich became Prime Minister for the first time and pushed for the GTS consortium. Despite the increased pressure from Russia,[109] the new government publicly claimed that it would not hand over Ukraine's GTS to Russia.[110] However, once the distribution of power changed in favour of Yanukovich,[111] the government suddenly changed its position. On 1 February 2007, Yanukovich supported the statement of the Russian president, who said that Ukraine could transfer the GTS in exchange for equal participation in the gas production in Russia.[112] The claim immediately mobilised the political opposition in Ukraine. Thus, Bloc Yulia Timoshenko (BYuT) blocked the work

---

106 In November 2005, the Khartsyzsk Pipe Plant (Donetsk) won the tender to supply pipes for the construction of the Bogorodchany–Uzhgorod pipeline. In December 2005, the Ukrainian government allocated land for the pipeline's construction. In April 2006, it chose a Ukrainian company as a contractor for the construction of the Bogorodchany–Uzhgorod pipeline. See 'Khartsyzskij trubnyy zavod pobedil v tendere na postavku trub dlya gazoprovoda "Bogorodchany–Uzhgorod"', in: UNIAN, 22 November 2005; 'Mezhdunarodnyy gazotransportnyy konsortsium opredelil genpodryadchika stroitel'stva gazoprovoda "Bogorodchany–Uzhgorod"', in: UNIAN, 19 April 2006.
107 'Gazprom gotov reshit' gazovuyu problemu s Ukrainoj pri uslovii sozdaniya polnocennogo gazotransportnogo kontsorciuma', in: UNIAN, 8 December 2005.
108 Grib, Natal'ya: Gaz za provod. Rossiya ob'yavila, chego ona khochet ot Ukrainy, in: Kommersant, no. 235, 14 December 2005, http://www.kommersant.ru/doc/635142.
109 Thus, Gazprom linked the question of gas prices to the GTS consortium in its 'complete' form. It also began blocking the convening of the consortium in its agreed ('incomplete') form ('Mintopehnergo Ukrainy obvinyaet rossijskuyu storonu v zatyagivanii s provedeniem sobraniya uchastnikov gazovogo konsortsiuma', in: UNIAN, 27 July 2006).
110 'Gazotransportnaya sistema ne stanet razmennoj monetoj v gazovykh peregovorakh s Rossiey – Boyko', in: UNIAN, 21 August 2006; 'Yanukovich zaveryaet, chto Ukraina ne otdast Rossii pravo kontrolya nad gazotransportnoj sistemoj', in: UNIAN, 6 September 2006.
111 For more details on the situation, see Malygina, Katerina: Ukraine as a neo-patrimonial state: understanding political change in Ukraine in 2005–2010, in: South East Europe Review, 2010 (vol. 13), no. 1, pp. 7–27.
112 'Yanukovich dopuskaet peredachu aktivov ukrainskoy GTS v obmen na paritetnoe uchastie v gazodobyvayushchikh aktivakh Rossii', in: UNIAN, 1 February 2007.

11. The Struggle Over Ukraine's Gas Transit Pipeline Network 299

of the parliament, calling for the adoption of a bill prohibiting any stripping of the assets of the state-owned Naftogaz Ukrainy, be it privatisation, concession, leasing, or the creation of joint ventures. Although the GTS privatisation was already banned by law in 1995, the Verkhovna Rada passed the proposed bill almost unanimously, with 430 out of 450 votes in favour.[113]

Nevertheless, the Yanukovich government continued non-public negotiations with Russia.[114] Consequently, the topic of GTS privatisation disappeared from the public discourse for some time, but only until August 2007, when Ukraine was in the middle of the election campaign due to pre-term parliamentary elections. On 22 August 2007, Naftogaz's CEO, Evgenii Bakulin, announced that Ukraine and Russia had been considering different options to exchange assets in the gas sector.[115] President Viktor Yushchenko called the negotiations 'unauthorised' and demanded that they be stopped.[116] Bakulin's statement also provoked securitising moves by the opposition. The Yushchenko party, known as 'Our Ukraine–People's Self-Defence' (NU-NS), released a statement in which Yanukovich's government was accused of trying to hand over an important strategic asset to Russia against the law and in exchange for dubious political support before the elections. According to NU-NS, the GTS handover would be the first step towards Ukraine losing its state sovereignty and would have a negative impact on the economy and people's well-being.[117] The reaction of Yulia Timoshenko was even fiercer: she warned the Prime Minister Viktor Yanukovich, the Minister of Energy Yuri Boiko, and the Naftogaz CEO Evgenii Bakulin that they would be held criminally liable for 'betrayal of national interests'.[118] The conflict ended in early September, when Yushchenko issued a presidential decree that transferred negotia-

---

113 Zakon Ukrainy no. 605-V 'Pro vnesennya zmin do Zakonu Ukrainy "Pro truboprovidnyj transport" shchodo pidpryyemstv mahistral'noho truboprovidnoho transportu', 6 February 2007, http://zakon4.rada.gov.ua/laws/show/605-v.
114 For instance, a few news reports stated that the government was negotiating Ukraine's participation in the Russian upstream sector ('Bojko obsudil s glavoj "Gazproma" Millerom uchastie Ukrainy v proektakh po dobyche gaza na territorii Rossii', in: UNIAN, 16 February 2007; 'Bojko obsudil s glavoj "Gazproma" Millerom detali uchastiya Ukrainy v proektakh po dobyche gaza v Rossii', in: UNIAN, 26 February 2007). It also attempted to revive the work of the GTS consortium in its 'small' form, which encountered opposition from the Russian side ('Predsedatelem OOO "Mezhdunarodnyj konsortsium po upravleniyu i razvitiyu gazotransportnoy sistemy Ukrainy" izbran Yuriy Boyko', in: UNIAN, 28 April 2007; 'Prem'er RF zayavlyaet, chto po proektu stroitel'stva gazoprovoda "Bogorodchany–Uzhgorod" sushchestvuyut problemy', in: UNIAN, 9 October 2007).
115 'Ukraina i Rossiya obsuzhdayut vozmozhnye varianty obmena aktivami v neftegazovoj sfere – glava "Naftogaza"', in: UNIAN, 21 August 2007.
116 'Prezident zayavlyaet, chto ne daval mandat na peregovory mezhdu RF i Ukrainoj v gazotransportnoy sfere', in: UNIAN, 22 August 2007.
117 'NU-NS prosit Prezidenta i silovye vedomstva ne dopustit' peredachi Rossii pravitel'stvom Yanukovicha chasti GTS Ukrainy', in: UNIAN, 22 August 2007.
118 'Timoshenko preduprezhdaet Yanukovicha, Boyko i Bakulina ob ugolovnoj otvetstvennosti za izmenu natsional'nykh interesov', in: UNIAN, 22 August 2007.

tions on gas supplies from the commercial to interstate level[119] and defined the officials authorised to comment on their progress.[120]

In 2008, the topic of GTS privatisation played a smaller role in Ukrainian politics than ever before. The reason for this may be attributed to a change in the dominant discourses. On the one hand, after becoming the Prime Minister in December 2007, Yulia Timoshenko started a campaign to exclude dubious intermediaries from the gas trade. Accompanied by a harsh rhetoric, her attempts quickly became the top issue in the Ukrainian media. At the same time, she curbed all speculations around GTS by repeatedly stating that the latter would remain in public ownership and would not become subject to negotiations with Russia.[121] Timoshenko kept her word, and even during the January 2009 gas conflict, Ukraine rejected Russia's idea of a trilateral consortium with the sole purpose of financing the supply of the so-called 'technical gas' needed to pump gas through Ukraine's GTS.[122]

The topic of the GTS consortium was replaced in 2008 with a relatively new discourse on GTS modernisation. Although the issue of GTS modernisation appeared in the Ukrainian public discourse before 2008, it remained rather technical and thus non-politicised. In 2008, however, the EU initiated a new framework of cooperation within its 'Eastern Partnership' programme that placed energy security high on the agenda. In this context, the EU began planning an International Investment Conference on the Modernisation of Ukraine's GTS, which had yet to take place in 2008. The conference that finally occurred in March 2009 ended with the signing of a joint EU–Ukraine Declaration that laid the framework for future investments in Ukraine's gas transit network.[123] Although the Declaration was rather vague and did not contain any binding

---

119 The agreement from 3 January 2006 that introduced the intermediary company Rosukrenergo was signed at the commercial level between Gazprom and Naftogaz Ukrainy and cancelled all previous intergovernmental agreements in this sphere.
120 Ukaz Prezydenta Ukrainy no. 842/2007 'Pro deyaki zakhody shchodo zabezpechennya staloho funkcionuvannya i rozvytku rynku pryrodnoho hazu', 7 September 2007, http://www.president.gov.ua/documents/6668.html.
121 Throughout the year, the presidential secretariat tried to stigmatize Timoshenko as a traitor to Ukraine and its national interest because she allegedly promised the Kremlin to meet all its requirements, including handing over Ukraine's GTS, in exchange for her support during the presidential elections. There were at least two public statements of this type: after the armed conflict between Russia and Georgia in August 2008 and after signing a memorandum with Russia in early October 2008 ('Otsutstvie edinoy pozitsii v Ukraine otnositel'no situatsii v Gruzii mozhet svidetel'stvovat' ob opredelennykh dogovorennostyakh s Rossiey o podderzhke lidera BYuT na vyborakh Prezidenta – SP', in: UNIAN, 14 August 2008; 'V Sekretariate Prezidenta schitayut, chto vcherashnij vizit Timoshenko v Moskvu stal unizheniem dostoinstva Ukrainy', in: UNIAN, 3 October 2008).
122 'Timoshenko zayavlyaet, chto ne podderzhivaet ideyu sozdaniya konsortsiumov na baze GTS Ukrainy', in: UNIAN, 21 January 2009.
123 Joint Declaration between the Government of Ukraine, the European Commission, the European Investment Bank, the European Bank for Reconstruction and Development, and

commitments from the EU's side, Russia's reaction to it was extremely negative.[124] Russian officials interpreted the move as ignoring Russia's interests as the main gas supplier and stated that the best way to modernise the Ukrainian GTS was by the creation of a trilateral GTS consortium. Nevertheless, Russia did not politicise this question any further, acknowledging poor chances for success in the 2009 pre-election year and seeking other ways to influence Ukraine, such as attempting to negotiate a new global energy treaty with transit as its focal point.[125] As a result, the GTS question did not become a central issue in the election campaign in 2009 compared to 2007, with only 'minor' presidential candidates addressing it on their programmes.[126]

Table 11-1 overleaf summarises the events described above in the form of confrontation between the government and the opposition. The case study shows that all privatisation suggestions emanated from the executive branch, either the president or the government. This is surprising given the social and stabilising function of the state's control over the gas transit in Ukraine. One would expect, on the contrary, that the executive branch, guided by the instinct of self-preservation, would resist any attempts at GTS privatisation. However, as the analysis suggests, the officials—primarily Leonid Kuchma—had always pursued this policy at times of changed geopolitical situations, which supports Kuchma's multi-vector foreign policy. Moreover, the analysis suggests that Kuchma's balancing act was a means to overcome the domestic and external constraints. In this sense, one can speak about an attempt to expand the scope of action rather than a serious policy course. This conclusion is also supported by inadequate public debates on the issue. Thus, the government discussed GTS privatisation only in terms of further expansion of gas transit and never considered the possible negative economic consequences indicated by the opposition. If GTS privatisation attempts were sincere, the government would have taken a more proactive stance.

The opposition reacted to attempts of the executive branch to improve its position through GTS privatisation by instrumentalising the debate around it. In the 1990s, the communists framed the GTS as the main strategic asset and the main source of state revenues. In the 2000s, the securitisation process reached its peak and was driven

---

the World Bank, 23 March 2009, http://eeas.europa.eu/energy/events/eu_ukraine_2009/joint_declaration_en.pdf.

124 The Russian delegation even demonstratively left the conference when it became clear that the final declaration was to be signed solely between the EU and Ukraine.

125 Cf. e.g., statements by the Russian Energy Minister Sergei Shmatko and the Gazprom CEO Alexei Miller ('Ministr energetiki RF schitaet, chto cherez polgoda budet vozmozhnost' vernut'sya k idee sozdaniya gazotransportnogo konsortsiuma', in: UNIAN, 18 May 2009; 'Miller schitaet, chto vopros o modernizatsii gazovoy otrasli Ukrainy vozniknet posle prezidentskikh vyborov', in: UNIAN, 11 October 2009).

126 For instance, Sergii Tigipko and Arsenii Yatsenyuk spoke repeatedly in favour of a tripartite GTS consortium.

primarily by the opposition leader Yulia Timoshenko as the main securitising actor. In the next section, I will explain how securitisation proceeded and why it was successful.

Table 11-1: Ukraine's Debate over GTS Privatisation in 1990–2009

	Government	Opposition
1995	In March, President Kuchma and the acting Prime Minister Marchuk support the creation of a joint venture, Gaztransit, which Russia made the primary condition for gas debt restructuring.	In September, Communists block the creation of Gaztransit by adding the pipeline maintenance companies to the list of enterprises not eligible for privatisation.
2000	Yushchenko's government advocates the GTS concession, whereas President Kuchma is in favour of the partial privatisation. Ultimately, President Kuchma agrees at the Sochi summit in October to sell Russia a stake in its GTS in exchange for gas debt restructuring and transit of Turkmen gas via Russian territory.	The Socialist Party and some non-parliamentary parties make a stand against GTS privatisation. In general, the Sochi deal was never realised due to 'Kuchmagate' scandal, in which 'oppositional four' played the primary role.*
2002	In June, President Kuchma signs the St. Petersburg Declaration envisaging the creation of the trilateral GTS consortium between Ukraine, Russia, and Germany. However, in October, Kinakh's government creates the GTS consortium on a parity basis with Russia. The subsequent Yanukovich government limits the operation of the GTS consortium to building new pipelines.	The opposition (primarily the Socialist Party and Yushchenko's bloc 'Our Ukraine') fails in its attempts to put the question of consortium's creation progress on parliamentary agenda. Yulia Timoshenko makes the first attempt to securitise the GTS in public.
2007	In February, Prime Minister Yanukovich announces that Ukraine might hand over its GTS to Russia. The non-public negotiations continue despite the oppositional campaign.	Timoshenko's BYuT secures almost unanimous adoption of the second bill prohibiting the GTS privatisation. During the pre-term parliamentary elections, Timoshenko's BYuT and Yushchenko's NU-NS further securitise the issue of the GTS privatisation.

* Namely, the socialist Oleksandr Moroz, the communist Petro Symonenko, the former Prime Minister Viktor Yushchenko, and the former Vice-Minister of Energy Yulia Timoshenko. By the time of Kuchmagate, neither Yushchenko nor Timoshenko had their electoral blocs, which were formed only a year later in 2001.
Source: Author's own analysis.

11.6. Struggle over Ukraine's GTS Through the Securitisation Lens

The process tracing conducted in the previous section showed that the government's attempts to change the GTS ownership status failed due to an active campaign by the

11. The Struggle Over Ukraine's Gas Transit Pipeline Network

political opposition. In this section, I will analyse the single securitisation move typical of this campaign.

Following the signing of the St. Petersburg Declaration, Yulia Timoshenko published an article in June 2002 in which she summarised the basic arguments against the creation of the GTS consortium. The following excerpts from her article illustrate how the securitisation is performed:

> [...] Thanks to the GTS, the Ukrainian state meets 50% of its needs for natural gas, reduces its price to domestic consumers, helping ultimately not only our miserable grandmothers who receive meagre pension but also major industries to stay afloat. I shall remind you, Ukraine's state budget might receive annually nearly a quarter of its revenues thanks to transit [fees—K.M.].
>
> There is no way around it, the GTS is the main breadwinner of Ukraine. Moreover, under current management system the country would have long been 'run upon the rocks' and come apart, suffocating in the energy crisis if not for 'native' GTS. Thus, one can confidently say that there would be no Ukrainian independence without GTS. And I have serious doubts whether it [the independence—K.M.] can be saved by agreeing to a gas consortium.
>
> Moreover, we must be aware that in addition to the economic benefits, GTS provides also significant political advantages, which, by the way, are constantly growing. Thanks to it [GTS—K.M.], Ukraine occupies a key position in the European energy system, and hence the overall security of the European Union.
>
> Only thanks to the strategic character of the Ukrainian GTS, the world leaders have to constantly maintain a dialogue with the official Kyiv, despite its kleptocratic and deeply anti-popular government and economic mess. [...]
>
> Without any exaggeration, the GTS is a major guarantee of our national security because neither the EU nor Russia nor the US will ever agree with such a development that could put into question the stable operation of the continent's main energy artery. [...]
>
> Given the striking asymmetry of Ukrainian–Russian trade relations, the expansion of Russian capital into our economy, handing over our gas pipelines to 'Big Brother' [Russia—K.M.] is, as I already said, a real adjudication to Ukrainian independence both economically and politically. [...]
>
> Ukraine will very soon feel the negative economic effects of the gas consortium. Experts' conclusions are univocal: first we will have a serious decrease in budget revenues. And it means (it is never superfluous to note) pensions, welfare payments, public sector wages, spending on science, medicine, national defence, in a word, everything that the President, the government, deputies promised to raise in their programmes, declarations and speeches. [...]
>
> Second, we should expect a significant increase in energy prices for domestic consumers. Given the extreme energy intensity of the Ukrainian major industry branches, the decisive role of energy prices in cost of the metallurgy products—our main export position, hope and support—Ukraine should be afraid of a new industrial collapse following the creation of the gas transit consortium. Next as a chain reaction—inflation, financial collapse, default, social and political crisis, the sale of state property for debts [...].[127]

In her article, Timoshenko clearly makes a securitising move, framing the GTS as an indispensable part of Ukraine's independence. Using the 'grammar of security', she also portrays the terrible situation that would occur if Ukrainian GTS were privatised.

---

127 Timoshenko, Yulia: Truba ukrayins'kij ekonomichnij nezalezhnosti, in: Dzerkalo tyzhnya, 29 June 2002, http://gazeta.dt.ua/POLITICS/truba__ukrayinskiy_ekonomichniy_nezalezhnosti.html.

The GTS privatisation is thus presented as an 'existential threat' and 'point of no return'. The referent object here is the Ukrainian state, which makes the securitisation frame powerful.

This frame created by Timoshenko in 2002 was strategically exploited by such oppositional parties as BYuT and NU-NS throughout the 2000s. Due to their active campaigns, the GTS privatisation became a taboo topic over time. Moreover, the government never attempted to debunk the myth that 'GTS is part of Ukraine's independence'. This may be a direct effect of the securitisation, which used an extremely powerful frame that, in my view, could be discursively bested only by yet another securitisation.[128]

Furthermore, the logic behind securitisation in the case of Ukraine's GTS was not 'rule bending', as the classical securitisation theory posits. Leaning on Vuori, one can attribute the GTS case to the 'deterrence' logic because the result of securitisation is no-action—the GTS was not privatised at the end.[129] I argue, however, that the securitisation logic is much subtler here. The message that Ukrainian oppositional forces tried to convey is that in contrast to those in power, they guarded the country's national interests, and they would not succumb to external pressure. This is the classical 'we vs. them' frame that helps mobilise people in many other areas. The claim's ultimate aim is therefore to discredit the government and undermine its legitimacy. Accordingly, the audience of the securitisation was not only Russia and the domestic elites but also the general public. That the securitisation was generally successful shows the results of an opinion poll conducted by the Razumkov Center in April 2013, which showed that approximately 60% of Ukrainians believed that the Ukrainian GTS should remain in public ownership, 12% thought that the state could lease its GTS to Russia in exchange for lower gas prices, another 15% were in favour of the trilateral GTS consortium between Ukraine, Russia, and the EU, and only 3% thought that the GTS should be privatised.[130] Unfortunately, there are no earlier polls on this issue that would allow comparison over time. Another important factor in the success of securitisation was its resonance with the existing macrosecutitisation, namely, the discourse of fear that Russia could colonise Ukraine once again and deprive the latter of

---

128   I should add that there may be many other possible explanations. As already argued, the Ukrainian governments never really wanted to privatize GTS, even at times when they acted as if they did. It may also be said that in a closed political context such as the Ukrainian one, governments do not like to debate in public, preferring other means (such as cooptation or coercion) to achieve their ends.
129   Vuori, Juha A.: Illocutionary logic and strands of securitization: applying the theory of securitization to the study of non-democratic political orders, in: European Journal of International Relations, 2008 (vol. 14), no. 1, pp. 65–99.
130   Sociological Service of the Razumkov Center, http://www.razumkov.org.ua/ukr/poll.php?poll_id=889.

its statehood. This discourse dominated Ukrainian politics in the 1990s and became rather stable throughout time.

11.7. Conclusion

The case of gas transit in Ukraine shows the limits of Russia's energy leverage and rent seeking in Ukraine. For more than twenty years, Ukraine's gas transit network has remained under the state's control despite numerous attempts to change its status. Undoubtedly, the role of the parliament in the GTS privatisation blockade was and remains large. In addition to legislative means, the Ukrainian opposition relied heavily on public discourse in its power struggles with the government. This article showed that the securitisation of GTS as an indispensable part of Ukraine's independence hindered its ultimate privatisation and gave Ukraine the strength to successfully resist Russia's pressure, leading, in some cases, to irrational and ineffective decisions with detrimental consequences for Ukraine's economy.

Furthermore, the struggle over GTS privatisation has been complex, involving many actors, each with his or her own goal. In this article, I have argued that the true meaning behind the struggle is the subtle power game between the government attempting to increase its scope of actions and the opposition attempting to discredit the government and boost its own image. Whereas the former utilised the strategy of 'multipolarity', the latter used securitisation as its tactic. Thus, this article showed that there is a different logic behind securitisation in non-Western contexts than the one proposed by the original securitisation theory and further developed the securitisation theory by adding 'discredit' to the list of possible securitisation logics. This article contributes also to the literature on hybrid regimes, showing that discourses in such states could be an important resource in achieving instrumental goals, especially when utilised by the opposition.

Annex

Oil and Gas Reserves and Production

Table A-1: Proven Crude Oil Reserves (in Million Tonnes)

	Azerbaijan	Kazakhstan	Russia	Turkmenistan
1998	163.7	736.6	8,634.1	68.2
1999	163.7	736.6	9,261.6	68.2
2000	163.7	736.6	9,411.6	68.2
2001	163.7	736.6	9,889.0	68.2
2002	954.8	736.6	10,380.0	68.2
2003	954.8	1,227.6	10,775.6	68.2
2004	954.8	1,227.6	10,707.4	68.2
2005	954.8	1,227.6	11,116.6	68.2
2006	954.8	1,227.6	11,362.1	81.8
2007	954.8	4,092.0	11,525.8	81.8
2008	954.8	4,092.0	11,471.2	81.8
2009	954.8	4,092.0	11,621.3	81.8
2010	954.8	4,092.0	11,607.6	81.8
2011	954.8	4,092.0	11,880.4	81.8
2012	954.8	4,092.0	11,894.1	81.8

Source: BP Statistical Review of World Energy 2013, http://www.bp.com/content/dam/bp/excel/Statistical-Review/statistical_review_of_world_energy_2013_workbook.xlsx (accessed 2 December 2013).

Table A-2: Crude Oil Production (in Million Tonnes)

	Azerbaijan	Kazakhstan	Russia	Turkmenistan
1985	13.2	22.7	542.3	6.8
1990	12.5	25.8	515.9	5.7
1991	11.8	26.6	461.9	5.4
1992	11.2	25.8	398.8	5.2
1993	10.3	23.0	354.9	4.4
1994	9.6	20.3	317.6	4.2
1995	9.2	20.6	310.7	4.1
1996	9.1	23.0	302.9	4.4
1997	9.0	25.8	307.4	5.4
1998	11.4	25.9	304.3	6.4
1999	13.9	30.1	304.8	7.1
2000	14.1	35.3	326.7	7.2
2001	15.0	40.1	351.7	8.0
2002	15.3	48.2	383.7	9.0
2003	15.4	52.4	425.7	10.0
2004	15.5	60.6	463.3	9.6
2005	22.2	62.6	474.8	9.5
2006	32.2	66.1	485.6	9.2
2007	42.6	68.4	496.8	9.8
2008	44.5	72.0	493.7	10.3
2009	50.4	78.2	500.8	10.4
2010	50.8	81.6	511.8	10.7
2011	45.6	82.4	518.5	10.7
2012	43.4	81.3	526.2	11.0

Source: BP Statistical Review of World Energy 2013, http://www.bp.com/content/dam/bp/excel/Statistical-Review/statistical_review_of_world_energy_2013_workbook.xlsx (accessed 2 December 2013).

Table A-3: Proven Natural Gas Reserves (in Trillion Cubic Metres)

	Azerbaijan	Kazakhstan	Russia	Turkmenistan
1997	0.6	1.3	30.7	2.4
1998	0.6	1.3	30.5	2.3
1999	0.9	1.2	29.8	2.3
2000	0.9	1.2	29.6	2.3
2001	0.9	1.2	29.7	2.3
2002	0.9	1.3	29.8	2.3
2003	0.9	1.3	30.5	2.3
2004	0.9	1.3	30.3	2.3
2005	0.9	1.3	30.3	2.3
2006	0.9	1.3	30.3	2.3
2007	0.9	1.3	30.4	2.3
2008	0.9	1.3	30.4	7.3
2009	0.9	1.3	31.1	7.3
2010	0.9	1.3	31.1	10.2
2011	0.9	1.3	32.9	17.5
2012	0.9	1.3	32.9	17.5

Source: BP Statistical Review of World Energy 2013, http://www.bp.com/content/dam/bp/excel/Statistical-Review/statistical_review_of_world_energy_2013_workbook.xlsx (accessed 2 December 2013).

Table A-4: Natural Gas Production (in Billion Cubic Metres)

	Azerbaijan	Kazakhstan	Russia	Turkmenistan
1985	12.7	4.9	418.1	75.3
1990	9.0	6.4	590.0	79.5
1991	7.8	7.1	581.9	76.3
1992	7.1	7.3	582.8	54.4
1993	6.2	6.1	559.5	59.1
1994	5.8	4.1	549.6	32.3
1995	6.0	5.3	532.6	29.2
1996	5.7	4.2	543.5	31.9
1997	5.4	6.1	515.2	15.7
1998	5.1	4.9	532.7	12.0
1999	5.4	6.6	535.7	20.6
2000	5.1	8.2	528.5	42.5
2001	5.0	9.2	526.2	46.4
2002	4.7	9.1	538.8	48.4
2003	4.6	11.9	561.5	53.5
2004	4.5	13.1	573.7	52.8
2005	5.2	13.5	580.1	57.0
2006	6.1	13.9	595.2	60.4
2007	9.8	16.7	592.0	65.4
2008	14.8	18.7	601.7	66.1
2009	14.8	17.8	527.7	36.4
2010	15.1	17.6	588.9	42.4
2011	14.8	19.3	607.0	59.5
2012	15.6	19.7	592.3	64.4

Source: BP Statistical Review of World Energy 2013, http://www.bp.com/content/dam/bp/excel/Statistical-Review/statistical_review_of_world_energy_2013_workbook.xlsx (accessed 2 December 2013).

Oil and Gas Pipelines in Europe and Central Asia

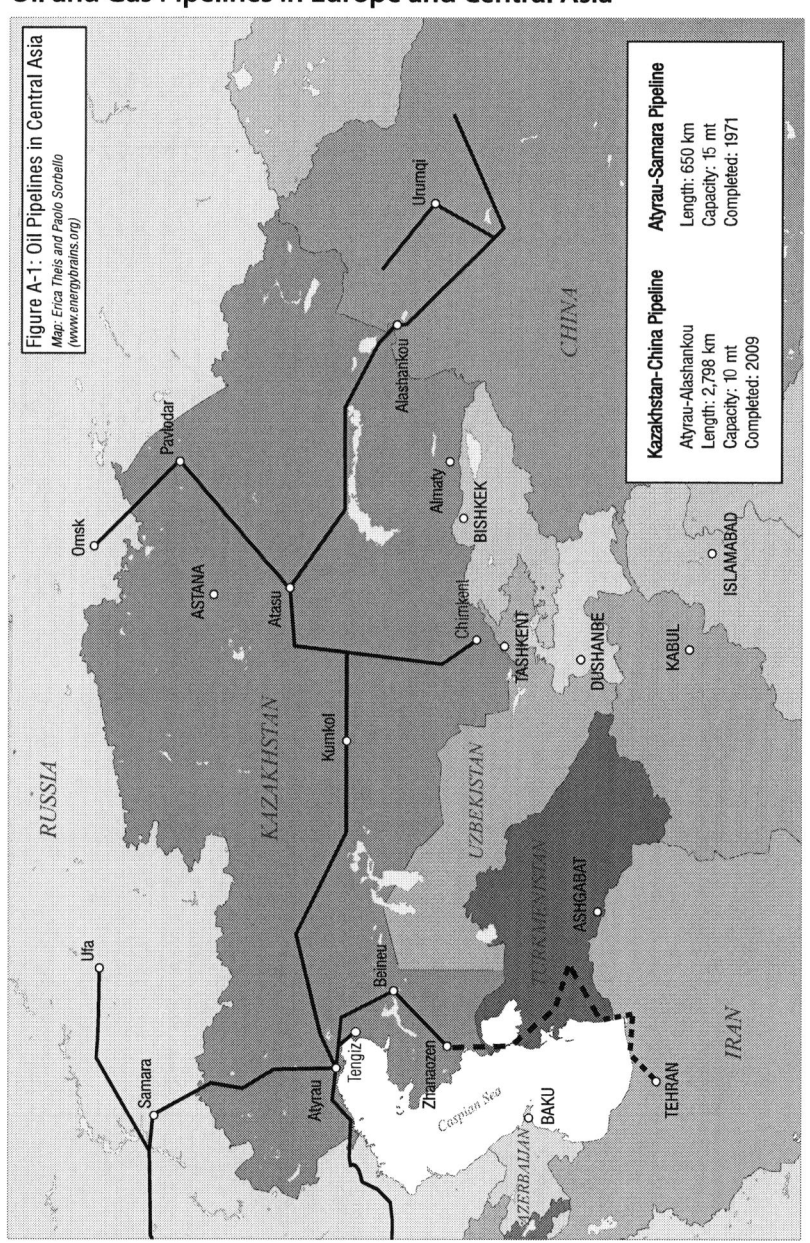

Figure A-1: Oil Pipelines in Central Asia
Map: Erica Theis and Paolo Sorbello
(www.energybrains.org)

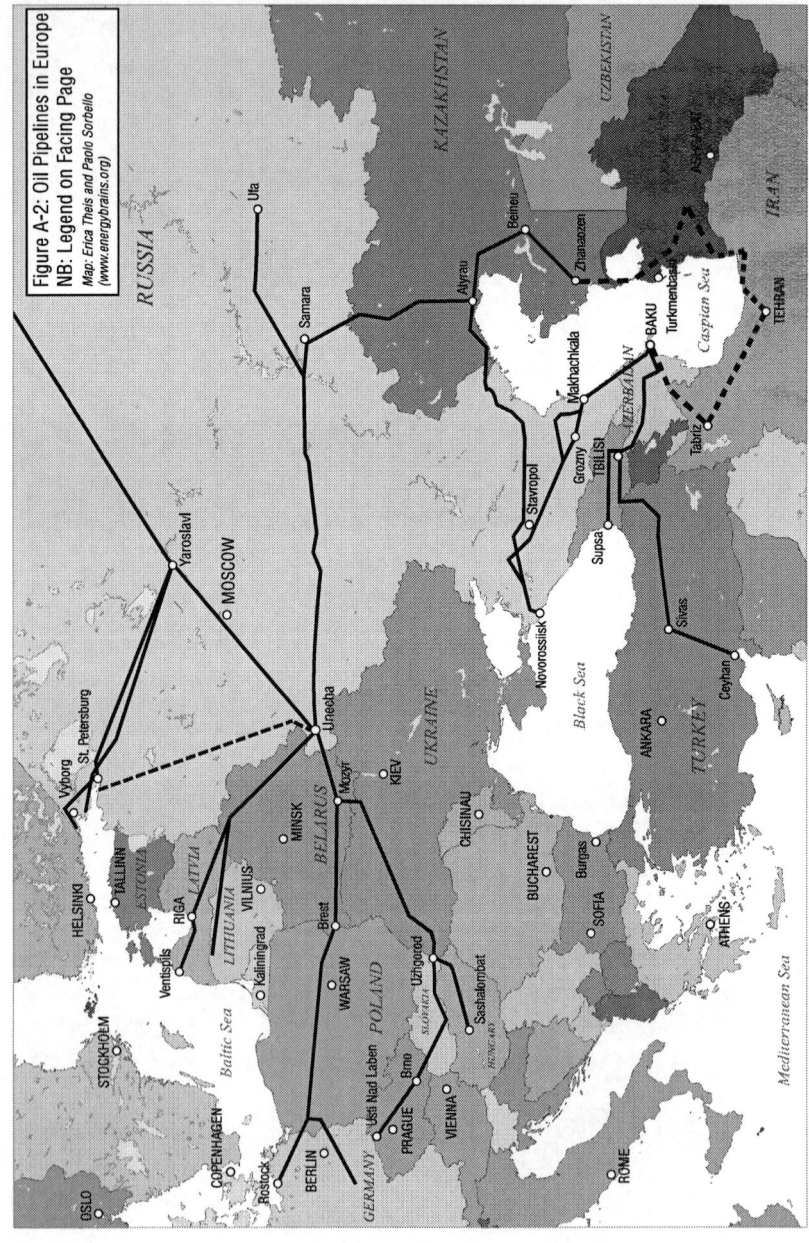

OIL PIPELINES - EUROPE

Baku - Supsa
Length: 833 km
Capacity: 6 mt
Completed: 1998
Commercial deliveries: 1999

Baku - Novorossiisk
Length: 1,330
Capacity: 7 mt
Commercial deliveries: 1997

Baku-Tbilisi-Ceyhan
Length: 1,768 km
Capacity: 60 mt
Completed: 2005
Commercial deliveries: 2006

Caspian Pipeline Consortium
Tengiz-Novorossiisk
Length: 1,511 km
Capacity: 35 mt (2012) 48 mt (2013)
Commercial deliveries: 2001

Baltic Pipeline System I
Yaroslavl-Primorsk
Length: 950 km
Capacity: 76 mt
Completed: 2001
Commercial deliveries: 2006

Baltic Pipeline System II
Unecha - Ust-Luga
Capacity: 50 mt

Druzhba
Almetyevsk-Schwedt/Sashalombat
Length: 8,900 km
Capacity: 81 mt
Completed: 1964

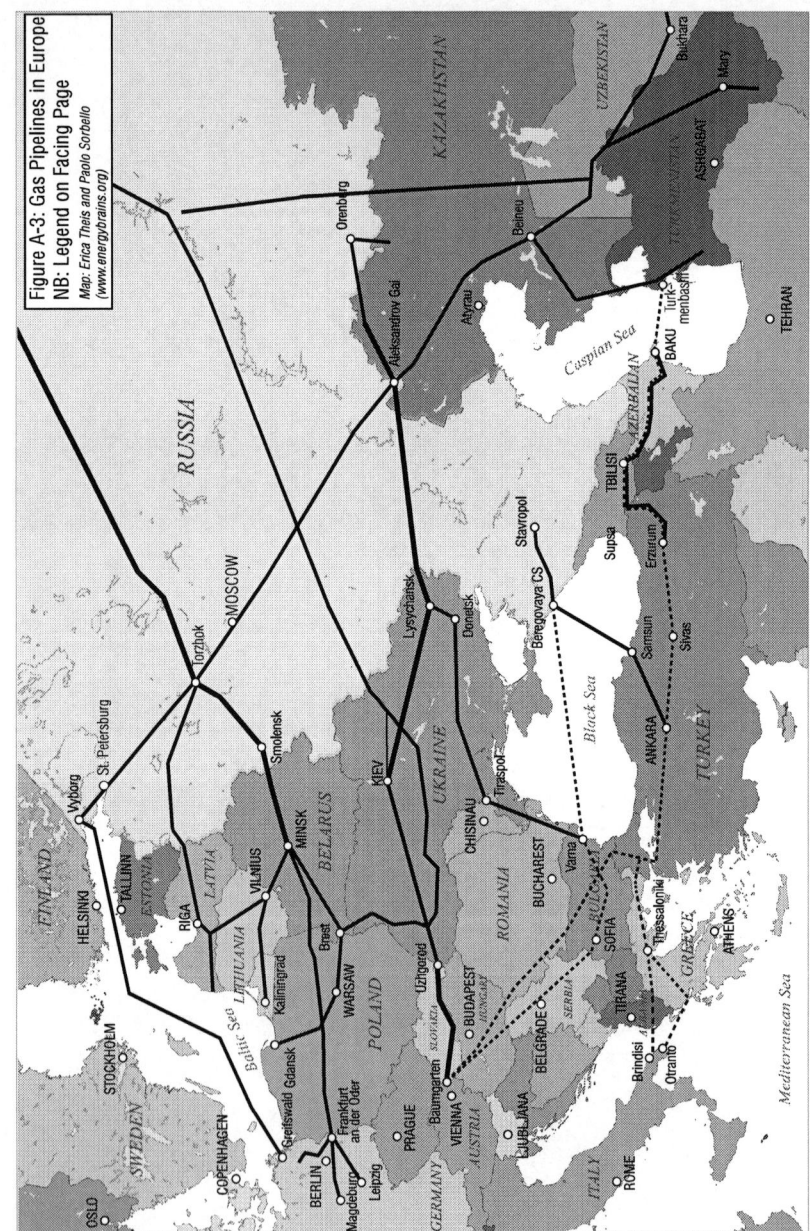

GAS PIPELINES - EUROPE

Yamal-Europe

Torzhok - Mallnow
Length: 1,998 km
Capacity: 33 bcm
Completed: 2005
Commercial deliveries: 2006

Bratstvo

Shebelinka-Pomary-Uzhgorod
Length: 4,500 km
Capacity (northernbound): 74 bcm
Completed: 1967-1980

Northern Lights

Urengoy-Minsk
Length: 7,377 km
Capacity: 55 bcm
Completed: 1969

Nord Stream

Vyborg-Greifswald
Length: 1,222 km
Capacity: 55 bcm
Operational since 2012 (2nd line)

South Stream

Beregovaya-Baumgarten
Length: 2,380 km
Capacity: 63 bcm
Proposed: 2007
Construction started in 2012

Blue Stream

Beregovaya-Samsun
Length: 1,213
Capacity: 16 bcm
Completed: 2003
Commercial deliveries: 2005

Central Asia - Centre

Central Asia-Aleksandrov Gai
Length: 2,000 km
Capacity: 90 bcm
Completed: 1985

Baku-Tbilisi-Erzurum

Baku-Tbilisi-Erzurum
Length: 692 km
Capacity: 6 bcm
Operational since 2006

Soyuz

Orenburg-Uzhgorod
Length: 2,750 km
Capacity: 26 bcm
Completed: 1978

Southern Gas Corridor

Trans-Caspian Pipeline

Turkmenbashi-Baku
Capacity: 30 bcm
Proposed: 1999

Trans-Adriatic Pipeline

Greece-Albania-Italy
Length: 867 km
Capacity: 10 bcm
Proposed: 2003

Nabucco West

Bulgaria-Austria
Length: 1,329 km
Capacity: 10 bcm
Proposed: 2012

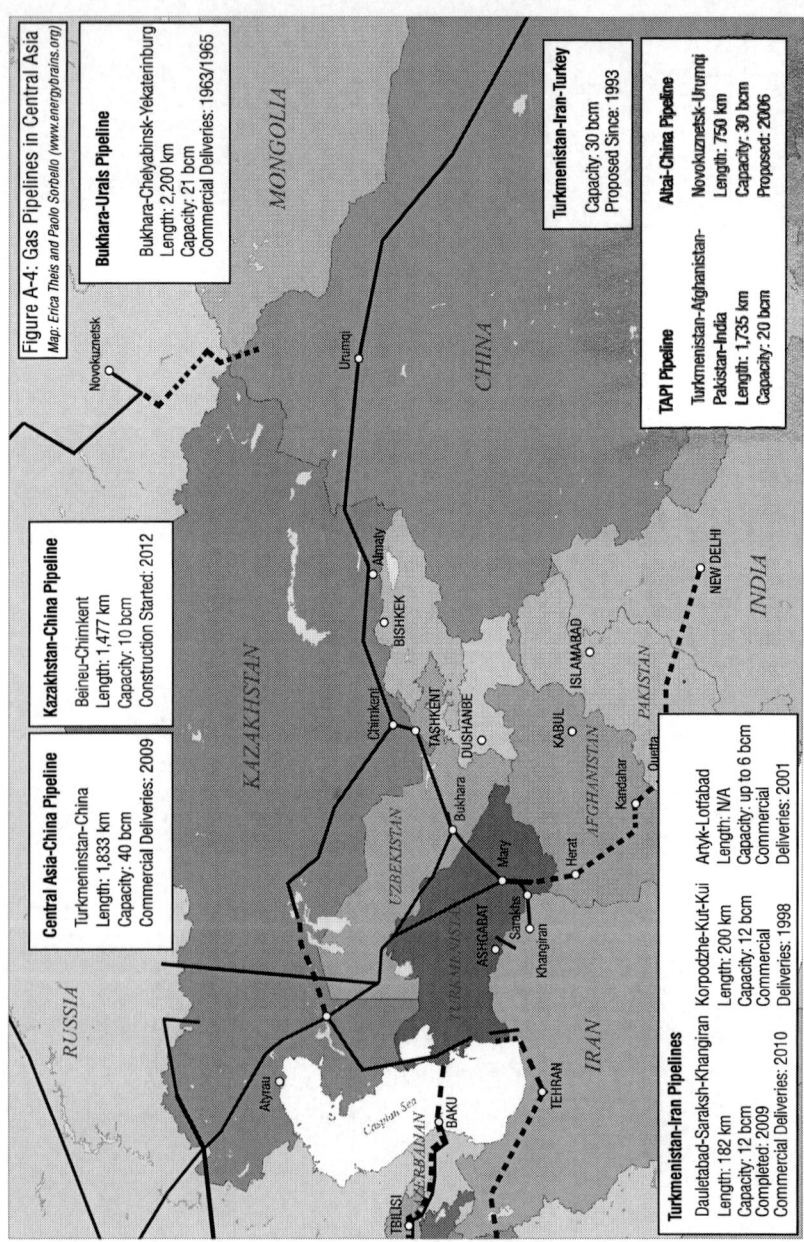

Figure A-4: Gas Pipelines in Central Asia
Map: Erica Theis and Paolo Sorbello (www.energybrains.org)

Selected Bibliography on Eurasian Export Pipelines

Aalto, Pami: Russia's energy policies: National, interregional and global levels, Cheltenham: Edward Elgar, 2012.

Aalto, Pami/ Dusseault, David/ Kennedy, Michael D./ Kivinen, Markku: Russia's energy relations in Europe and the Far East: towards a social structurationist approach to energy policy formation, in: Journal of International Relations and Development (forthcoming, advance online publication,18 January 2013), pp. 1–29, http://www.palgrave-journals.com/jird/journal/vaop/ncurrent/abs/jird201229a.html.

Adams, Jan S.: Pipelines and pipedreams: can Russia continue to dominate Caspian Basis energy?, in: Problems of Post-Communism, 1998 (vol. 45), no. 5, pp. 26–36.

Adams, Jan S.: Russia's gas diplomacy, in: Problems of Post-Socialism, 2002 (vol. 49), no. 3, pp. 14–22.

Afifi, S.N./ Hassan, M.G./ Zobaa, A.F.: The impacts of the proposed Nabucco gas pipeline on EU common energy policy, in: Energy Sources, Part B: Economics, Planning, and Policy, 2013 (vol. 8), no. 1, pp. 14–27.

Akiner, Shirin (ed.): The Caspian: politics, energy and security, London: Routledge, 2004.

Allakhverdieva, Aysel: The Baku–Tbilisi–Ceyhan pipeline in the context of human rights, in: The Caucasus and Globalization, 2007 (vol. 1), no. 4, pp. 95–100, http://www.ca-c.org/c-g/2007/journal_eng/c-g-3/09.shtml.

Aseeva, Anna: Re-thinking Europe's gas supplies after the 2009 Russia–Ukraine crisis, in: The China and Eurasia Forum Quarterly, 2010 (vol. 8), no. 1, pp. 127–138.

Babali, Tuncay: Prospects of export routes for Kashagan oil, in: Energy Policy, 2009 (vol. 37), no. 4, pp. 1298–1308.

Babali, Tuncay: Turkey at the energy crossroads, in: Middle East Quarterly, 2009 (vol. 16), no. 2, pp. 25–33.

Babayan, David: The Northern Caspian on China's geopolitical agenda, in: Central Asia and the Caucasus, 2012 (vol. 13), no. 3, pp. 56–67, http://www.ca-c.org/journal/2012-03-eng/06.shtml.

Babayan, Nelli: The geek, the bully, and the freaks: diversifying EU energy sources through and exercising influence in the South Caucasus, in: Boening, Astrid/ Kremer, Jan-Frederik/ Loon, Aukje van (eds): Global power Europe, Berlin: Springer, 2013, pp. 147–164.

Bacik, Gokhan/ Aras, Bulent: The impediments to establish an energy-regime in Caucasia, in: Hetland, Jens/ Gochitashvili, Teimuraz (eds.): Security of natural gas supply through transit countries, Boston: Kluwer Academic Publishers, 2004, pp. 343–354.

Badalyan, Lusine: Interlinked energy supply and security challenges in the South Caucasus, in: Caucasus Analytical Digest, 12 December 2011, no. 33, pp. 2–5, http://www.isn.ethz.ch/Digital-Library/Publications/Detail/?ots591=0c54e3b3-1e9c-be1e-2c24-a6a8c7060233&lng=en&id=135318.

Baev, Pavel K.: Russian energy policy and military power: Putin's quest for greatness, London: Routledge, 2009.

Baev, Pavel K.: Energy intrigues on the EU's southern flank: applying game theory, in: Problems of Post-Communism, 2010 (vol. 57), no. 3, pp. 11–22.

Baev, Pavel K./ Øverland, Indra: The South Stream versus Nabucco pipeline race: geopolitical and economic (ir)rationales and political stakes in mega-projects, in: International Affairs, 2010 (vol. 86), no. 5, pp. 1075–1090.

Bagirov, Sabit (ed.): Trans-Anatolian gas pipeline: challenges and prospects for the Black Sea countries and the Balkans, Baku: Qanun Publishing, 2012, http://edf.az/ts_general/download/The_Conference_Papers.pdf.

Bahgat, Gawdat: Pipeline diplomacy: the geopolitics of the Caspian Sea region, International Studies Perspectives, 2002 (vol. 3), no. 3, pp. 310–327.

Bahgat, Gawdat: Prospects for energy cooperation in the Caspian Sea, in: Communist and Post-Communist Studies, 2007 (vol. 40), no. 2, pp. 157–168.

Baizakova, Kuralai I.: Energy security issues in the foreign policy of the Republic of Kazakhstan, in: American Foreign Policy Interests, 2010 (vol. 32), no. 2, pp. 103–109.

Balmaceda, Margarita M.: Gas, oil and the linkages between domestic and foreign policies: the case of Ukraine, in: Europe-Asia Studies, 1998 (vol. 50), no. 2, pp. 257–286.

Balmaceda, Margarita M.: Ukraine's persistent energy crisis, in: Problems of Post-Communism, 2004 (vol. 51), no. 4, pp. 40–50.

Balmaceda, Margarita M.: Explaining the management of energy dependency in Ukraine. Possibilities and limits of a domestic-centered perspective, Mannheim: Mannheimer Zentrum für Europäische Sozialforschung, 2004, http://www.uni-mannheim.de/fkks/fkks30.pdf.

Balmaceda, Margarita M.: Belarus: oil, gas, transit pipelines and Russian foreign energy policy, London: GMB Publishing, 2006.

Balmaceda, Margarita M.: Energy dependency, politics and corruption in the former Soviet Union: Russia's power, oligarchs' profits and Ukraine's missing energy policy, 1995–2006, London: Routledge, 2008.

Balmaceda, Margarita M.: Russia's central and eastern European energy transit corridor: Ukraine and Belarus, in: Aalto, Pami (ed.): Russia's energy policies: national, interregional and global levels, Cheltenham: Edward Elgar, 2012, pp. 136–155.

Balmaceda, Margarita M.: The politics of energy dependency: Ukraine, Belarus, and Lithuania between domestic oligarchs and Russian pressure, Toronto: University of Toronto Press, 2013.

Balzer, Harley: The Putin thesis and Russian energy policy, in: Post-Soviet Affairs, 2005 (vol. 21), no. 3, pp. 210–225.

Becker, Abraham S.: Russia and Caspian oil: Moscow loses control, in: Post-Soviet Affairs, 2000 (vol. 16), no. 2, pp. 91–132.

Bedeski, Robert E./ Swanström, Niklas (eds): Eurasia's ascent in energy and geopolitics: rivalry or partnership for China, Russia and Central Asia?, New York: Routledge, 2012.

Begoyan, Anush: United States policy in the South Caucasus: securitisation of the Baku Ceyhan project, in: Iran and the Caucasus, 2004 (vol. 8), no. 1, pp. 141–155.

Bilgin, Mert: New prospects in the political economy of inner-Caspian hydrocarbons and western energy corridor through Turkey, in: Energy Policy, 2007 (vol. 35), no. 12, pp. 6383–6394.

Bilgin, Mert: Geopolitics of European natural gas demand: supplies from Russia, Caspian and the Middle East, in: Energy Policy, 2009 (vol. 37), no. 11, pp. 4482–4492.

Blank, Stephen: Russia's new gas deal with China: background and implications, in: Northeast Asia Energy Focus, 2009 (vol. 6), no. 4, pp. 16–30.

Blinken, Antony J.: Ally versus ally: America, Europe, and the Siberian pipeline crisis, New York: Praeger, 1987.

Bosse, Giselle: The EU's geopolitical vision of a European energy space: when 'Gulliver' meets 'white elephants' and Verdi's Babylonian king, in: Geopolitics, 2011 (vol. 16), no. 3, pp. 512–535.

Bradshaw, Michael J.: Global energy dilemmas: a geographical perspective, in: Geographical Journal, 2010 (vol. 176), no. 4, pp. 275–290.

Bruce, Chloë: Fraternal friction or fraternal fiction? The gas factor in Russian–Belarusian relations, Oxford: Oxford Institute for Energy Studies, 2005, http://www.oxforden ergy.org/wpcms/wp-content/uploads/2010/11/NG8-FraternalFrictionOrFraternal fictionTheGasFactorInRussianBelarusianRelations-ChloeBruce-2005.pdf.

Carroll, Toby: The cutting edge of accumulation: neoliberal risk mitigation, the Baku–Tbilisi–Ceyhan pipeline and its impact, in: Antipode, 2012 (vol. 44), no. 2, pp. 281–302.

Chadwick, Margaret/ Long, David/ Nissanke, Machiko: Soviet oil exports: trade adjustments, refining constraints and market behaviour, Oxford: Oxford Institute for Energy Studies, 1987.

Chang, Chun-Ping/ Berdiev, Aziz N./ Lee, Chien-Chiang: Energy exports, globalization and economic growth: the case of South Caucasus, in: Economic Modelling, 2013 (vol. 33), pp. 333–346.

Chen, Xiaoqin: Central Asian factors in energy relationship between China and Russia, in: Asian Social Science, 2012 (vol. 8), no. 7, pp. 33–39, http://ccsenet.org/journal/index.php/ass/article/view/17600/11783.

Chow, Edward C./ Hendrix, Leigh E.: Central Asia pipelines: field of dreams and reality, in: NBR: Pipeline politics in Asia: the intersection of demand, energy markets, and supply routes, Seattle, WA: The National Bureau of Asian Research, 2010, pp. 29–42 (Special Report no. 23), http://nbr.org/publications/specialreport/pdf/Free/123113/SR23_Pipeline_Politics.pdf.

Ciarreta, Aitor/ Nasirov, Shahriyar: Development trends in the Azerbaijan oil and gas sector: Achievements and challenges, in: Energy Policy, 2012 (vol. 40), pp. 282–292.

Cohen, Ariel: Kazakhstan: the road to independence. Energy policy and the birth of a nation, Stockholm/ Washington, DC: Central Asia–Caucasus Institute & Silk Road Studies Program, 2008, http://www.isdp.eu/files/publications/books/08/ac08kazakhstanindependence.pdf.

Cooley, Alexander: Principles in the pipeline: managing transatlantic values and interests in Central Asia, in: International Affairs, 2008 (vol. 84), no. 6, pp. 1173–1188.

Cornell, Svante E./ Nilsson, Niklas (eds): Europe's energy security: Gazprom's dominance and Caspian supply alternatives, Stockholm/ Washington, DC: Central Asia–Caucasus Institute & Silk Road Studies Program, 2008, http://www.silkroadstudies.org/new/docs/publications/EnergySecurity.pdf.

Cornell, Svante E./ Tsereteli, Mamuka/ Socor, Vladimir: Geostrategic implications of the Baku–Tbilisi–Ceyhan pipeline, in: Starr, S. Frederick/ Cornell, Svante E. (eds): The Baku–Tbilisi–Ceyhan pipeline: oil window to the West, Stockholm/ Washington, DC: Central Asia–Caucasus Institute & Silk Road Studies Program, 2005, pp. 17–38.

Coşkun, Bezen Balamir/ Carlson, Richard: New energy geopolitics: why does Turkey matter?, in: Insight Turkey, 2010 (vol. 12), no. 3, pp. 205–220.

Crandell, Maureen: The role of Central Asian gas: is it possible to bypass Russia?, in: Dellecker, Adrian/ Gomart, Thomas (eds): Russian energy security and foreign policy, London: Routledge, 2011, pp. 74–85.

Cutler, Robert M.: Turkey and the geopolitics of Turkmenistan's natural gas, in: Review of International Affairs, 2001 (vol. 1), no. 2, pp. 20–33.

Cutler, Robert M.: The central Eurasian hydrocarbon energy complex: from Central Asia to Central Europe, in: Amineh, Mehdi Parvizi/ Guang, Yang (eds): Secure oil and alternative energy: the geopolitics of energy paths of China and the European Union, Leiden: Koninklijke Brill, 2012, pp. pp. 41–74.

D'Anieri, Paul: Economic interdependence in Ukrainian-Russian relations, Albany, NY: State University of New York Press, 1999.

Dellecker, Adrian: Caspian Pipeline Consortium: bellwether of Russia's investment climate?, Paris: Institut français des relations internationales, 2008, http://www.ifri.org/downloads/ifrirnvdelleckercpcengjuin2008.pdf.

Demidova, Ksenia: The deal of the century: the Reagan administration and the Soviet pipeline, in: Patel, Kiran Klaus/ Weisbrode, Kenneth (eds): European integration and the Atlantic community in the 1980s, Cambridge: Cambridge University Press, 2013, pp. 59-82.

Dienes, Leslie/ Shabad, Theodore: The Soviet energy system: resource use and policies, Washington DC: Winston & Sons, 1979.

Dobronravina, N.A.: Neft', gaz i transportnoe proklyatie: Kazakhstan, Turkmenistan, Azerbaidzhan, in: Dobronravina, N.A./ Marganiya, O.L. (eds): Resursnoe proklyatie: neft', gaz: modernizatsia obshchestva, St. Petersburg: GU VSHE, 2008, pp. 441-501.

Dulambayeva, R.T./ Boluspayev, S.A./ Daribayeva, M.Z.: Energy sector of Kazakhstan: current state and prospects of development, in: World Applied Sciences Journal, 2013 (vol. 21), no. 7, pp. 968-974, http://idosi.org/wasj/wasj21(7)13/3.pdf.

Egizarian, Ashot: Iran, Turkey, and Russia: semi-peripheral strategies in Central Asia and the Southern Caucasus, in: Central Asia and the Caucasus, 2013 (vol. 14), no. 1, pp. 57-70, http://www.ca-c.org/journal/2013/journal_eng/cac-01/04.shtml.

Energy Charter Secretariat: Oil flows and export capacity in the Caspian Sea and Black Sea regions, Brussels: Energy Charter Secretariat, 2008, http://www.encharter.org/fileadmin/user_upload/document/Oil_Flows.pdf.

Energy Charter Secretariat: Bringing gas to the market. Gas transit and transmission tariffs in Energy Charter Treaty countries: regulatory aspects and tariff methodologies, Brussels: Energy Charter Secretariat, 2012, http://www.encharter.org/fileadmin/user_upload/Publications/Gas_Tariffs_2012_ENG.pdf.

Energy Charter Secretariat: Bringing oil to the market: transport tariffs and underlying methodologies for cross-border crude oil and products pipelines, Brussels: Energy Charter Secretariat, 2012, http://www.encharter.org/fileadmin/user_upload/Publications/Oil_Pipeline_Tariffs_2012_ENG.pdf.

Erdogdu, Erkan: Bypassing Russia: Nabucco project and its implications for the European gas security, in: Renewable and Sustainable Energy Reviews, 2010 (vol. 14), no. 9, pp. 2936-2945.

Ericson, Richard E.: Eurasian natural gas pipelines: the political economy of network interdependence, in: Eurasian Geography and Economics, 2009 (vol. 50), no. 1, pp. 28-57.

Ericson, Richard E.: Eurasian natural gas: significance and recent developments, in: Eurasian Geography and Economics, 2012 (vol. 53), no. 5, pp. 615-648.

Estrada, Javier/ Bergesen, Helge Ole/ Moe, Arild/ Sydnes, Anne Kristin: Natural gas in Europe: markets, organisation and politics, London: Pinter, 1988.

Evangelista, Matthew: From each according to its abilities: competing theoretical approaches to the post-Soviet energy sector, in: Wallander, Celeste A. (ed.): The sources of Russian foreign policy after the Cold War, Boulder, CO: Westview Press, 1996, pp. 173–205.

Eyyubov, Kamran: Factors determining the differences in FDI inflows into Azerbaijan's and Russia's oil sectors after independence, in: The Caucasus & Globalization, 2011 (vol. 5), no. 3–4, pp. 30–43.

Faber van der Meulen, Evert: Gas supply and EU–Russia relations, in: Europe-Asia Studies, 2009 (vol. 61), no. 5, pp. 833–856.

Faraji Rad, Abdol Reza/ Moradi, Heydar: Examining the TAPI pipeline and its impact on regional and cross-regional rivalry, in: Central Asia and the Caucasus, 2012 (vol. 13), no. 2, pp. 83–97, http://www.ca-c.org/online/2012/journal_eng/cac-02/08.shtml.

Finon, Dominique: The EU foreign gas policy of transit corridors: autopsy of the stillborn Nabucco project, in: OPEC Energy Review, 2011 (vol. 35), no. 1, pp. 47–69.

Finon, Dominique/ Locatelli, Catherine/ Mima, Silvana: Long-term competition between gas infrastructures developments in Asia: the construction on the Siberia and Caspian export development by cross-border pipelines, Grenoble: Institut d'Economie et de Politique de l'Energie, 2000.

Fredholm, Michael: The Russian energy strategy & energy policy: pipeline diplomacy or mutual dependence?, Watchfield: Defence Academy of the United Kingdom, Conflict Studies Research Centre, 2005.

Garrison, Jean A./ Abdurahmonov, Ahad: Explaining the Central Asian energy game: complex interdependence and how small states influence their big neighbors, in: Asian Perspective, 2011 (vol. 35), no. 3, pp. 381–405.

Ghaleb, Alexander: Natural gas as an instrument of Russian state power, Carlisle, PA: Strategic Studies Institute, U.S. Army War College, 2011.

Ginzburg, Veniamin/ Troschke, Martina: The exports of Turkmenistan's energy resources: stabilisation without a market economy, in: Central Asia and the Caucasus, 2003, no. 6(24), pp. 108–117, http://www.ca-c.org/journal/2003/journal_eng/cac-06/15.ginen.shtml.

Giuli, Marco: Nabucco pipeline and the Turmenistan connumdrum, in: Caucasian Review of International Affairs, 2008 (vol. 2), no. 3, pp. 124–132, http://www.cria-online.org/Journal/4/Nabucco_pipeline_-_the_Turkmenistan_conundrum_done.pdf.

Global Witness: It's a gas: funny business in the Turkmen–Ukraine gas trade, Washington, DC: Global Witness, 2006, http://www.globalwitness.org/library/its-gas-funny-business-turkmen-ukraine-gas-trade.

Goldman, Marshall I.: Moscow's new economic imperialism, in: Current History, 2008 (vol. 107), no. 711, pp. 322-329.

Goldman, Marshall I.: Petrostate: Putin, power, and the new Russia, Oxford: Oxford University Press, 2008.

Gonchar, Michael/ Martynyuk, Vitalii/ Chubyk, Andriy: The impact of Nord Stream, South Stream on the gas transit via Ukraine and security of gas supplies to Ukraine and the EU, in: Liuhto, Kari (ed.): EU-Russia gas connection: pipes, politics and problems, Turku: Pan-European Institute, 2009, pp. 49-69, http://www.utu.fi/fi/yksikot/tse/yksikot/PEI/raportit-ja-tietopaketit/Documents/Liuhto%200809%20web.pdf.

Götz, Roland: Mythos Diversifizierung. Europa und das Erdgas des Kaspiraums, in: Osteuropa, 2007 (vol. 57), no. 8-9, pp. 449-462.

Götz, Roland: Die Nabucco-Gaspipeline: Probleme und Alternativen, in: Osteuropa-Wirtschaft, 2009 (vol. 54), no. 1-2, pp. 1-9.

Green, Milford B./ Sagers, Matthew J.: Changes in Soviet natural gas flows: 1970-1985, in: Professional Geographer, 1985 (vol. 37), no. 3, pp. 310-319.

Grigoriev, Leonid: Russia, Gazprom and the CAC, in: Dellecker, Adrian/ Gomart, Thomas (eds): Russian energy security and foreign policy, London: Routledge, 2011, pp. 147-169.

Guillet, Jérôme: How to get a pipeline built: myth and reality, in: Dellecker, Adrian/ Gomart, Thomas (eds.): Russian energy security and foreign policy, London: Routledge, 2011, pp. 58-73.

Guliyev, Farid/ Akhrarkhodjaeva, Nozima: Transportation of Kazakhstani oil via the Caspian Sea (TKOC): arrangements, actors and interests, Lysaker: Fridtjof Nansen Institute 2008 (RussCasp Working Paper), http://www.fni.no/russcasp/Kazakh_Azeri_%20oil_transport_RussCasp_Working_Paper.pdf.

Guliyev, Farid/ Akhrarkhodjaeva, Nozima: The Trans-Caspian energy route: cronyism, competition and cooperation in Kazakh oil export, in: Energy Policy, 2009 (vol. 37), no. 8, pp. 3171-3182.

Gustafson, Thane: The Soviet gas campaign: politics and policy in Soviet decisonmaking, Santa Monica, CA: RAND, 1983.

Gustafson, Thane: Soviet negotiating strategy: the East-West gas pipeline deal, 1980-1984, Santa Monica, CA: RAND, 1985.

Haase, Nadine: Die Pipelineinfrastruktur in der Kaspischen Region, Berlin: SWP, 2002 (Aktuelle SWP-Dokumentation, Reihe D Nr. 25).

Hadfield, Amelia: Energy and foreign policy: EU-Russia energy dynamics, in: Smith, Steve/ Hadfield, Amelia/ Dunne, Tim (eds.): Foreign policy: theories, actors, cases, Oxford: Oxford University Press, 2007, pp. 441-462.

Hanson, Philip/ Teague, Elizabeth: Big business and the state in Russia, in: Europe-Asia Studies, 2005 (vol. 57), no. 5, pp. 657–680.

Hardt John P.: Soviet energy policy in Eastern Europe, in: Meiklejohn-Terry, Sarah (ed.): Soviet policy in Eastern Europe, New Haven, CT: Yale University Press, 1984, pp. 189–220.

Hassmann, Heinrich: Erdöl in der Sowjetunion. Geschichte, Gebiete, Probleme, Hamburg: Indstrieverlag von Hernhaussen KG, 1951.

Heinrich Andreas: Transit pipelines, in: Pleines, Heiko: Energy in Ukraine, London: Financial Times, 1998, pp. 141–172 (Financial Times Management Report).

Heinrich Andreas: Rußlands Gazprom. Teil II: Gazprom als Akteur auf internationaler Ebene, Cologne: BIOst, 1999 (Bericht des Bundesinstitut für ostwissenschaftliche und internationale Studien no. 34).

Heinrich, Andreas: Poland as a transit country for Russian natural gas: potential for conflict, Koszalin: KICES, 2007.

Heinrich, Andreas: Russlands Exportpipelines: Diversifizierung oder Bestandssicherung?, Russland-Analysen, no. 217, 25 March 2011, pp. 18–23, http://www.laender-analysen.de/russland/pdf/Russlandanalysen217.pdf.

Hetland, Jens/ Gochitashvili, Teimuraz (eds): Security of natural gas supply through transit countries, Dordrecht: Kluwer academic Publishers, 2003.

Hill, Fiona: Energy empire: oil, gas and Russia's revival, London: The Foreign Policy Centre, 2004.

Hodgkins, Jordan A.: Soviet power: energy resources, production and potentials, Englewood Cliffs, NJ: Prentice-Hall, 1961.

Högselius, Per: Red gas: Russia and the origins of European energy dependence, New York: Palgrave Macmillan, 2013.

Horák, Slavomír: Turkmenistan's shifting energy geopolitics in 2009–2011: European perspectives, in: Problems of Post-Communism, 2012 (vol. 59), no. 2, pp 18–30.

Huiron, Zhao/ Hongwei, Wu: China's energy foreign policy towards the Caspian region: the case of Kazakhstan, in: Amineh, Mehdi Parvizi/ Guang, Yang (eds): Secure oil and alternative energy: the geopolitics of energy paths of China and the European Union, Leiden: Koninklijke Brill, 2012, pp. 167–196.

Ibrahimov, Rovshan: Energy resource transportation by countries with no access to the open sea (an Azerbaijan case study), in: The Caucasus and Globalization, 2011 (vol. 5), no. 1–2, pp. 63–69, http://www.ca-c.org/c-g/2011/journal_eng/c-g-1-2/c-gE-1-2.2011.pdf.

Idan, Avinoam/ Shaffer, Brenda: The foreign policies of post-Soviet landlocked states, in: Post-Soviet Affairs, 2011, (vol. 27), no. 3, pp. 241–268.

International Energy Agency: Caspian oil and gas: the supply potential of Central Asia and Transcaucasus, Paris: IEA, 1998.

Ipek, Pinar: The role of oil and gas in Kazakhstan's foreign policy: looking east or west?, in: Europe-Asia Studies, 2007 (vol. 59), no. 7, pp. 1179–1199.

Ipek, Pinar: Azerbaijan's foreign policy and challenges for energy security, in: Middle East Journal, 2009 (vol. 63), no. 2, pp. 227–239.

Janardhanan, Nandakumar: China's search for energy and its strategy towards Central Asia, in: International Journal of Energy Sector Management, 2009 (vol. 3), no. 2, pp. 102–107.

Jentleson, Bruce W.: Pipeline politics: the complex political economy of East–West energy trade, Ithaca, NY: Cornell University Press, 1986.

Kandiyoti, Rafael: Pipelines: oil flows and crude politics, London: IB Tauris, 2008.

Kandiyoti, Rafael: What price access to the open seas? The geopolitics of oil and gas transmission from the Trans-Caspian republics, in: Central Asian Survey, 2008 (vol. 27), no. 1, pp. 75–93.

Kassenova, Nargis: Kazakhstan and the South Caucasus corridor in the wake of the Georgia–Russia war, Brussels: EUCAM, 2009 (EUCAM Policy Brief no. 3), http://www.fride.org/download/PB3_EUCAM_Kajastan_Caucasus_ENG_en09.pdf.

Kivinen, Markku: Public and business actors in Russia's energy policy, in: Aalto, Pami (ed.): Russia's energy policies: national, interregional and global levels, Cheltenham: Edward Elgar, 2012, pp. 45–62.

Kjærnet, Heidi: Azerbaijani–Russian relations and the economization of foreign policy, in: Øverland, Indra/ Kjærnet, Heidi/ Kendall-Taylor, Andrea (eds): Caspian energy policy: Azerbaijan, Kazakhstan and Turkmenistan, London: Routledge, 2010, pp. 150–161.

Korchemkin, Mikhail: Russia's oil and gas exports to the former Soviet Union, in: Kaminski, Bartlomiej (ed.): Economic transition in Russia and the new states of Eurasia, New York: M.E. Sharp, 1996, pp. 121–137.

Kovalev, Felix N.: Transportation of Caspian oil through Russia, in: Amirahmadi, Hooshang (ed.): The Caspian region at a crossroad: challenges of a new frontier of energy and development, New York: St. Martin's Press, 200, pp. 155–162.

Kramer, John: Soviet-CEMA energy ties, in: Problems of Communism, 1985 (vol. 34), no. 4, pp. 32–47.

Krauer-Pacheco, Ksenia: Turkey as a transit country and energy hub: the link to its foreign policy aims, Bremen: Research Centre for East European Studies, 2011, http://www.forschungsstelle.uni-bremen.de/UserFiles/file/fsoAP118.pdf.

Kropatcheva, Elena: Playing both ends against the middle: Russia's geopolitical energy games with the EU and Ukraine, in: Geopolitics, 2011 (vol. 16), no. 3, pp. 553–573.

Kubicek, Paul: Energy politics and geopolitical competition in the Caspian Basin, in: Journal of Eurasian Studies, 2013 (vol. 4), no. 2, pp. 171–180.

Küçük, Zeki Furkan: China's energy policy toward Central Asia and the importance of Kazakhstan, in: Central Asia and the Caucasus, 2009, no. 3(57), pp. 33–46, http://www.ca-c.org/journal/2009-03-eng/03.shtml.

Kusznir, Julia: TAP, Nabucco West, and South Stream: the pipeline dilemma in the Caspian Sea basin and its consequences for the development of the Southern Gas Corridor, in: Caucasus Analytical Digest, 18 February 2013, no. 47, pp. 2–8, http://www.isn.ethz.ch/Digital-Library/Publications/Detail/?ots591=0c54e3b3-1e9c-be1e-2c24-a6a8c7060233&lng=en&id=160678.

Kusznir, Julia: The Nabucco gas pipeline project and its impact on EU energy policy in the South Caucasus, in: Caucasus Analytical Digest, 12 December 2011, no. 33, pp. 9–13, http://www.isn.ethz.ch/Digital-Library/Publications/Detail/?ots591=0c54e3b3-1e9c-be1e-2c24-a6a8c7060233&lng=en&id=135318.

Kyj, Myroslaw J./ Kyj, Larissa: An institution-stakeholder framework for examining business relationship dynamics in a transforming Eastern Europe, in: Journal of World Business, 2009 (vol. 44), no. 3, pp. 300–310 [Odessa–Brody oil pipeline].

Laldjebaev, Murodbek: The water-energy puzzle in Central Asia: the Tajikistan perspective, in: Water Resources Development, 2010 (vol. 26), no. 1, pp. 23–36.

Lane, David (ed.): The political economy of Russian oil, Lanham, MD: Rowman and Littlefield, 1999.

Lateigne, Marc: China, energy security and Central Asian diplomacy: bilateral and multilateral approaches, in: Øverland, Indra/ Kjærnet, Heidi/ Kendall-Taylor, Andrea (eds): Caspian energy policy: Azerbaijan, Kazakhstan and Turkmenistan, London: Routledge, 2010, pp. 101–115.

Laurila, Juhani: Transit transport between the European Union and Russia in light of Russian geopolitics and economics, in: Emerging Markets Finance and Trade, 2003 (vol. 39), no. 5, pp. 27–57.

Liutho, Kari: Energy in Russia's foreign policy, Turku: Pan-European Institute, 2010, http://www.utu.fi/fi/yksikot/tse/yksikot/PEI/raportit-ja-tietopaketit/Documents/Liuhto_final_netti.pdf.

Locatelli, Catherine: Russian and Caspian hydrocarbons: energy supply stakes for the European Union, in: Europe-Asia Studies, 2010 (vol. 62), no. 6, pp. 959–971.

Lough, John: Russia's energy diplomacy, London: Chatham House, 2011 (Briefing Paper), http://www.chathamhouse.org/sites/default/files/19352_0511bp_lough.pdf.

Lussac, Samuel J.: Ensuring European energy security in Russian 'Near Abroad': the case of the South Caucasus, in: European Security, 2010 (vol. 19), no. 4, pp. 607–625.

Mabro, Robert/ Wybrew-Bond, Ian (eds): Gas to Europe: the strategies of four major suppliers, Oxford: Oxford University Press, 1999.

Magomedov, Arbakhan/ Nikerov, Ruslan: The South Stream vs. Nabucco: the 'pipeline war' gains momentum, in: The Caucasus and Globalization, 2011 (vol. 5), no. 1–2, pp. 15–23, http://www.ca-c.org/c-g/2011/journal_eng/c-g-1-2/c-gE-1-2.2011.pdf.

Malyhina, Kateryna: Die Erdgasversorgung der EU unter besonderer Berücksichtigung der Ukraine als Transitland, Bremen: Research Centre for East European Studies, 2009, http://www.forschungsstelle.uni-bremen.de/UserFiles/file/06-Publikationen/Arbeitspapiere/fsoap105.pdf.

Mangott, Gerhard: Gestörte Leitungen: Russland und die Gasversorgung der EU, in: Erler, Gernot/ Schulze, Peter W. (eds): Die Europäisierung Russlands: Moskau zwischen Modernisierungspartnerschaft und Großmachtrolle, Frankfurt/ Main: Campus, pp. 189–213.

Marketos, Thrassy N.: China's energy geopolitics: the Shanghai Cooperation Organization and Central Asia, London: Routledge, 2009.

Marketos, Thrassy N.: Eastern Caspian Sea energy geopolitics: a litmus test for the U.S. – Russia – China struggle for the geostrategic control of Eurasia, in: Caucasian Review of International Affairs, 2009 (vol. 3), no. 1, pp. 2–19, http://www.cria-online.org/6_2.html.

Marketos, Thrassyvoulos (Thrassy) N.: Chinese strategy toward Central and South Asia: energy interests and energy security, in: Central Asia and the Caucasus, 2012 (vol. 13), no. 1, pp. 43–52, http://www.ca-c.org/online/2012/journal_eng/cac-01/03.shtml.

Matveev, Vladimir: China's gas policy in Central Asia, in: Central Asia and the Caucasus, 2008, no. 5(53), pp. 77–88.

Merciai, Patrizio: The Euro-Siberian gas pipeline dispute: a compelling case for the adoption of jurisdictional codes of conduct, in: Maryland Journal of International Law and Trade, 1984 (vol. 8), no. 1, pp. 1–52, http://digitalcommons.law.umaryland.edu/mjil/vol8/iss1/3.

Miles, Carolyn: The Caspian pipeline debate continues: why not Iran?, in: Journal of International Affairs, 1999 (vol. 53), no. 1, pp. 325–346.

Mitchell, John V./ Beck, Peter/ Grubb, Michael: The new geopolitics of energy, London: Royal Institute of International Affairs, 1996.

Mitchell, John V./ Morita, Koji/ Selley, Norman/ Stern Jonathan: The new economy of oil: impacts on business, geopolitics and society, London: Royal Institute of International Affairs/ Earthscan, 2001.

Mitrova, Tatiana/ Pirani, Simon/ Stern, Jonathan P.: Russia, the CIS and Europe: gas trade and transit, in: Pirani, Simon (ed.): Russian and CIS gas markets and their impact on Europe, Oxford: Oxford University Press, 2009, pp. 395–441.

Movkebaeva, Galiia A.: Energy cooperation among Kazakhstan, Russia, and China within the Shanghai Cooperation Organization, in: Russian Politics and Law, 2013 (vol. 51), no. 1, pp. 80–87.

Nanay, Julia: Russia's role in the Eurasian energy market: seeking control in the face of growing challenges, in Perovic, Jeronim/ Orttung, Robert/ Wenger, Andreas (eds): Russian energy power and foreign relations: implications for conflict and cooperation, London: Routledge, 2009, pp. 109–131.

Nanay, Julia/ Smith Stegen, KarenL: Russia and the Caspian region: challenges for transatlantic energy security?, in Journal for Transatlantic Studies, 2012 (vol. 10), no. 4, pp. 343–357.

Nathan, Hippu S.K./ Kulkarni, Sanket S./ Ahuja, Dilip R.: Pipeline politics: a study of India's proposed cross-border gas projects, in: Energy Policy, 2013 (vol. 62), pp. 145–156.

Newnham, Randall: Soviet oil and gas trade with Eastern Europe: developments in the 1980s, Santa Monica, CA: California Seminar on International Security and Foreign Policy, 1990 (Discussion Papers 112).

Newnham, Randall: Oil, carrots, and sticks: Russia's energy resources as a foreign policy tool, in: Journal of Eurasian Studies, 2011 (vol. 2), no. 2, pp. 134–143.

Newnham, Randall: Pipeline politics: Russian energy sanctions and the 2010 Ukrainian elections, in: Journal of Eurasian Studies, 2013 (vol. 4), no. 2, pp. 115–122.

Nies, Susanne: Oil and gas delivery to Europe: an overview of existing and planned infrastructures, Paris: Institut français des relations internationales, 2011.

Nuriyev, Elkhan: The South Caucasus at the crossroads: conflicts, Caspian oil and great power conflicts, Münster: Lit Verlag.

Nygren, Bertil: Putin's use of natural gas to reintegrate the CIS region, in: Problems of Post-Communism, 2008 (vol. 55), no. 4, pp. 3–15.

Öğütçü, Mehmet/ Ma, Xin: Growing links in energy and geopolitics: China, Russia, and Central Asia, CEPMLP Internet Journal, March 2008, http://www.dundee.ac.uk/cepmlp/gateway/files.php?file=CEPMLP_IJ_Mar08-Growing_links_916741751.pdf.

Olcott, Martha B.: International gas trade in Central Asia: Turkmenistan, Iran, Russia and Afghanistan, Stanford, CA: Stanford Institute for International Studies, 2004 (Working Paper 28), http://iis-db.stanford.edu/pubs/20605/Turkmenistan_final.pdf.

Olcott, Martha B.: The energy dimension in Russian global strategy: Vladimir Putin and the geopolitics of oil, Houston, TX: Rice University, The James A. Baker III Institute for Public Policy, 2004, http://carnegieendowment.org/files/wp-2005-01_olcott_english1.pdf.

Olcott, Martha B.: International gas trade in Central Asia: Turkmenistan, Iran, Russia, and Afghanistan, in: Victor, David G./ Jaffe, Amy M./ Hayes, Mark H. (eds): Natural gas and geopolitics: from 1970 to 2040, Cambridge: Cambridge University Press, 2006, pp. 202–233.

Olcott, Martha B.: Kazmunaigaz: Kazakhstan's national oil and gas company, Houston, TX: Rice University, James A. Baker III Institute for Public Policy, 2007, http://carnegieendowment.org/files/Kaz_Olcott.pdf.

Omonbude, Ekpen J.: The economics of transit oil and gas pipelines: a review of the fundamentals, in: OPEC Energy Review, 2009 (vol. 33), no. 2, pp. 125–139.

Omonbude, Ekpen J.: Cross-border oil and gas pipelines and the role of the transit country: economics, challenges, and solutions, Basingstoke: Palgrave Macmillan, 2013.

Orban, Anita: Power, energy and the new Russian imperialism, Westport, CT: Praeger Security International, 2008.

Orttung, Robert: Energy and state-society relation: socio-political aspects of Russia's energy wealth, in Perovic, Jeronim/ Orttung, Robert/ Wenger, Andreas (eds): Russian energy power and foreign relations: implications for conflict and cooperation, London: Routledge, 2009, pp. 51–71.

Orttung, Robert W./ Øverland, Indra: A limited toolbox: explaining the constraints on Russia's foreign energy policy, in: Journal of Eurasian Studies, 2011 (vol. 2),no. 1, pp. 74–85.

Orttung, Robert W./ Perović, Jeronim/ Pleines, Heiko/ Schröder, Hans-Henning (eds): Russia's energy sector between politics and business, Bremen: Research Centre for East European Studies, 2008, http://www.forschungsstelle.uni-bremen.de/UserFiles/file/06-Publikationen/Arbeitspapiere/fsoAP92.pdf.

Ostrowski, Wojciech: Politics and oil in Kazakhstan, London: Routledge, 2009.

Øverland, Indra/ Torjesen, Stina: Just good friends: Kazakhstan's and Turkmenistan's energy relations with Russia, in: Øverland, Indra/ Kjærnet, Heidi/ Kendall-Taylor, Andrea (eds): Caspian energy policy: Azerbaijan, Kazakhstan and Turkmenistan, London: Routledge, 2010, pp. 136–149.

Øverland, Indra/ Kjærnet, Heidi/ Kendall-Taylor, Andrea (eds): Caspian energy politics: Azerbaijan, Kazakhstan and Turkmenistan, London: Routledge, 2010.

Palkin, S.E.: An energy triangle: China, Russia, and Kazakhstan, in: Problems of Economic Transition, 2012 (vol. 55), no. 1, pp. 36–50.

Panyushkin, Valerii V./ Zygar', Michail V.: Gazprom: novoe russkoe oruzhie, Moscow: Zakharov, 2008.

Paramonov, Vladimir/ Strokov, Aleksey: Russia and Central Asia: existing and potential oil and gas trade, Shrivenham: Defence Academy of the UK, Advanced Research and Assessment Group, 2008, http://www.da.mod.uk/colleges/arag/document-listings/ca/08%2803%29VP%20English.pdf.

Paramonov, Vladimir/ Strokov, Aleksey: Russian oil and gas projects and investments in Central Asia, Shrivenham: Defence Academy of the UK, Advanced Research and Assessment Group, 2008, http://www.da.mod.uk/colleges/arag/document-listings/ca/08%2819%29VP%20English.pdf.

Pardo Sierra, Oscar: a corridor through thorns: EU energy security and the Southern Energy Corridor, in: European Security, 2010 (vol. 19), no. 4, pp. 643–660.

Pasukeviciute, Irma/ Roe, Michael: Strategic policy and the logistics of crude oil transit in Lithuania, in: Energy Policy, 2005 (vol. 33), no. 7, pp. 857–866.

Peimani, Hooman: The Caspian pipeline dilemma: political games and economic losses, Westport, CT: Praeger Publishers, 2001.

Perovic, Jeronim: Farce ums Gas: Russland, die Ukraine und die EU-Energiepolitik, in: Osteuropa, 2009 (vol. 59), no. 1, pp. 19–35.

Peyrouse, Sébastien: Economic aspects of the Chinese–Central Asia rapprochement, Stockholm/ Washington, DC: Central Asia-Caucasus Institute & Silk Road Studies Program, 2007, http://www.silkroadstudies.org/new/docs/Silkroadpapers/2007/0709China-Central_Asia.pdf.

Pirani, Simon: Ukraine's energy sector, Oxford: Oxford Institute for Energy Studies, 2007, http://www.oxfordenergy.org/wpcms/wp-content/uploads/2010/11/NG21-UkrainesGasSector-SimonPirani-2007.pdf.

Pirani, Simon: a gas dependent state, in: Pirani, Simon (ed.): Russian and CIS gas markets and their impact on Europe, Oxford: Oxford University Press, 2009, pp. 93–132.

Pirani, Simon: Am Tropf. Die Ukraine, Russland und das Erdgas, in: Osteuropa, 2010 (vol. 60), no. 2, pp. 237–256.

Pirani, Simon/ Stern, Jonathan/ Yafimava, Katja: The Russo–Ukrainian gas dispute of January 2009: a comprehensive assessment, Oxford: Oxford Institute for Energy Studies, 2009, http://www.oxfordenergy.org/wpcms/wp-content/uploads/2010/11/NG27-TheRussoUkrainianGasDisputeofJanuary2009AComprehensiveAssessment-JonathanSternSimonPiraniKatjaYafimava-2009.pdf.

Pleines, Heiko: Developing Russia's oil and gas industry, in Perovic, Jeronim/ Orttung, Robert/ Wenger, Andreas (eds): Russian energy power and foreign relations: implications for conflict and cooperation, London: Routledge, 2009, pp. 71–87.

Roberts, John: Caspian pipelines, London: Royal Institute for International Affairs, 1996.

Rogozhina, N.: The Caspian: oil transit and problems of ecology, in: Problems of Economic Transition, 2010 (vol. 53), no. 5, pp. 86–93.

Rousseau, Richard: Competing geopolitical interests of China, Russia and the United States in Central Asia and the Caspian region, in: Khazar Journal of Humanities and Social Sciences, 2011 (vol. 14), no. 3, pp. 13–29, http://jhss-khazar.org/wp-content/uploads/2011/11/02R-Rousseau-Competing-Geopolitical-Interests-of-China-Russia-Khazar-Journal1.pdf.

Ruseckas, Laurent: a tale of two seas: the pipeline politics of the Caspian and Black Seas, Cambridge, MA: Cambridge Energy Research Associates, 1994.

Rutland, Peter: Oil, politics and foreign policy, in: Lane, David (ed.): The political economy of Russian oil, Lanham, MD: Rowman & Littlefield, 1999, pp. 163–188.

Rutland, Peter: US energy policy and the former Soviet Union, in: Perovic, Jeronim/ Orttung, Robert W./ Wegner, Andreas (eds): Russian energy power and foreign relations: implications for conflict and cooperation, London: Routledge, 2009, pp. 181–200.

Rzayeva, Gulmira/ Tsakiris, Theodoros G.R.: Azerbaijani gas strategy and the EU's Southern Corridor, Baku: Strategic Imperative, 2012, http://www.eliamep.gr/wp-content/uploads/2012/08/tsakiris1.pdf.

Sagers, Matthew J./ Nicoud, Jennifer: Development of East Siberian gas field and pipeline to China: a research communication, in: Post-Soviet Geography and Economics, 1997 (vol. 38), no. 5, pp. 288–295.

Sampson, Paul: Pipeline to power: a battle of two gas companies leaves Turkmenistan the winner, in: Transitions, 1998 (vol. 5), no. 10, pp. 50–55.

Sander, Michael: Deutsch-russische Beziehungen im Gassektor: wirtschaftliche Rahmenbedingungen, Interorganisationsnetzwerke und die Verhandlungen zur Nord Stream Pipeline, Baden-Baden: Nomos, 2012.

Sartori, Nicolò: The Southern Gas Corridor: needs, opportunities and constraints, Rome: Istituto Affari Internazionali, 2011, http://www.iai.it/pdf/DoclAI/iai1108.pdf.

Saurbek, Zhanibek: Kazakh–Chinese energy relations: economic pragmatism or political cooperation?, in: China and Eurasia Forum Quarterly, 2008 (vol. 6), no. 1, pp. 79–93.

Sayari, Sabri: Turkey's Caspian interests: economic and security opportunities, in: Ebel, Robert/ Menon, Rajan (eds): Energy and conflict in Central Asia and the Caucasus, Lanham, MD: Rowman and Littlefield, 2000, pp. 225–246.

Schmidt-Felzmann, Anke: EU member states' energy relations with Russia: conflicting approaches to securing natural gas supplies, in: Geopolitics, 2011 (vol. 16), no. 3, pp. 574–599.

Shadrina, Elena: Russia's foreign energy policy: norms, ideas and driving dynamics, Turku: Pan-European Institute, 2010, http://www.utu.fi/fi/yksikot/tse/yksikot/PEI/raportit-ja-tietopaketit/Documents/Shadrina_final_netti.pdf.

Shaffer, Brenda: Energy politics, Philadelphia, PA: University of Pennsylvania Press, 2009.

Shaffer, Brenda: Caspian energy phase II: beyond 2005, in: Energy Policy, 2010 (vol. 38), no. 11, pp. 7209-7215.

Shelest, Anna: The Georgian-Russian Conflict, Energy Projects, and Security of the Black Sea-Caspian Region, in: Central Asia and the Caucasus, 2009, no. 4-5(58-59), pp. 192-199, http://www.ca-c.org/journal/2009-04-05-eng/16.shtml.

Shen, Simon: Great power politics: India's absence from ideological energy diplomacy in Central Asia, in: The China and Eurasia Forum Quarterly, 2010 (vol. 8), no. 1, pp. 95-110.

Sidar, Cenk/ Winrow, Gareth: Turkey and South Stream: Turco-Russian rapprochement and the future of the Southern Corridor, in: Turkish Policy Quarterly, 2011 (vol. 10), no. 2, pp. 51-61.

Siddiky, Ishrak Ahmed: Towards a new framework for cross-border pipelines: the International Pipeline Agency, in: China and Eurasia Forum Quarterly, 2010 (vol. 8), no. 3, pp. 121-127.

Siddiky, Ishrak Ahmed: Towards a new framework for cross-border pipelines: the International Pipeline Agency, in: Energy Policy, 2011 (vol. 39), no. 9, pp. 5344-5346.

Smirnov, Sergey: The gas pipelines: a game of Caspian patience, in: Central Asia and the Caucasus, 2007, no. 6(48), pp. 77-84, http://www.ca-c.org/journal/2007-06-eng/07.shtml.

Smith, Hanna: Russian foreign policy and energy: the case of the Nord Stream gas pipeline, in: Aalto, Pami (ed.): Russia's energy policies: national, interregional and global levels, Cheltenham: Edward Elgar, 2012, pp. 117-135.

Smith Stegen, Karen: Deconstructing the 'energy weapon': Russia's threat to Europe as case study, in: Energy Policy, 2011 (vol. 39), no. 10, pp. 6505-6513.

Smits, Robert-Jan H. M.: The Siberian gas pipeline dispute: U.S. policy and European interests, Genève: Institut universitaire de hautes études internationales, 1985.

Smolansky, Oles M.: Ukraine's quest for independence: the fuel factor, in: Europe-Asia Studies, 1995 (vol. 47), no. 1, pp. 67-90.

Sobjak, Anita/ Zasztowt, Konrad: Nabucco West—perspectives and relevance: the reconfigured scenario, in: PISM Policy Papers (The Polish Institute of International Affairs, Warsaw), 2012, no. 44, http://www.pism.pl/files/?id_plik=12268.

Sorbello, Paolo/ Grandi, Ludovico: From concentration to competition: the struggle for power between the Kremlin and Gazprom through the study of TNK-BP and South-Stream, in: Irish Slavonic Studies, 2013 (vol. 25), pp. 106-119.

Souleimanov, Emil/ Černý, Filip: The Southern Caucasus pipelines and the Caspian 'oil diplomacy': the issue of transporting Caspian oil and natural gas to world markets, in: RIPS: Revista de Investigaciones Políticas y Sociológicas, 2012 (vol. 11), no. 4, pp. 77-105, http://www.usc.es/revistas/index.php/rips/article/download/696/690.

Sovacool, Benjamin K.: The interpretive flexibility of oil and gas pipelines: case studies from Southeast Asia and the Caspian Sea, in: Technological Forecasting & Social Change, 2011 (vol. 78), no. 4, pp. 610–620.

Sovacool, Benjamin K.: Reconfiguring territoriality and energy security: global production networks and the Baku–Tbilisi–Ceyhan (BTC) pipeline, in: Journal of Cleaner Production, 2012 (vol. 32), no. 9, pp. 210–218.

Starr, S. Frederick/ Cornell, Svante E. (eds): The Baku–Tbilisi–Ceyhan pipeline: oil window to the West, Stockholm/ Washington, DC: Central Asia – Caucasus Institute & Silk Road Studies Program, 2005.

Stent, Angela E.: Restoration and revolution in Putin's foreign policy, in: Europe-Asia Studies, 2008 (vol. 60), no. 9, pp. 1089–1106.

Stern, Jonathan P.: Specters and pipe dreams, in: Foreign Policy, 1982, No 48, pp. 21–36.

Stern, Jonathan P.: International gas trade in Europe: the politics of exporting and importing countries, reprint, Aldershot: Gower, 1986.

Stern, Jonathan P.: Soviet oil and gas exports to the West: commercial transaction or security threat?, Aldershot: Gower, 1989.

Stern, Jonathan P.: Soviet and Russian gas: the origins and evolution of Gazprom's export strategy, in: Mabro, Robert/ Wybrew-Bond, Ian (eds): Gas to Europe: the strategies of four major suppliers, Oxford: Oxford University Press, 1999, pp. 135–199.

Stern, Jonathan P.: The Russian–Ukrainian gas crisis of January 2006, Oxford: Oxford Institute for Energy Studies, 2006, http://www.oxfordenergy.org/wpcms/wp-content/uploads/2011/01/Jan2006-RussiaUkraineGasCrisis-JonathanStern.pdf.

Stern, Jonathan P.: Natural gas security problems in Europe: the Russian–Ukrainian crisis of 2006, in: Asia-Pacific Review, 2006 (vol. 13), no. 1, pp. 32–59.

Stulberg, Adam N.: Eurasia's pipeline tangle: practical lessons from cross-border pipeline operations, in: Russia in Global Affairs, 24 September 2011, http://eng.globalaffairs.ru/number/Eurasias-Pipeline-Tangle--15337.

Stulberg, Adam N.: Strategic bargaining and pipeline politics: confronting the credible commitment problem in Eurasian energy transit, in: Review of International Political Economy, 2012 (vol. 19), no. 5, pp. 808–836.

Syroezhkin, Konstantin: China's Presence in the Energy Sector of Central Asia, in: Central Asia and the Caucasus, 2012 (vol. 13), no. 1, pp. 20–43, http://www.ca-c.org/journal/2012-01-eng/02.shtml.

Tayfur, M. Fatih/ Göyman, Korel: Decision making in Turkish foreign policy: the Caspian oil pipeline issue, in: Middle Eastern Studies, 2002 (vol. 38), no. 2, pp. 101–122.

Tekin, Ali/ Williams, Paul A.: EU–Russian relations and Turkey's role as an energy corridor, in: Europe-Asia Studies, 2009 (vol. 61), no. 2, pp. 337–356.

Tkachenko, Stanislav L.: Actors in Russia's energy policy towards the EU, in: Aalto, Pami (ed.): The EU-Russian energy dialogue: Europe's future energy security, Aldershot: Ashgate, 2008, pp. 163-192.

Tomberg, Igor: Energy policy and energy projects in Central Eurasia, in: Central Asia and the Caucasus, 2007, no. 6(48), pp. 38-50, http://ca-c.org/journal/2007-06-eng/04.shtml.

Tomberg, Igor: Central Asian gas in the changing geopolitical context, in: Central Asia and the Caucasus, 2012 (vol. 13), no. 1, pp. 7-19, http://www.ca-c.org/journal/2012-01-eng/01.shtml.

Tsereteli, Mamuka: Economic and energy security: connecting Europe and the Black Sea-Caspian region, Stockholm/ Washington, DC: Central Asia-Caucasus Institute &Silk Road Studies Program, 2008, http://www.isdp.eu/images/stories/isdp-main-pdf/2008_tsereteli_economic-and-energy-security.pdf.

Tsygankov, Andrey P.: Vladimir Putin's vision for Russia as a normal great power, in: Post Soviet Affairs, 2008 (vol. 21), no. 2, pp. 132-158.

Umbach, Frank: Energy security in Eurasia: clashing interests, in: Dellecker, Adrian/ Gomart, Thomas (eds): Russian energy security and foreign policy, London: Routledge, 2011, pp. 23-38.

Umbach, Frank: Competing for Caspian energy resources: Russia's and China's energy (foreign) policies and the implications for the EU's energy security, in: Amineh, Mehdi Parvizi/ Guang, Yang (eds): Secure oil and alternative energy: the geopolitics of energy paths of China and the European Union, Leiden: Koninklijke Brill, 2012, pp. 75-114.

United Nations Development Programme/ World Bank: Cross-border oil and gas pipelines: problems and prospects, Washington DC: Energy Sector Management Assistance Programme, 2003, http://siteresources.worldbank.org/INTOGMC/Resources/crossborderoilandgaspipelines.pdf.

Van Oudenaren, John: The Urengoi pipeline: prospects for Soviet leverage, Santa Monica, CA: RAND, 1984.

Victor, Nadejda M./ Victor, David G.: Bypassing Ukraine: exporting Russian gas to Poland and Germany, in: Victor, David G./ Jaffe, Amy M./ Hayes, Mark H. (eds): Natural gas and geopolitics: from 1970 to 2040, Cambridge: Cambridge University Press, 2006, pp. 122-168.

Viter, Olena/ Pavlenko, Rostyslav/ Honchar, Mykhaylo: Ukraine: post-revolution energy policy and relations with Russia, London: GMB Publishing, 2006.

Vitkus, Gediminas: Russian pipeline diplomacy: a Lithuanian response, in: Acta Slavica Iaponica, 2009 (vol. 26), pp. 25-46.

Webb, John/ Reteyum, Aleksey: Russia's Baltic pipeline system: a state priority with commercial significance? Cambridge, MA: Cambridge Energy Research Associates, 2001.

Westphal, Kirsten: Russian gas, Ukrainian pipelines, and European supply security: lessons of the 2009 controversies, Berlin: SWP, 2009 (SWP Research Paper), http://www.swp-berlin.org/fileadmin/contents/products/research_papers/2009_RP11_wep_ks.pdf.

Woods, Stan: Pipeline politics: the allies at odds, Aberdeen: Centre for Defence Studies, 1983.

Xu, Xiaojie: The oil and gas links between Central Asia and China: a geopolitical perspective, in: OPEC Review, 1999 (vol. 23), no. 1, pp. 33–54.

Yafimava, Katja: The June 2010 Russian–Belarusian gas transit dispute: a surprise that was to be expected, Oxford, Oxford Institute for Energy Studies, 2010, http://www.oxfordenergy.org/wpcms/wp-content/uploads/2010/11/NG43-TheJune2010RussianBelarusianGasTransitDisputeASurpriseThatWasToBeExpected-KatjaYafimava-2010.pdf.

Yafimava, Katja: The transit dimension of EU energy security: Russian gas transit across Ukraine, Belarus, and Moldova. Oxford: Oxford University Press, 2011.

Yenikeyeff, Shamil M.: Kazakhstan's gas: export markets and export routes, Oxford: Oxford Institute for Energy Studies, 2008, http://www.oxfordenergy.org/wpcms/wp-content/uploads/2010/11/NG25-KazakhstansgasExportMarketsandExportRoutes-ShamilYenikeyeff-2008.pdf.

Yenikeyeff, Shamil M.: Energy interests of the 'great powers' in Central Asia: cooperation or conflict?, in: The International Spectator, 2011 (vol. 46), no. 3, pp. 61–78.

Youngs, Richard: Energy security: Europe's new foreign policy challenge, London: Routledge, 2009.

Zhukov, Stanislav: Turkmenistan: an exporter in transition, in: Pirani, Simon (ed.): Russian and CIS gas markets and their impact on Europe, Oxford: Oxford University Press, 2009, pp. 271–315.

Ziegler, Charles E.: The energy factor in China's foreign policy, in: Journal of Chinese Political Science, 2006 (vol. 11), no. 1, pp. 1–24.

About the Authors

Farkhod Aminjonov is a PhD candidate at the Balsillie School of International Affairs, Waterloo, Canada. He has a BA degree in International Law from the University of World Economy and Diplomacy in Tashkent, Uzbekistan. He has also completed two Master's programs: an MA in Political Science at the OSCE Academy in Bishkek, Kyrgyzstan, and an MA in International Studies at the University of Tsukuba, Japan.

Lusine Badalyan is a PhD student at the Research Centre for East European Studies at the University of Bremen, Germany, and an associate fellow at the Bremen International Graduate School of Social Sciences (BIGSS). She received a BA in International Relations from Yerevan State University, Armenia, and an MA in International Relations and Global Governance from Jacobs University Bremen, Germany. Her PhD project is conducted within the research project 'Domestic debates and foreign policy-making in the Caspian region: the case of export pipelines from Azerbaijan, Kazakhstan and Turkmenistan' with financial support from the Volkswagen Foundation.

Boris Barkanov is a visiting assistant professor in the Political Science Department at West Virginia University, USA. After receiving his PhD from UC Berkeley in 2010, he was a postdoctoral fellow at the Davis Center for Russian and Eurasian Studies at Harvard University. His research is at the intersection of international relations and comparative politics, and focuses substantively on Eurasian energy markets, Russian politics and foreign policy, and Russia's relations with its neighbours.

Inna Chuvychkina is a PhD candidate at the University of Bremen, Research Centre for East European Studies, Germany. She received her BA in International Relations and her MA in Political Science from the Peoples' Friendship University in Moscow, Russia. Her research focuses on International Relations, Russian energy policy and Russia–EU relations.

Andreas Heinrich is a researcher at the Research Centre for East European Studies, University of Bremen, Germany. He received his PhD in political science from the Free University in Berlin, Germany. His research focuses on energy relations in the former Soviet Union and Eastern Europe.

Irina Kustova is a PhD candidate in the School of International Studies at the University of Trento, Italy. She holds an MA in European Studies from Maastricht University, specialist degrees in International Economics from the Ural Federal University, Russia) and in International Law from the Ural State Law Academy, Russia. In 2012, she was a visiting scholar at the Energy Research Institute of the Russian Academy of Sciences and at the Centre for Energy and Resource Security, King's College London. Her research interests include EU–Russia relations in the gas sector, European energy markets, energy security, and European integration.

Katerina Malygina is a researcher and a PhD candidate at the Research Centre for East European Studies at the University of Bremen, Germany, and the editor of the online-journal 'Ukraine-Analysen'. She holds an MA degree in International Relations from the Catholic University of Eichstätt-Ingolstadt and was a DAAD/OSI and Friedrich Ebert Foundation scholarship holder. Her research interests include post-Soviet hybrid regimes, the energy sector of the CIS states, politics and foreign policy of Ukraine.

András Molnár is a second year PhD student in the field of International Relations at Corvinus University of Budapest, Hungary. He received a BA degree in International Relations from Széchenyi István University and an MA degree in Diplomacy from Corvinus University. His research topic is the Russian foreign energy policy in the light of neo-classical game theory.

Eric Pardo Sauvageot is a PhD candidate at the Department of International Studies of the Universidad Complutense de Madrid (UCM), Spain, and editorial coordinator of the 'UNISCI Discussion Papers'. He has been a research fellow at the Davis Center for Russian and Eurasian Studies, Harvard University, the Aleksanteri Center for Russian Studies at Helsinki University and the Center for European and Russian Studies (CEURUS) at Tartu University. His main research interests are the post-Soviet space, energy and Russia–Japan relations.

Rufat Rustamov is a PhD student at the Gandja State University, Azerbaijan. He received his BA degree from Baku State University and his MA from the Baku Slavic University, both in the field of International Relations. The author participated in the Erasmus Mundus Programme 'Partnership for Azerbaijan, Georgia, and Armenia'. His research interests include energy supplies from the Caspian region to Europe and export pipelines.

Niels Smeets is a PhD candidate and a research fellow at the Leuven International and European Studies Institute (LINES) and a member of the Research Group Russia and Eurasia at the University of Leuven, Belgium. He holds MA degrees in Comparative and International Politics, Management Studies, and Slavonic Studies. He studied at the St. Petersburg State University on an Erasmus Mundus scholarship. His research focuses on energy security in the Eurasian space, with special attention to the Russian Energy Strategy.

Paolo Sorbello received an MA (honours) in International and Diplomatic Studies from the University of Bologna, Italy, in 2011. Since 2012, he is enrolled in a 2-year Erasmus Mundus International Masters in Russian, Central and East European studies organised by the University of Glasgow, UK. In 2013, he spent his second year doing research in Almaty, Kazakhstan, in affiliation with KIMEP University.

CHANGING EUROPE

Edited by Dr. Sabine Fischer, Dr. Heiko Pleines and
Prof. Dr. Hans-Henning Schröder

ISSN 1863-8716

1   Sabine Fischer, Heiko Pleines, Hans-Henning Schröder (eds.)
    Movements, Migrants, Marginalisation
    Challenges of societal and political participation in Eastern Europe and the enlarged EU
    ISBN 978-3-89821-733-0

2   David Lane, György Lengyel, Jochen Tholen (eds.)
    Restructuring of the Economic Elites after State Socialism
    Recruitment, Institutions and Attitudes
    ISBN 978-3-89821-754-5

3   Daniela Obradovic, Heiko Pleines (eds.)
    The Capacity of Central and East European Interest Groups to Participate in EU Governance
    ISBN 978-3-89821-750-7

4   Sabine Fischer, Heiko Pleines (eds.)
    Crises and Conflicts in Post-Socialist Societies
    The Role of Ethnic, Political and Social Identities
    ISBN 978-3-89821-855-9

5   Julia Kusznir, Heiko Pleines (eds.)
    Trade Unions from Post-Socialist Member States in EU Governance
    ISBN 978-3-89821-857-3

6   Sabine Fischer, Heiko Pleines (eds.)
    The EU and Central & Eastern Europe
    Successes and Failures of Europeanization in Politics and Society
    ISBN 978-3-89821-948-8

7   Sabine Fischer, Heiko Pleines (eds.)
    Civil Society in Central and Eastern Europe
    ISBN 978-3-8382-0041-5

8   Zdenka Mansfeldová, Heiko Pleines (eds.)
    Informal Relations from Democratic Representation to Corruption
    Case Studies from Central and Eastern Europe
    ISBN 978-3-8382-0173-3

ibidem-Verlag

Melchiorstr. 15

D-70439 Stuttgart

info@ibidem-verlag.de

www.ibidem-verlag.de
www.ibidem.eu
www.edition-noema.de
www.autorenbetreuung.de